THE SPECIAL THEORY OF RELATIVITY

THE SPECIAL THEORY OF RELATIVITY

BY

J. AHARONI

DOVER PUBLICATIONS, INC.
NEW YORK

Published in Canada by General Publishing Company, Ltd., 30 Lesmill Road, Don Mills, Toronto, Ontario.
Published in the United Kingdom by Constable and Company, Ltd., 10 Orange Street, London WC2H 7EG.

This Dover edition, first published in 1985, is an unabridged and unaltered republication of the second, revised edition (1965) of the work first published by Oxford University Press in 1959. It is reprinted by special arrangement with Oxford University Press, 200 Madison Avenue, New York, N.Y. 10016.

Manufactured in the United States of America
Dover Publications, Inc., 31 East 2nd Street, Mineola, N.Y. 11501

Library of Congress Cataloging in Publication Data

Aharoni, J.
The special theory of relativity.

Reprint. Originally published: 2nd ed. London: Oxford University Press, 1965.
Bibliography: p.
Includes index.
1. Relativity (Physics) 2. Calculus of tensors. 3. Spinor analysis. I. Title.
QC173.55.A37 1985 530.1′1 84-25922
ISBN 0-486-64870-2

PREFACE TO THE SECOND EDITION

SOON after the publication of the first edition an important paper, entitled 'Self-energy and stability of the classical electron', was published by F. Rohrlich. It clarified the old puzzle of the 4/3 factor appearing in the expression for the electromagnetic mass m_0 of an electron when expressed in terms of its electrostatic energy, as derived by M. Abraham and H. A. Lorentz. Rohrlich proved that in a consistent application of the special theory of relativity, the 4/3 factor does not occur and must be replaced by 1. Rohrlich's paper initiated new interest in the problem and it turned out that actually a similar solution had already been proposed by B. Kwal in 1949 and the same result obtained as far back as 1922 by E. Fermi who used a different method. It can now be stated that the abolition of the 4/3 factor is also implicit in Dirac's paper on the classical theory of the electron (1938). It is difficult to explain why all the earlier papers passed unnoticed. Possibly this was due to Poincaré's idea to link the 4/3 factor with the instability of an electric charge on purely electrostatic forces. This new situation required a revision of the part of the book dealing with electrodynamics and to give a clear exposition of this development I found it necessary to enlarge this part considerably and also to include a paragraph on the retarded and advanced potentials as well as a brief account of the Wheeler–Feynman absorber theory of radiation. On revising the book, I also made a number of improvements, particularly in the part dealing with spinors.

To Dr. G. A. Perkins I am indebted for some criticism in connexion with the Proca field.

J. A.

Imperial College
London, S.W. 7

PREFACE TO FIRST EDITION

IT is generally agreed that the mathematical tool for dealing with the theory of relativity is the tensor calculus or, in view of quantum-mechanics, more generally the spinor calculus. However, tensors have a repelling effect on many physicists and spinors have only made the situation worse. In writing this book I have been guided by the idea that this attitude arises from the way tensors are usually introduced in lectures and textbooks. They are so often defined as entities which transform in a certain way without any explanation being given why we should consider one particular mode of transformation and not another. It has been my experience in teaching the theory of relativity, that when a student is taken into confidence and the fact that (to take the three-dimensional case) tensor transformations constitute representations of the group of rotations which result from the isotropy of space is disclosed, tensors become attractive and lively. No doubt the reluctance of lecturers and textbooks to deal with this relationship between tensors and the group of rotations, or between the four-dimensional tensors and the Lorentz group, is due to the fact that group-theory itself appears to many physicists as an obstacle. It is true that group-theory, although essentially algebraic in nature, constitutes a difficult branch of mathematics. However, for our purpose, only very elementary knowledge of group-theory is required, essentially only a few fundamental concepts. These are explained in the text as much as possible, but the consultation of any elementary textbook on matrices and groups may prove helpful to the reader who has no knowledge at all of these subjects.

This book deals with the theory of relativity in both its tensor and spinor aspects. The latter are closely linked with quantum-theory. Although I am not dealing with quantized fields, I hope that I have elucidated sufficiently the reasons why spinors come to play a role in quantum-theory. In a sense, one of the main objects of the book is to provide an introduction to the application of the theory of relativity in quantum-theory, particularly in field theories.

In writing the book I had in mind, in the first instance, the student of Physics, but I feel that it should be of help also to the student of Applied Mathematics who wishes to acquaint himself with the physical background of tensor and spinor calculus.

I have not found it practical to give references in the text, and the reader will only find a list of references at the end of the book. Of these, I should like to draw attention in particular to Corson's book and the one by O. Costa de Beauregard, which I have found very helpful. I have adopted Corson's notation for tensors, two-spinors, and four-spinors, but in a number of definitions with respect to covariant and contravariant indices and in some other ways I have adopted a somewhat different procedure.

Finally, I should like to mention that when dealing with the non-conservation of parity and its repercussion on the full Lorentz group, I had the advantage of some helpful discussion on the matter with my colleagues Professor A. Salam, Dr. P. T. Matthews, and Dr. J. C. Taylor, to whom I wish to express my thanks.

J. A.

Imperial College
London, S.W. 7

CONTENTS

THE SPECIAL THEORY OF RELATIVITY

CHAPTER 1

SPACE AND TIME IN INERTIAL SYSTEMS

1. Object and Main Principle of the Theory

The theory of relativity deals with the relations which exist between physical entities as they appear to different observers who are in motion relative to each other. Originally the theory dealt mainly with entities of electromagnetic nature. Einstein's first paper on the subject (1905, *Ann. Physik*, **17**) was entitled: 'On Electrodynamics of Moving Bodies.' However, from the outset it was recognized that the theory is of fundamental importance in the whole of physics.

The theory of relativity developed in two main stages, both essentially the work of Einstein. In the first stage the so-called 'Special Theory of Relativity' was established, in which only one class of relative motion of different observers is considered, a motion uniform in direction and velocity. In the second stage the foundation was laid for a more general theory in which no restriction is made with regard to relative motion. It turned out that the general theory is closely linked up with a theory of gravitation. As such it is therefore perhaps of a more restricted interest. On the other hand, some very general features of physical laws already follow from the special theory of relativity and it is primarily sufficient to put any suggested physical law to a relativistic test in the special sense only.† In this book we deal with the special theory only.

The special theory of relativity is based essentially on one principle. Before this is stated it is, however, necessary to introduce the concept of an inertial frame of reference. Let x, y, z form three orthogonal axes and let us assume that any free body, upon which no external force is acting, is found to be either at rest or moving along a straight line relative to x, y, z irrespective of position and direction in which the body may be moving. Such a system of coordinates is said to be inertial. In an inertial system it is possible to distribute identical clocks and adjust their 'zeros', or phases, so that each free body appears to be moving not only along a straight line but also with constant velocity.‡

† In other words we shall ignore the effect which matter has on the metrical properties of space-time. The four-dimensional space-time is regarded as pseudo-Euclidean.

‡ If the moving bodies consist of clocks, these clocks will be found to move with some constant velocity as obtainable from the clocks themselves.

This adjustment is not unique, for if we introduce a phase change in each clock which is a linear function of its position x, y, z it will be found again that every free body moves with constant velocity. To achieve a further restriction on the definition of time we can introduce a method of synchronization of clocks at different places which makes further physical experiences assume a simple form. One important experience in this respect can be summed up by saying that the space is homogeneous and isotropic, at least as far as any dynamical observations are considered. The mechanical properties of any physical system are independent of its position in space or orientation, if all external interference with the system is eliminated. Thus, from dynamical observation of internal properties of a given system it is impossible to tell where it is situated in space or what orientation it possesses. Position and orientation have a meaning only in a relative sense. We can speak of position and orientation only relative to other bodies, or relative to some system of coordinates which is also just a body of suitable shape.

Let us now consider a gun which fires bullets in different directions. Then to conform with the isotropy and homogeneity of the space we have to define time at all the space points in such a way that the velocity of the bullets will be found to have the same value irrespective of the direction along which they move. This is achieved by using identical clocks at all space points and by synchronizing them in such a way that two bullets which are fired by identical guns from any point A to any point B and vice versa, will collide at mid-point if the firing takes place at the same time. The above definition of time makes use therefore of the inherent isotropy of space. It has the simple property of transitivity. If a clock at A is synchronous with a clock at B and a clock at C, then B and C are also synchronous. (See Fig. 1 where bullets fired from A and B and colliding at O will also collide with a bullet fired from C, if it is fired at the same time as the bullet from A.)

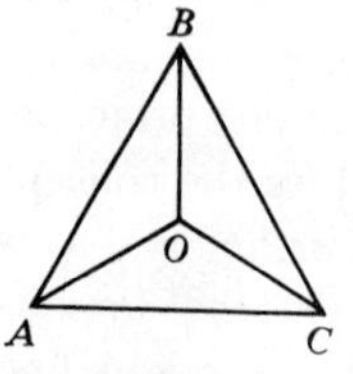

FIG. 1

Any time definition which does not possess this property would be highly artificial and make the laws of physics look unduly complicated. The synchronization test of clocks as described above takes advantage of the inherent homogeneity and isotropy of space. In this connexion it may be added that to synchronize clocks by bringing them to a master clock (say of A) and then taking them back to their respective locations need not necessarily lead to a time definition which is identical with the former. In other words, it does not necessarily follow from the isotropy of space that clocks which have been transported away from a master clock through different distances will remain synchronized, even if they are transported with the same speed. (The speed can be checked with the help of the clocks themselves.) We shall see later that, according to the theory of relativity, if two synchronized clocks are transported from one position in space to another position via different routes they will in general cease to be synchronous, although in practice the effect is too small to be noticed.

We call a set of axes, x, y, z as considered above, together with clocks which have been synchronized in the manner described, an inertial system. The name arises from the fact that in such a frame of reference a body upon which no force is acting will either stay at rest or move in a straight line with constant velocity. Hence the law of inertia holds with respect to such a frame of reference.

We can now formulate the special principle of relativity upon which the whole theory is based: Let K be an inertial frame of reference, then, according to the principle of relativity, in every set of axes x', y', z' which moves along a straight line and with constant velocity relative to K it is equally possible to adjust the clocks as in K, and all the frames of reference so obtained are inertial and are completely equivalent to each other not only with respect to the law of inertia, but also with respect to any other law or physical relation. In other words, according to the special principle of relativity there exist an infinite number of inertial frames of reference which are equivalent as far as the formulation of the laws of physics is concerned. In general, two inertial frames of reference move relative to each other along a straight line and with constant velocity. This is the only possible motion between inertial frames.

The principle of relativity as far as dynamics is concerned was acknowledged even before the modern theory of relativity developed. It was admitted that both from kinematical and from dynamical observations it is impossible to decide between two inertial frames of reference which 'moves' and which is 'at rest' or which is more privileged. All dynamical laws are identical in inertial frames of reference and it is impossible to devise a dynamical experiment which will enable us to detect absolute uniform motion.

For a long time physicists maintained, however, that the principle of relativity does not hold in optics and wherever electromagnetic phenomena are involved. The prevailing idea was that there exists one frame of reference in which the laws of optics and of the electromagnetic field assume a particularly simple form. In other frames, which move relative to this exceptional one, the laws are of a different and necessarily more complicated form because the velocity relative to the 'absolute' frame finds its way into their formulation. It was thought that the absolute space is at rest relative to the ether, a medium which was supposed to fill the universe and without which it appeared difficult to account for the wave properties of light and for the existence of an electromagnetic field.

However, all the attempts to determine any motion relative to the hypothetical ether proved vain, and nothing could therefore appear more natural than to extend the principle of relativity from dynamics to the whole of physics. The most important experiment in this connexion is the Michelson–Morley experiment described in the Appendix. Nevertheless, this step with regard to the principle of relativity could not be taken light-heartedly, because it turned out that it necessitated the abandonment of certain preconceived ideas about space and time which were tacitly assumed to hold from the early days of physics. In particular, as we shall see later, the idea of absolute simultaneity had to be abandoned. Although simultaneity has a unique and definite meaning in each inertial frame of reference, this no longer holds when the times of two inertial frames in motion relative to each other are compared. Simultaneous events in one system are non-simultaneous in the other.

Since it is impossible to distinguish between inertial frames of reference, both when dynamical relations as well as those pertaining to other branches of physics are considered, it follows that we can also make use of light signals to synchronize clocks. A source of light exhibits the same properties irrespective of its position in space or orientation and this holds in every inertial frame of reference. Consequently we can synchronize clocks by replacing the guns and bullets by sources of light and light signals. This will make no difference for the resulting synchronization, both methods yielding the same time. In one respect, however, light signals differ from bullets fired from a gun. In the latter case the velocity depends on the state of motion of the gun, and therefore it was tacitly assumed when we defined the synchronization method, that the guns were at rest in each considered frame. In the case of light we know from experience that the velocity of light is not only independent of the particular nature of the source and the colour of the emitted light, but also independent of the state of motion of the source in the considered frame. Thus if two sources, one at rest and the other in motion, emit at the moment when the moving source passes near the source at rest, two light signals in a given direction, these signals travel with the same velocity as if they were one signal. This constitutes the law of constancy of light velocity. Experimental evidence for this law is given in the Appendix. Hence, when we use light signals to synchronize clocks, we need not insist on identical sources, nor do we have to insist upon their being at rest in the considered space frame. Although this is of no consequence as regards the synchronization of clocks in a given inertial frame of reference, it has a decisive effect on the interrelations which exist between the coordinates x, y, z, t of events relative to one frame of reference and x', y', z', t' of the same events when viewed from a second inertial frame of reference. The homogeneous and isotropic behaviour of light propagation on the one hand, and the independence of the velocity of light of the particular properties and state of motion of the source on the other hand, indicate that light can travel with only one velocity in any considered frame. From the principle of relativity, it then follows further that only one value for the velocity of light can exist, no matter which inertial frame is used. In section 4 we shall prove, following Einstein, that because of this property of invariance of light velocity, simultaneity cannot have absolute meaning with respect to inertial systems which move relative to each other. At this stage, we note that since light travels with the same velocity in all inertial frames it follows that the simplest method of synchronizing clocks and indeed of defining time in any frame is to send light signals from a master clock. If a signal is sent at time t to a point at distance l from the master clock, the time on arrival of the signal at this point is $t+l/c$. The time obtained in this way is identical with the one obtained by the methods described previously, for instance by mechanical methods. In this matter of using a master clock and sending light signals to determine the time at distant points we must be on our guard against forming the opinion that the invariance of light velocity is nothing but a result of our definition of time. The invariance of the velocity of light is not a consequence of an adjustment of clocks which is so arranged as to yield the same value c in any inertial frame. The invariance of light velocity is a consequence of two laws, both based on experiments: (1) The principle of relativity, which in our present context tells us that uniform motion of a source of light can only be relative. Hence, we cannot say that the frame of reference is at rest and the source moves or vice versa. Only relative motion is observable.

The experimental evidence for this relativity in optics was first provided by the Michelson–Morley experiment (see Appendix). (2) The law of constancy of light velocity. According to this law the velocity is independent of the motion of the source. (For experimental evidence see Appendix.) This law, combined with the principle of relativity, leads us immediately to the result that the velocity of light will turn out to have the same value in all inertial frames provided we use clocks and rods of identical manufacture and synchronize them in the same way in all inertial frames (for instance mechanically) and in accordance with the isotropy of space.

2. The General Law of Transformation between Inertial Systems

In this section we shall derive a general relation which must hold between the space and time coordinates of inertial systems, if the principle of relativity is to be satisfied. Let K and K' be two inertial frames of reference, both using the same unit of time and the same unit of length. Then if the velocity of K' relative to K is given by a vector $\mathbf{v}$ (v_x, v_y, v_z), the velocity of K relative to K' is given by

$$\mathbf{v}' = -\mathbf{v} = -(v_{x'}, v_{y'}, v_{z'}).$$

v_x, v_y, v_z are the components of $\mathbf{v}$ relative to K and $v_{x'}$, $v_{y'}$, $v_{z'}$ are the components of $\mathbf{v}$ relative to K'. K and K' are in general inclined to each other. Rotations from K to K' may be characterized by three angles, α, β, γ, and a rotation from K' to K by α', β', γ'. The inverse of α, β, γ is $-\gamma$, $-\beta$, $-\alpha$.

Each physical event at x, y, z at time t is observed in K' at x', y', z', t' and vice versa. For this reason there must exist a relation between these two sets of variables, which depends on the relative velocity and orientation of the two sets. We call this relation the transformation equations and write them as follows:

$$\begin{aligned} x' &= \phi_1(x, y, z, t;\quad \alpha, \beta, \gamma;\quad v_x, v_y, v_z), \\ y' &= \phi_2(x, y, z, t;\quad \alpha, \beta, \gamma;\quad v_x, v_y, v_z), \\ z' &= \phi_3(x, y, z, t;\quad \alpha, \beta, \gamma;\quad v_x, v_y, v_z), \\ t' &= \phi_4(x, y, z, t;\quad \alpha, \beta, \gamma;\quad v_x, v_y, v_z). \end{aligned} \tag{2.1}$$

Now from the principle of relativity it follows that the inverse of equations (2.1) must be identical with the original equations, provided v_x, v_y, v_z are replaced by $v'_{x'} = -v_{x'}$, $v'_{y'} = -v_{y'}$, $v'_{z'} = -v_{z'}$, and α, β, γ

by $\alpha' = -\gamma$, $\beta' = -\beta$, $\gamma' = -\alpha$. Hence the functions $\phi_1,..., \phi_4$ are such that if we solve eqns. (2.1), we should find:

$$\begin{aligned} x &= \phi_1(x',y',z',t'; \quad \alpha',\beta',\gamma'; \quad v'_{x'}, v'_{y'}, v'_{z'}), \\ y &= \phi_2(x',y',z',t'; \quad \alpha',\beta',\gamma'; \quad v'_{x'}, v'_{y'}, v'_{z'}), \\ z &= \phi_3(x',y',z',t'; \quad \alpha',\beta',\gamma'; \quad v'_{x'}, v'_{y'}, v'_{z'}), \\ t &= \phi_4(x',y',z',t'; \quad \alpha',\beta',\gamma'; \quad v'_{x'}, v'_{y'}, v'_{z'}). \end{aligned} \tag{2.2}$$

Both frames of reference are inertial. Hence a uniform motion of a body along a straight line in K must appear as a uniform motion along a straight line in K'. From eqns. (2.1) we get

$$dx' = \frac{\partial \phi_1}{\partial x} dx + \frac{\partial \phi_1}{\partial y} dy + \frac{\partial \phi_1}{\partial z} dz + \frac{\partial \phi_1}{\partial t} dt, \tag{2.3}$$

with similar relations for dy', dz', and dt'. This leads to the following relations between velocity components u_x, u_y, u_z in K and corresponding components $u_{x'}$, $u_{y'}$, $u_{z'}$ in K':†

$$u_{x'} = \frac{dx'}{dt'} = \frac{(\partial\phi_1/\partial x)dx + (\partial\phi_1/\partial y)dy + (\partial\phi_1/\partial z)dz + (\partial\phi_1/\partial t)dt}{(\partial\phi_4/\partial x)dx + (\partial\phi_4/\partial y)dy + (\partial\phi_4/\partial z)dz + (\partial\phi_4/\partial t)dt}. \tag{2.4}$$

Or, dividing top and bottom by dt, we get

$$u_{x'} = \frac{dx'}{dt'} = \frac{u_x(\partial\phi_1/\partial x) + u_y(\partial\phi_1/\partial y) + u_z(\partial\phi_1/\partial z) + (\partial\phi_1/\partial t)}{u_x(\partial\phi_4/\partial x) + u_y(\partial\phi_4/\partial y) + u_z(\partial\phi_4/\partial z) + (\partial\phi_4/\partial t)}. \tag{2.5}$$

Similar relations hold for $u_{y'}$, $u_{z'}$.

We shall be mainly interested in Cartesian coordinates x, y, z and the simplest procedure with regard to t is to consider it also as a fourth Cartesian coordinate. Thus we introduce a four-dimensional space with Cartesian coordinates x, y, z, t. In such a space a uniform motion along a straight line in the three-dimensional space x, y, z appears as a straight line. And so we are now seeking those transformations which transform a straight line in $K(x,y,z,t)$ into a straight line in $K'(x',y',z',t')$. The most general transformations which achieve this are of the form

$$x' = \frac{a_{11}x + a_{12}y + a_{13}z + a_{14}t + k_1}{\alpha x + \beta y + \gamma z + \delta t + \epsilon}, \tag{2.6}$$

with similar relations for y', z', and t', where the denominator is the same for all the coordinates. However, in general such transformations transform an infinite point x, y, z, $t \to \infty$ into a finite point and this cannot hold for two observers who move relative to each other. In both frames of reference an infinite point is an infinite point. Consequently, we must limit ourselves to those transformations which leave the infinite plane

† To simplify the notation we shall from now on use $u_{x'}$ for $u'_{x'}$, etc.

invariant. This can be achieved by replacing the common linear denominator $\alpha x+\beta y+\gamma z+\delta t+\epsilon$ by a constant which is different from zero. We thus come to the conclusion that the most general transformation between two inertial frames $K(x,y,z,t)$ and $K'(x',y',z',t')$ is a linear transformation of the form

$$\begin{aligned} x' &= a_{11}x+a_{12}y+a_{13}z+a_{14}t+k_1, \\ y' &= a_{21}x+a_{22}y+a_{23}z+a_{24}t+k_2, \\ z' &= a_{31}x+a_{32}y+a_{33}z+a_{34}t+k_3, \\ t' &= a_{41}x+a_{42}y+a_{43}z+a_{44}t+k_4. \end{aligned} \tag{2.7}$$

In these equations the a_{ik} depend on the relative velocity between the two observers and the relative inclination of the axes, but are independent of x, y, z, t. The k_i are constants of integration and without much loss of generality we may equate them to zero. It will then simply follow that at time $t = t' = 0$ the point $x = y = z = 0$ coincides with $x' = y' = z' = 0$.

We should now apply the principle of relativity to eqns. (2.7) i.e. solve them and subject them to the condition which is expressed in (2.2). This task can be greatly simplified without loss of generality at this stage, if we let K and K' use axes which are parallel to each other and if we let the relative motion proceed along the x and x' axes with a velocity v and $-v$ respectively. Then for both systems $\alpha = \beta = \gamma = 0$ and we need not use these variables explicitly. It follows further that for a given value of y (or z) the corresponding y' (or z') will be constant independent of x, t, and z (or y). Also, the relationship between x' and x cannot depend on y and z:

$$\begin{aligned} x' &= a_{11}x + a_{14}t \\ y' &= a_{22}y \\ z' &= a_{33}z \\ t' &= a_{41}x+a_{42}y+a_{43}z+a_{44}t. \end{aligned} \tag{2.8}$$

From the principle of relativity it follows that if we reverse these equations and express x, y, z, t in terms of x', y', z', t', we should get, to begin with, equations which are of the same form as (2.8). Actually we find by solving (2.8):

$$\begin{aligned} x &= (a_{22}a_{33}a_{44}x'+a_{14}a_{33}a_{42}y'+a_{14}a_{22}a_{43}z'-a_{14}a_{22}a_{33}t')\Delta^{-1} \\ y &= a_{22}^{-1}y' \\ z &= a_{33}^{-1}z' \\ t &= (-a_{22}a_{33}a_{41}x'-a_{11}a_{33}a_{42}y'-a_{11}a_{22}a_{43}z'+a_{11}a_{22}a_{33}t')\Delta^{-1} \end{aligned} \tag{2.9}$$

where Δ is the determinant of eqns. (2.8) and is given by

$$\Delta = a_{22}a_{33}(a_{11}a_{44}-a_{41}a_{14}). \tag{2.10}$$

In these equations all the a_{ik} are functions of $+v$, the same functions as in (2.8).

Eqns. (2.9) are not of the same form as eqns. (2.8) unless we make the following coefficients vanish:

$$a_{42} = a_{43} = 0. \tag{2.11}$$

We then get, instead of (2.8),

$$\begin{aligned} x' &= a_{11}x && && +a_{14}t \\ y' &= && a_{22}y && \\ z' &= && && a_{33}z \\ t' &= a_{41}x && && +a_{44}t \end{aligned} \tag{2.12}$$

and, for the reverse equations,

$$\begin{aligned} x &= a_{22}a_{33}a_{44}\Delta^{-1}x' && && -a_{14}a_{22}a_{33}\Delta^{-1}t' \\ y &= && a_{22}^{-1}y' && \\ z & && a_{33}^{-1}z' && \\ t &= -a_{22}a_{33}a_{41}\Delta^{-1}x' && && +a_{11}a_{22}a_{33}\Delta^{-1}t', \end{aligned} \tag{2.13}$$

where Δ is the same as before (2.10). Both sets of equations are now of the same form and thus conform to the principle of relativity. However, this principle makes the further demand that the coefficients in both sets of equations have the same value provided we replace v by $-v$ in (2.12). For this purpose we must discuss the dependence of the various coefficients on the relative velocity. Consider any point $x' =$ constant which moves with a velocity $-v$ along the x-axis. In the K-system we shall have

$$a_{11}x+a_{14}t = x' = \text{constant}. \tag{2.14}$$

This is the equation of motion of the point x'. The velocity of this point is obtained from

$$a_{11}dx+a_{14}dt = 0, \tag{2.15}$$

$$\frac{dx}{dt} = -\frac{a_{14}(v)}{a_{11}(v)} = -v. \tag{2.16}$$

This is also the velocity of any other point x' relative to K, in fact the velocity of the whole frame of K' relative to K. Now from the principle of relativity it follows that a similar relation must hold for the corresponding coefficients in (2.13) provided we replace v by $-v$ in the right-hand side of (2.16):

$$-\frac{a_{14}(-v)}{a_{11}(-v)} = \frac{a_{14}\,a_{22}\,a_{33}}{a_{22}\,a_{33}\,a_{44}} = +v; \tag{2.17}$$

therefore
$$\frac{a_{14}(v)}{a_{44}(v)} = v. \tag{2.18}$$

From eqns. (2.16) and (2.18) we get

$$a_{11}(v) = a_{44}(v). \tag{2.19}$$

These coefficients will be denoted by 'a'. Finally, we can also conclude that the following relation must hold:

$$a_{22} = a_{33} = b \quad \text{(say)}. \tag{2.20}$$

This follows from the isotropy of space. In our present case the motion is along the x-axis and the behaviour of both the y- and z-axes, which are perpendicular to the x-axis, must be identical. We are thus left with three constants a, b, a_{41} all depending on the velocity. The transformation equations are:

$$\begin{aligned} x' &= ax && && +avt \\ y' &= && by && \\ z' &= && && bz \\ t' &= a_{41}x && && +at. \end{aligned} \tag{2.21}$$

For the determinant, eqn. (2.10), we now get

$$\Delta = ab^2(a-a_{41}v), \tag{2.22}$$

and (2.13) becomes

$$\begin{aligned} x &= (a-a_{41}v)^{-1}x' && && -(a-a_{41}v)^{-1}vt' \\ y &= && b^{-1}y' && \\ z &= && b^{-1}z' && \\ t &= -a^{-1}(a-a_{41}v)^{-1}a_{41}x' && && +(a-a_{41}v)^{-1}t'. \end{aligned} \tag{2.23}$$

As already mentioned, the coefficients of (2.23) should be obtainable from those of (2.21) by replacing $a(v)$, $b(v)$, $a_{41}(v)$, and $a(v)b$ by $a(-v)$, $b(-v)$, $a_{41}(-v)$, and $-a(-v)v$. This leads to the following relations:

$$a(-v) = (a(v)-a_{41}(v)v)^{-1}, \tag{2.24}$$

$$a_{41}(-v) = -a_{41}(v)(a(v)-a_{41}(v)v)^{-1}a^{-1}(v), \tag{2.25}$$

$$b(-v) = b^{-1}(v). \tag{2.26}$$

From these equations we get

$$\frac{a_{41}(-v)}{a_{41}(v)} = -\frac{a(-v)}{a(v)}. \tag{2.27}$$

We shall now prove that $a(v)$ must be an even function of the velocity. For let us assume that an observer in K measures the length of a rod $A'B'$ which is at rest in K'. K performs this measurement by marking the positions of A' and B' along the x-axis at a given time t. These markings are two events which take place at two positions x_1 and x_2 and from the first of eqns. (2.21) we get

$$x_1' - x_2' = a(x_1 - x_2). \tag{2.28}$$

$x_1' - x_2'$ is the length of the rod relative to K', in which frame the rod rests. Denoting this length by l_0 we get

$$l_0 = a(v)(x_1 - x_2). \tag{2.29}$$

Let us assume that the rod is moved in the opposite direction. In this case K may observe the ends A' and B' at positions $\bar{x}_1$ and $\bar{x}_2$ and we must have

$$l_0 = a(-v)(\bar{x}_1 - \bar{x}_2). \tag{2.30}$$

Since the length of a rod as measured by K cannot depend on the direction in which the rod moves (isotropy of space), it follows that $x_1 - x_2 = \bar{x}_1 - \bar{x}_2$, and we thus get

$$a(v) = a(-v). \tag{2.31}$$

Using this result, it follows from (2.24) that

$$a_{41} = v^{-1}(a(v) - a^{-1}(v)). \tag{2.32}$$

From now on we can again write just 'a' for $a(v)$.

It only remains to discuss (2.26). Here again we may argue that $b(v)$ is an even function of the velocity, because such relations as $y' = by$, $z' = bz$ cannot depend on the direction of the velocity. Hence

$$b(v) = b(-v). \tag{2.33}$$

From this equation and (2.26) we get

$$b(v) = \pm 1. \tag{2.34}$$

The minus sign may be ruled out, because for $v = 0$ we should have $y' = y$, $z' = z$.

In conclusion, we thus find that the transformation equations contain only one unknown function $a(v)$:

$$\begin{aligned} x' &= ax && && +avt \\ y' &= && y && \\ z' &= && && z \\ t' &= v^{-1}(a - a^{-1})x && && +at. \end{aligned} \tag{2.35}$$

The determinant of these equations is

$$\Delta = a^2 - a(a - a^{-1}) = 1, \tag{2.36}$$

and if we solve them with respect to x, y, z, t we get the same equation again with the exception of the sign of v:

$$\begin{aligned} x &= ax' && -avt' \\ y &= y' && \\ z &= z' && \\ t &= -v^{-1}(a - a^{-1})x' && +at'. \end{aligned} \tag{2.37}$$

To sum up the foregoing calculation we can say that if K and K' are inertial frames of reference with a relative velocity v, eqns. (2.35) must necessarily hold between the respective space and time coordinates. To derive this result we have assumed that both K and K' use the same units of length and time and that in both systems the clocks are synchronized in the same self-consistent way so that any relative velocity between K and K' has the same value in the appropriate direction, no matter at which points in K and K' it is measured. We have also made explicit use of the principle of relativity. In the next section we shall derive the functional dependence of 'a' upon v.

3. The Existence of a Universal and Invariant Velocity

From a dimensional consideration it follows that the coefficient $a(v)$ in eqn. (2.35) must be a pure number. Therefore $a(v)$ is either a constant number which is independent of v, or else it is equal to $f(v^2/c^2)$ where c^2 has the dimension of the square of a velocity and is independent of v. From the principle of relativity it then follows that c^2 must have the same value whether observed in K or K'.

First let us consider the consequence which follows from the assumption that a is after all independent of v. Then, since for $v = 0$ the transformation eqns. (2.35) should lead to the identical transformation $x' = x$, etc., a must be equal to 1. Thus eqns. (2.35) become:

$$\begin{aligned} x' &= x && +vt \\ y' &= y && \\ z' &= z && \\ t' &= t && \end{aligned} \tag{3.1}$$

These represent the classical pre-relativistic transformation known as the Galilei transformation. It is in full accord with the principle of relativity.

The second possibility arises if

$$a = f(v^2/c^2), \tag{3.2}$$

where c^2 is a universal constant having the same value no matter in which inertial frame it is observed. The existence of such a constant indicates that whenever universal constants yield on multiplication a quantity of the dimension of c^2 they must also yield the same value. Different values would involve different modes of transformation from x, y, z, t to x', y', z', t', and this is impossible for given sets of clocks and standards of lengths. Nor can c^2 be equal to $c_1 c_2$ where $c_1 \neq c_2$ and both are invariant velocities, for c_1^2 and c_2^2 would also be invariant and these are necessarily equal. Hence there exists only one invariant velocity to be denoted by c.

For convenience let us call anything which moves with a velocity c a signal. The velocity of a signal is independent of the direction in which it moves and must appear with the same value irrespective of the inertial frame used. This actually enables us to evaluate the function $a = f(v^2/c^2)$. From (2.35) we can derive the law of transformation of a velocity in the same way as we derived (2.5) from (2.1). We get for a velocity $\mathbf{u} = (u_x, u_y, u_z)$,

$$u_{x'} = \frac{au_x + av}{v^{-1}(a - a^{-1})u_x + a},$$

$$u_{y'} = \frac{u_y}{v^{-1}(a - a^{-1})u_x + a},$$

$$u_{z'} = \frac{u_z}{v^{-1}(a - a^{-1})u_x + a}. \tag{3.3}$$

Hence,
$$u_{x'}^2 + u_{y'}^2 + u_{z'}^2 = \frac{a^2u_x^2 + u_y^2 + u_z^2 + 2a^2u_x v + a^2v^2}{(v^{-1}(a - a^{-1})u_x + a)^2}. \tag{3.4}$$

In the case of a signal we must have

$$c^2 = c_{x'}^2 + c_{y'}^2 + c_{z'}^2 = c_x^2 + c_y^2 + c_z^2. \tag{3.5}$$

Consequently we get from (3.4) and (3.5):

$$c^2 = \frac{c^2 + (a^2 - 1)c_x^2 + 2a^3c_x v + a^2v^2}{(v^{-1}(a - a^{-1})c_x + a)^2}. \tag{3.6}$$

This yields

$$[c^2(a^2 - 1) - a^2v^2] + 2v^{-1}[c^2(a^2 - 1) - a^2v^2]c_x + \\ + (a^2 - 1)v^{-2}a^{-2}[c^2(a^2 - 1) - a^2v^2]c_x^2 = 0. \tag{3.7}$$

Since c_x is arbitrary and variable within a wide range each of the coefficients in the square brackets must vanish. They all yield the relation:

$$a(v) = \left(1-\frac{v^2}{c^2}\right)^{-\frac{1}{2}}. \tag{3.8}$$

With this value for $a(v)$, (2.35) becomes

$$\begin{aligned} x' &= ax \qquad\qquad +avt \\ y' &= \quad y \\ z' &= \qquad z \\ t' &= \frac{av}{c^2}x \qquad +at. \end{aligned} \tag{3.9}$$

We now see that the case $a(v) = 1$, which leads to a Galilei transformation, can be looked upon as a special case of the more general relation (3.8). If the signal velocity c happens to be infinite (3.9) will be identical with (3.1). Hence we can say quite generally that the principle of relativity points to the existence of signals which travel with the same velocity no matter in what inertial frame their motion is observed. From this principle we cannot derive the value of the velocity nor even decide whether it is finite or infinite. The classical transformation of Galilei implies that signals travel with infinite velocities. However, we now see that signals travelling with finite velocity c are equally compatible with the principle of relativity. The value of c may be found only through observations.

4. The Lorentz Transformation and the Relativity of Simultaneity

In the Appendix some experimental evidence is discussed which proves (as mentioned already) that light travels with an invariant velocity in all inertial frames of reference, so that the velocity of light has to be identified with the 'c' which follows from the principle of relativity. The best value of c at present is given by†

$$c = 2{\cdot}997929\times 10^5 \pm 0{\cdot}2 \text{ km/sec}, \tag{4.1}$$

which is, however, usually approximated to 3×10^5 km/sec.

It is usual to denote $a(v)$ by γ and vc^{-1} by β so that

$$a(v) = \gamma = \frac{1}{(1-\beta^2)^{\frac{1}{2}}}. \tag{4.2}$$

† E. Bergestrand, *Suppl. Nuovo Cimento*, No. 1, 1957, p. 224.

Eqns. (3.9) become

$$\begin{aligned} x' &= \gamma x && +\gamma vt \\ y' &= \quad y && \\ z' &= \quad z && \\ t' &= \frac{\gamma v}{c^2}x && +\gamma t. \end{aligned} \tag{4.3}$$

These equations express the Lorentz transformation which forms the basis of the special theory of relativity. As the name indicates, Lorentz was the first to derive them. However, his derivation was not based on the principle of relativity (in conjunction with the law of constancy of light velocity) but was obtained through attempting to derive rules of transformation which would make Maxwell's equations assume the same form both in a frame which is at rest relative to the ether (in which Lorentz believed) as well as in a frame which moves relative to it. Lorentz attempted to prove in this way that no experiment of an electromagnetic nature could demonstrate motion relative to the ether. To some extent he was opposed to the theory of relativity. Independently of Lorentz and of each other, H. Poincaré and A. Einstein also derived eqns. (4.3). Einstein was the first to approach the problem in a general way and on the basis of the principle of relativity. On the other hand, Poincaré's approach, like that of Lorentz, was based on the properties of Maxwell's equations. The derivation given here differs somewhat from that of Einstein. Einstein makes use of the invariance of light velocity right from the outset, thus using the relation

$$x^2+y^2+z^2+c^2t^2 = x'^2+y'^2+z'^2-c^2t'^2. \tag{4.4}$$

This equation expresses the fact that a spherical light wave which is emitted from the origin of K at $t = 0$ will also appear relative to K' as a spherical wave which spreads with the velocity c, the same as in K.

In this section we shall discuss some of the main features of the Lorentz transformation. Its most striking property concerns the concept of simultaneity. We have already seen that simultaneity has a definite physical meaning in each frame of reference. However, according to (4.3) this does not apply when interrelations between K and K' are considered. Events which are simultaneous in K are not simultaneous in K' and vice versa. This is caused by the term which makes t' vary along the x-axis for a given value of t, so that if two events at x_1 and x_2 occur at the same time t, they are registered as occurring at two different times relative to K' : $t'_1 = \gamma vc^{-2}x_1+\gamma t, t'_2 = \gamma vc^{-2}x_2+\gamma t$. Before the theory of relativity was first announced, it was accepted without criticism that simultaneity is absolute and two events are either simultaneous or not, no matter in what frame of reference they are observed. Einstein was the first to question this statement and pointed out that simultaneity, like any other physical property, must be subject to experimental test. Indeed, if one accepts the laws of constancy of light velocity and the principle of relativity as experimentally well founded, then one is necessarily led to the conclusion that simultaneity can have no absolute meaning for observers who move relative to each other. Einstein argued somewhat as follows: Suppose we take two rulers AB and $A'B'$ and mark their mid-points by M and M'. Let x_1, x_2 be the coordinates of A, B and x_1', x_2' those of A', B'. Let us now assume that $A'B'$ is made to move relative to AB with constant velocity v and that when x_1' passes near x_1 a certain event takes place at these points (which now coincide); and let us also assume that when x_2' coincides with x_2 another event takes place at these points. Suppose it turns out that two light signals which are emitted from x_1 and x_2 towards M when the events take place, collide at the mid-point M. Then K will be justified in concluding that the events were simultaneous. Apart from concluding that the events are simultaneous, K will also conclude that the length of the ruler $A'B'$ while moving, is equal to the length of AB and that when the events took place M' also coincided with M thus confirming that M' is the middle-point of $A'B'$. Let us now examine the events from the point of view of K'. Because of the law of constancy of light velocity K' can use the same light signals which were used by K. If he chooses to send his own signals this will make no difference because two light signals, which are emitted from two sources at coincidence, travel like one signal irrespective of the relative motion between the sources. Hence K' can use the above signals for finding out whether the events are simultaneous or not. If they collide at M', they are simultaneous, otherwise they are not. However, when M coincided with M' the signals had not yet collided at M and consequently when at some time later they did collide at M (this is the assumption which was made above) M' was no longer in coincidence with M and the signals therefore collided relative to K' at some point M'' which is not the mid-point of $A'B'$. K' will therefore conclude that the signals were not emitted at the same time. Hence the sending of the signals from A' and B' could not have been simultaneous with respect to K' if they were simultaneous with respect to K.

It is instructive to consider the above test also by using mechanical devices instead of light signals. Suppose we make use of guns which fire bullets towards M and M' when the events at A and A', B and B' occur (Fig. 2). In this case K and K' will have to use their own guns at rest in their respective systems, so that instead of the single pair of light signals as used before two pairs of bullets must be used now. The problem we have to solve is, will the bullets of K' collide at M', if we know that those of K collide at M, and if we also know that the bullets $m_{A'}$ and m_A were fired when A and A' coincided, and that m_B, $m_{B'}$ were fired when B and B' coincided? In this case, since we are dealing with four bullets (instead of two light signals) we cannot tell as before, whether $m_{A'}$ and $m_{B'}$ will collide at M' or whether it will be some other point, if we know that m_A and m_B collide at M. This can only be decided experimentally. In pre-relativistic physics it was assumed that if m_A and m_B collide at M, $m_{A'}$ and $m_{B'}$ will collide at M', and this result was justified theoretically as follows:

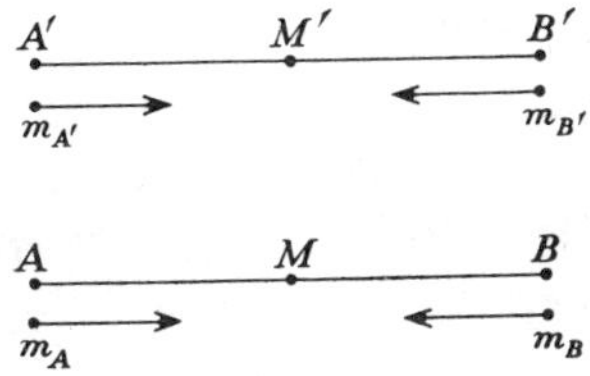

FIG. 2

From the point of view of K the point M' has at any time t a distance

$$\tfrac{1}{2}l+vt \quad \text{from } A$$

and

$$\tfrac{1}{2}l-vt \quad \text{from } B.$$

The bullets $m_{A'}$ and $m_{B'}$ have relative to K velocities $u+v$ and $u-v$. Consequently $m_{A'}$ reaches M' at a time t_1 which is a solution of

$$\tfrac{1}{2}l+vt_1 = (u+v)t_1, \tag{4.5}$$

and $m_{B'}$ reaches M' at a time t_2 which is a solution of

$$\tfrac{1}{2}l-vt_2 = (u-v)t_2. \tag{4.6}$$

From (4.5) and (4.6) we get

$$t_1 = t_2 = \frac{l}{2u}. \tag{4.7}$$

Hence, the bullets $m_{A'}$ and $m_{B'}$ collide at M' and the firing of the bullets is simultaneous both in K and K'. The fallacy of this argument lies in the assumed law of composition of velocities ($u+v$ and $u-v$). This in itself must be checked experimentally, a task which is not easier than the original one of determining where the bullets collide. If the classical law of composition holds, simultaneity is absolute. The theory of relativity establishes, however, a different law of composition of velocities which is in harmony with the law of invariance of the velocity of light, and in this way simultaneity loses its absolute significance.

The relativistic law of composition of velocities is directly obtainable from (4.3):

$$dx' = \gamma\,dx + \gamma v\,dt = \gamma(dx/dt + v)\,dt,$$

$$dt' = (\gamma v/c^2)\,dx + \gamma\,dt = \gamma\{(v/c^2)\,dx/dt + 1\}\,dt. \tag{4.8}$$

Consequently, if $u_x = dx/dt$, $u_{x'} = dx'/dt'$, we get

$$u_{x'} = \frac{u_x + v}{1 + u_x v/c^2}. \tag{4.9 a}$$

In the same way we also get

$$u_{y'} = \frac{u_y}{\gamma(1 + u_x v/c^2)}, \tag{4.9 b}$$

$$u_{z'} = \frac{u_z}{\gamma(1 + u_x v/c^2)}. \tag{4.9 c}$$

As will be seen, the classical law is obtained from the new law by making $c = \infty$. The classical law in itself does not contradict the principle of relativity, but it is based on an invariant velocity of infinite value.

From (4.9) we confirm that c is an invariant. Let us assume that a light signal travels along the x-axis ($c_x = c$). We then have

$$c_{x'} = \frac{c_x + v}{1 + c_x v/c^2} = \frac{c_x + v}{1 + v/c_x} = c_x. \tag{4.10}$$

Similarly, if it travels along the y-axis ($c_x = c_z = 0$, $c_y = c$),

$$c_{x'} = v,$$

$$c_{y'} = \frac{c}{\gamma},$$

$$c_{x'}^2 + c_{y'}^2 = \frac{c^2}{\gamma^2} + v^2 = c^2\left(\frac{1}{\gamma^2} + \frac{v^2}{c^2}\right) = c^2\left(1 - \frac{v^2}{c^2} + \frac{v^2}{c^2}\right) = c^2, \tag{4.11}$$

so that the magnitude of the velocity is again c.

Ordinary mechanical velocities are too small when compared with c and cannot therefore provide a suitable test of the relativistic law of composition. An experimental verification is obtained, however, through Fizeau's experiment, in which essentially the velocity of light in moving water is compared with the velocity in water at rest. If the velocity in water at rest is c/n where n is the refractive index, and the water is moved with a velocity v, we get for the combined velocity v_d

$$v_d = \frac{c/n + v}{1 + vc/c^2 n} \cong \frac{c}{n} + v\left(1 - \frac{1}{n^2}\right), \tag{4.12}$$

where the approximation is sufficiently accurate for small v/c. In the interferometer arrangement used by Fizeau, Fresnel's coefficient of drag $(1-n^{-2})$ has been accurately verified.

Composition of velocities or transformation of a velocity from one inertial frame to another, must be distinguished from addition of velocities which are all defined relative to a given frame. Thus, if relative to K two bodies move along the x-axis with velocities v_1 and v_2 respectively, then relative to K the velocity between these two bodies is $v_1 \pm v_2$ depending on whether they move in opposite directions or in the same direction. In particular, relative to K the velocity between two light signals which travel in opposite directions is $2c$. Each of the light signals covers in one second a distance c and the distance between the light signals after one second is therefore $2c$.

Another problem is involved if we wish to evaluate the velocity between two frames, say K_a and K_b, with reference to each other, when the velocities of K_a and K_b are given relative to some other frame of reference K (or K', K'', etc.). Let these velocities be $\mathbf{u}_a$ and $\mathbf{u}_b$ ($\mathbf{u}'_a$, $\mathbf{u}'_b$, etc.). Whether measured from K_a or K_b, the relative velocity between them has the same magnitude, while the components of the relative velocity as referred to K_a will be obtainable from those referred to K_b by a space rotation which depends on the mutual orientation between K_a and K_b. Since the magnitude of the velocity between K_a and K_b as measured in these systems cannot depend on some other system which is moving relative to them, we must always get the same value independently of the frame of reference which we may be using to describe the motion of K_a and K_b. Let the dependence of the relative velocity v between K_a and K_b on $\mathbf{u}_a$ and $\mathbf{u}_b$ be given by a function f, then we must have

$$v = f(\mathbf{u}_a, \mathbf{u}_b) = f(\mathbf{u}'_a, \mathbf{u}'_b), \quad \text{etc.} \tag{4.13}$$

so that $f(\mathbf{u}_a, \mathbf{u}_b)$ is an invariant. Under Galilean transformation the function f is given by

$$v^2 = (\mathbf{u}_a - \mathbf{u}_b)^2 \tag{4.14}$$

and the invariance with regard to Galilean transformation is directly evident. Under Lorentz transformations f is a more complicated function. There will be no loss of generality involved if we assume that K_a, K_b, and K have parallel axes. Let $\mathbf{x}_a$ be a radius vector describing the position of the origin of K_a relative to K and similarly for $\mathbf{x}_b$. Then

$$\mathbf{u}_a = \frac{d\mathbf{x}_a}{dt}, \qquad \mathbf{u}_b = \frac{d\mathbf{x}_b}{dt}. \tag{4.15}$$

Since v is independent of any particular choice of frame of reference, we now introduce one, K', which transforms one of the systems, say K_b, to rest. In this frame of reference, the velocities of K_a and K_b are given by

$$\mathbf{u}'_a = \frac{d\mathbf{x}'_a}{dt'}, \qquad \mathbf{u}'_b = \frac{d\mathbf{x}'_b}{dt'} = 0. \tag{4.16}$$

$\mathbf{u}'_a$ is identical with the relative velocity between K_a and K_b as referred to K_a which we are seeking. $(\mathbf{x}'_a, t')$ is obtainable from $(\mathbf{x}_a, t)$ by a Lorentz transformation from K to K' with a relative velocity $-\mathbf{u}_b$ (the one which transformed K_b to rest). Hence, all we have to do is to derive $\mathbf{u}'_a$ from $\mathbf{u}_a$, when $(\mathbf{x}'_a, t')$ is derived from $(\mathbf{x}_a, t)$ through a Lorentz transformation. For simplicity assume $\mathbf{u}_b = (u_b, 0, 0)$. Then with $\gamma = \gamma(u_b)$ we get, similarly to (4.8) and (4.9),

$$u_{ax'} = \frac{u_{ax}-u_{bx}}{1-u_{ax}u_{bx}/c^2}, \qquad u_{ay'} = \frac{u_{ay}}{(1-u_{ax}u_{bx}/c^2)\gamma},$$

$$u_{az'} = \frac{u_{az}}{(1-u_{ax}u_{bx}/c^2)\gamma}. \tag{4.17}$$

From these components we can obtain $u^2_{ax'}+u^2_{ay'}+u^2_{az'} = v^2$ which is the square of the relative velocity between K_a and K_b. Using the general Lorentz transformation (11.6), p. 50 (with $R = 1$, for parallel systems), the following general result is obtained:

$$v^2 = \frac{(\mathbf{u}_a-\mathbf{u}_b)^2-c^{-2}[\mathbf{u}_a, \mathbf{u}_b]^2}{(1-\mathbf{u}_a\,\mathbf{u}_b/c^2)^2}. \tag{4.18}$$

As can be seen, the Galilean case follows if we let c tend to infinity. Also if we let $\mathbf{u}_a = \mathbf{c}$ and $\mathbf{u}_b = -\mathbf{c}$, we get $v = c$, so that the velocity with which one light signal can travel relative to another, when they travel in opposite direction, as viewed from the signal, is c.

5. The Behaviour of Measuring Rods and Clocks in Motion

Consider a rod of length l_0 as measured in K', where we assume the rod to be at rest and lying along the x-axis. Let x'_A and x'_B be the coordinates of the ends A and B of the rod in K'. If K wishes to measure the length of this rod while in motion, he will have to observe the positions of A and B at the same time, say t. Let x_A and x_B be these positions. From (4.3) we get

$$x'_A = \gamma x_A+\gamma vt,$$
$$x'_B = \gamma x_B+\gamma vt. \tag{5.1}$$

Consequently,
$$x'_A-x'_B = \gamma(x_A-x_B). \tag{5.2}$$

Since $x'_A-x'_B$ is equal to the length of the rod at rest in K', and x_A-x_B is its length while in motion, we get the following relation between these quantities:
$$l = \gamma^{-1}l_0 = l_0(1-\beta^2)^{\frac{1}{2}}. \tag{5.3}$$

Thus we find that when a rod moves relative to an inertial frame of reference in the direction of its length it appears shorter. This effect can only be observed while the rod is in motion. When the rod is brought to rest it appears again with its rest length.

Next let us consider a moving clock. In this case it will be more convenient to assume the clock at rest in K. If K' wishes to compare this clock with his own clocks he must use at least two clocks separated by some distance. Let us assume that when the clock of K passes the first of the clocks of K', the indications of these clocks are t_1 and t'_1 respectively, and when the second clock of K' is passed the indications are t_2 and t'_2. All these events, which consist of comparisons of clocks, take place relative to K at the same point x. We have from the fourth of eqns. (4.3)

$$t'_1 = \frac{\gamma v}{c^2}x+\gamma t_1, \qquad t'_2 = \frac{\gamma v}{c^2}x+\gamma t_2. \tag{5.4}$$

Consequently,

$$t'_2-t'_1 = \gamma(t_2-t_1). \tag{5.5}$$

Let us denote t_2-t_1 by T_0 and $t'_2-t'_1$ by T. T_0 is a time period as indicated by a clock at rest while T corresponds to this period as indicated by clocks relative to which the first clock is moving. We get

$$T_0 = \gamma^{-1}T = T(1-\beta^2)^{\frac{1}{2}}. \tag{5.6}$$

This equation shows that the indication T of a clock which moves relative to other clocks appears as if the clock is slow ($T_0 < T$). The hands which indicate the time appear to move with a smaller angular velocity if a clock is moved and its readings compared with a set of stationary clocks. If the clock is brought to rest it regains its normal indication of flow of time, but it will now bear a fixed delay when compared with the clocks which were at rest. This delay has accumulated during the period in which the clock was slowed down, and cannot be made good automatically. Unlike a rod a clock is an integrating device which memorizes all the delays it underwent in the course of its motion.

It may be argued that at the moment of deceleration of the clock an effect might take place which would make it regain the time lost and thus bring it into agreement with the stationary clocks. However, the lag in time which accumulated during the uniform motion is proportional to the length of the distance travelled and can thus be made large for long distances. It is not possible for various time-lags to be compensated by the same deceleration. Consequently, if acceleration could have an additional effect on clocks it is entirely independent of the one obtained from the Lorentz transformation. However, there exists no experimental evidence for acceleration having any effect on the intrinsic properties of a clock. Thus, when a hydrogen atom, which may be looked upon as a clock, is falling acceleratedly towards the sun, the evidence is that it yields at each moment a Doppler effect which is accounted for by its instantaneous velocity and by assuming that it is identical with a hydrogen atom on earth (after correcting for the effect of the sun's and earth's gravitational fields on a hydrogen atom at the respective positions). We therefore assume that acceleration as such has no effect on clocks.

Suppose now we move a clock from A to B with a velocity u which is not necessarily constant. Let the clock indicate the correct time at A, when compared with the local clock. Owing to the motion to B every time element dT_0 as indicated by the moving clock is related to the corresponding time element dT in the inertial frame of reference K by

$$dT_0 = dT(1-u^2/c^2)^{\frac{1}{2}}. \tag{5.7}$$

On integrating from T_{01} to T_{02} we get

$$\int_{T_{01}}^{T_{02}} dT_0 = \int_{T_1}^{T_2} dT(1-u^2/c^2)^{\frac{1}{2}}, \tag{5.8}$$

or

$$T_{02}-T_{01} = \int_{T_1}^{T_2} dT(1-u^2/c^2)^{\frac{1}{2}}. \tag{5.9}$$

Because of this relation we have always

$$T_{02}-T_{01} < T_2-T_1. \tag{5.10}$$

In particular, if u is constant,

$$T_{02}-T_{01} = (1-u^2/c^2)^{\frac{1}{2}}(T_2-T_1). \tag{5.11}$$

This property of a clock to get slowed down when in motion, makes it impossible to define the synchronism of clocks by means of an auxiliary clock which is transported from one place to another, without corrections. If we send two synchronizing clocks with velocities v_1 and v_2, on arrival at their destination they will indicate different times. We could use their indications only by extrapolating to a velocity $v_1 = v_2 = 0$.

In deriving (5.9) we found it convenient to assume the clock to be at rest in K. Since the Lorentz transformation is symmetrical in K and K', we can also obtain the same result by assuming the clock to be at rest in K'; only in this case, because the Lorentz transformation (4.3) happens to be given in terms of x, y, z, t on the right-hand side, the calculation is slightly more lengthy. Since the clock is now always at a point x', we get for two different times t_1 and t_2

$$x' = \gamma x_1 + \gamma v t_1,$$

$$x' = \gamma x_2 + \gamma v t_2. \tag{5.12}$$

Consequently, $$(x_2 - x_1) = -v(t_2 - t_1). \tag{5.13}$$

From the fourth of eqns. (4.3) we get

$$t'_1 = (\gamma v/c^2)x_1 + \gamma t_1,$$

$$t'_2 = (\gamma v/c^2)x_2 + \gamma t_2. \tag{5.14}$$

Hence

$$t'_2 - t'_1 = (\gamma v/c^2)(x_2 - x_1) + \gamma(t_2 - t_1) = -(\gamma v^2/c^2)(t_2 - t_1) + \gamma(t_2 - t_1)$$

$$= \gamma(1 - v^2/c^2)(t_2 - t_1). \tag{5.15}$$

Or, as before, $$T_0 = \gamma^{-1} T. \tag{5.16}$$

6. The Clock Paradox

Suppose a clock C is moved with a velocity u from A to B (Fig. 3). Let us assume that at A, C was synchronized with a clock C_A and let the journey from A to B last $\frac{1}{2}T_C$ seconds as indicated by the moving clock. On arrival at B the clock C is compared with a clock C_B which is synchronous with C_A.† From the Lorentz transformation it follows, as we have seen in the previous section, that the local clock will indicate a more advanced time,

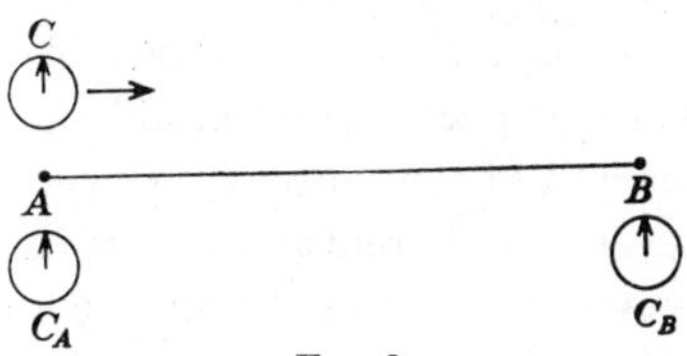

Fig. 3

$$T_B = \tfrac{1}{2}T_C(1 - u^2/c^2)^{-\frac{1}{2}}, \tag{6.1}$$

so that $$T_B > \tfrac{1}{2}T_C. \tag{6.2}$$

Let us now reverse the motion of C and bring it back to A, with the same velocity as before. Altogether the journey will have lasted T_C seconds (as indicated by C) and the total delay produced is obtained from the formula

$$T_A = T_C(1 - u^2/c^2)^{-\frac{1}{2}}, \tag{6.3}$$

so that $$T_A > T_C. \tag{6.4}$$

It has been argued occasionally that this deviation between C and C_A contradicts the principle of relativity and proves that the theory is not self-consistent. Suppose we attach a frame of reference to C.

† It is assumed that the initial quick acceleration until the velocity u is attained and the final quick deceleration are produced by some external force. For instance the clock C can be electrically charged and manipulated by other charged bodies.

In this frame C_A appears to move at first to the left and then back to the right until it meets C again. Applying the Lorentz transformation in the same way as before, we should now find

$$T_C > T_A \tag{6.5}$$

and this contradicts our previous result.

In reality the argument which leads to (6.5) is invalid. In our first experiment, when C is moved, C_A and C_B are resting in an inertial system K and we can apply the Lorentz transformation with regard to a clock C which moves with a velocity u relative to K. Actually, as already pointed out, it is immaterial whether C moves with a constant or a variable velocity u. If the latter is the case, each time-element dt_0 of C will appear as $dt = \gamma_u dt_0$ in K, and altogether the delay which will be found when C is brought back to A and compared with C_A will be obtained from

$$T_A = \int\limits_0^{T_C} \frac{dt_0}{(1-u^2/c^2)^{\frac{1}{2}}} = \int\limits_0^{T_C} \gamma_u \, dt_0. \tag{6.6}$$

On the other hand, the frame K' attached to C is not inertial. Whether moved with a variable velocity all the way or with a velocity which is necessarily variable at the beginning, at the reversal, and at the end of the journey, in every case K' is not inertial and we cannot apply the Lorentz transformation for an observer in K'. For this reason (6.5) does not hold. K and K' are not equivalent in our present case and the special principle of relativity is not applicable. From the principle of relativity we can only expect that if C_A rests in an inertial frame of reference and C is moved away and back, we shall find that C is lagging behind C_A, while if C is left at rest in an inertial frame of reference and C_A is moved away and back, we shall find that C_A is lagging behind C. It may be added, that in the general theory of relativity where all frames of reference are equivalent, the slowing of a clock which has been moved relative to a clock at rest in an inertial frame is accounted for in both cases, i.e. whether a frame K or K' is used. In the latter case there appear in K' inertial forces which are equivalent to gravitational forces.† Because of these forces, C_A moves in a non-uniform way (it 'falls' in the gravitational field which exists in K') while clock C has to be supported in a fixed position in K' by a suitable force. At some stage of the motion both clocks are in the

† Apart from these, the external (for instance electrical) forces which are required to cause the relative acceleration between C and C_A, C_B are also present as before. The experiment is identical with the former and we have only changed the frame of reference.

same neighbourhood in the gravitational potential and are affected by it in the same way. At another stage, when the reversal of motion takes place, the clocks are apart and are subject to different gravitational potentials, their difference being proportional to the distance between the clocks. The result of this is that the clock C_A is made to advance in its time indication at such a rate that when it is brought back to C, the latter is found to be retarded when compared with C_A.

7. The Invariant Interval

Let a spherical light wave be emitted at x_0, y_0, z_0 at time t_0. Independently of the motion of the source the wave front relative to an inertial frame will then satisfy the equation

$$|\mathbf{R}-\mathbf{R}_0| = c(t-t_0), \tag{7.1}$$

where $\mathbf{R}_0$ is the radius vector x_0, y_0, z_0, $\mathbf{R}$ the radius vector of any point on the wave front, and c is the velocity of light. Eqn. (7.1) may also be expressed as follows:

$$(x-x_0)^2+(y-y_0)^2+(z-z_0)^2 = c^2(t-t_0)^2. \tag{7.2}$$

From the principle of relativity it then follows that an identical equation will also hold in any other inertial frame K':

$$(x'-x_0')^2+(y'-y_0')^2+(z'-z_0')^2 = c^2(t'-t_0')^2. \tag{7.3}$$

Although it is hardly necessary to reaffirm that by using a Lorentz transformation (7.3) follows from (7.2) (or vice versa) it will be useful to perform this calculation. It will be sufficient to prove that this holds exactly for an infinitesimal progress of the wave, i.e. for the following differential form:

$$dx^2+dy^2+dz^2-c^2\,dt^2 = 0. \tag{7.4}$$

From the inverse of (4.3) we have

$$\begin{aligned} dx &= \gamma\,dx'-\gamma v\,dt', \\ dy &= dy', \\ dz &= dz', \\ dt &= -(\gamma v/c^2)dx'+\gamma\,dt'. \end{aligned} \tag{7.5}$$

Hence

$$\begin{aligned} dx^2+dy^2+dz^2-c^2\,dt^2 &= \gamma^2\,dx'^2+\gamma^2v^2\,dt'^2-2\gamma^2v\,dx'\,dt'+ \\ &\quad +dy'^2+dz'^2-(\gamma^2v^2/c^2)\,dx'^2+ \\ &\quad +2\gamma^2v\,dx'\,dt'-c^2\gamma^2\,dt'^2 \\ &= dx'^2+dy'^2+dz'^2-c^2\,dt'^2. \end{aligned} \tag{7.6}$$

Indeed, as we have already pointed out, Einstein derived the Lorentz transformation by applying the invariance of the law of propagation of light and the principle of relativity.†

The calculation which leads to (7.6) confirms not only the invariance of the law of propagation of light (the vanishing of the quadratic form) but also the invariance of the expression

$$ds^2 = dx^2+dy^2+dz^2-c^2\,dt^2. \tag{7.7}$$

Let two events in K be separated in space and time by dx, dy, dz, dt. Then ds is called the interval between these events. Hence from the invariance of ds it follows that a given pair of events are separated by the same interval no matter in which inertial frame they are observed.

The invariance of ds can also be proved in a more direct way from the principle of relativity and the law of propagation of light, viz. if ds is to vanish in K, ds' must vanish in K'. Hence, the following relation must hold:

$$ds = a(v)\,ds', \tag{7.8}$$

where $a(v)$ can only depend on the relative velocity between K and K'. From the isotropy of space, it follows that $a(v)$ must be an even function of v:

$$a(v) = a(-v). \tag{7.9}$$

Also, from the principle of relativity,

$$ds' = a(-v)\,ds. \tag{7.10}$$

Consequently $a(v) = \pm 1$. The negative value can be ruled out, because for $v = 0$, $a = 1$. Hence $ds = ds'$.

It may be added that we cannot apply a similar argument to the relation

$$dl' = (1-\beta^2)^{\frac{1}{2}}\,dl. \tag{7.11}$$

In this equation dl refers to the spatial distance between two events which K considers simultaneous (a simultaneous observation of the ends of a rod at rest), while dl' refers to a spatial distance between two events which are simultaneous in K'. The first pair of events is different from the second pair of events. The transformation of Lorentz leads to the relation (7.11) and all we can expect from the principle of relativity in this case is that a similar relation should hold if the rod is at rest in K',

$$dl = (1-\beta^2)^{\frac{1}{2}}\,dl'. \tag{7.12}$$

However, to avoid confusion between these two relations which refer

† From the invariance of this law, i.e. from $dx^2+dy^2+dz^2-c^2\,dt^2 = 0$ in all inertial frames of reference, it is not yet possible to derive the Lorentz transformation completely. Some additional arguments, all based on the symmetry between K and K', such as we have used in deriving the Lorentz transformation, had to be used by Einstein.

to two different physical experiments, we will do better to express them as follows:

$$dl_{k'} = (1-\beta^2)^{\frac{1}{2}}\, dl_{0k}, \tag{7.13}$$

$$dl_{k} = (1-\beta^2)^{\frac{1}{2}}\, dl_{0k'}. \tag{7.14}$$

In this way both equations express the same thing, namely, that a moving rod appears shorter when observed from a system relative to which the rod is moving. Even if dl_{0k} and $dl_{0k'}$ should be numerically equal (as is the case if the same rod is used throughout) we cannot eliminate them to get the relation $(1-\beta^2)^{\frac{1}{2}} = 1$, as appears possible if we use equations (7.11) and (7.12) on their face value.

An interval ds between two events can be either real, zero, or imaginary, depending on ds^2 being positive, zero, or negative:

$$c^2dt^2 \lesseqgtr dx^2+dy^2+dz^2. \tag{7.15}$$

In the first case the interval is called space-like, in the second case a null-interval, and in the third case time-like. From the invariance of ds it follows that the character of an interval is independent of the frame of reference.

In the case of a time-like interval the events are separated by such a length of time that the distance $c\,dt$ which a ray of light would travel in the time dt is greater than the space distance between the considered events. It follows that when the interval is time-like it is always possible to find a system of coordinates in which the events appear in the same place. This system will have to move with a velocity which is not greater than that of light relative to the original frame of reference in which the events happen to appear in separate places. The converse is also true: if the events appear in the same place, relative to some system, the interval is time-like relative to any other system. Since two events which are separated by a time-like interval can always be transformed to space coincidence, their essential separation is that of time. The time separation is smallest in a frame of reference in which events appear in the same place.

In a space-like interval, $dx^2+dy^2+dz^2 > c^2\,dt^2$. In this case the distance in space is such that even a ray of light cannot cover it in the available time dt. Any pair of events which are simultaneous relative to some frame of reference ($dt = 0$) are separated by a space-like interval. Conversely, it is always possible to find a frame of reference in which two events separated by a space-like interval appear simultaneous. For if we assume that the smallest possible separation in time is not zero but $dt_0 \neq 0$, we shall have $dx^2+dy^2+dz^2-c^2\,dt_0^2 > 0$.

Relative to this system the events occur at two points, say A and B, separated by a distance $(dx^2+dy^2+dz^2)^{\frac{1}{2}} = dl_{AB}$, whereby we assume for convenience that the event at B occurs after the event at A by dt_0 seconds. Now if we consider a system of coordinates which moves from A to B, the events in this system will take place in a smaller space separation. Consequently dt_0' will have to be smaller than dt_0 if ds is to retain its value. In the case of a time-like interval this process can be continued until the space distance is zero, and in the case of a space-like interval until the time separation is zero.

A null-interval is present if two events are separated in space and time in such a way that a ray of light starting its journey from the place and time of the first event can just reach the second event.

The invariance of ds is characteristic of all Lorentz transformations, also when the motion is not along the x-axis. In deriving (4.3) we made use of the assumption that the yz-plane is perpendicular to the x-axis, but we have not made use of the assumption that the y- and z-axes are perpendicular to each other. As the motion proceeds along the x-axis not only is ds an invariant but also dy and dz. However, in the more general case the motion is not along the x-axis, but it can be obtained from a combination of suitable rotations of axes in K and K' and the special Lorentz transformation which we have considered so far.

8. The Proper Time

Consider a clock C which moves relative to an inertial frame of reference K. If the momentary velocity is u, the relation between $d\tau$ as indicated by the moving clock and the corresponding time element dt as observed through the clocks of K is given by

$$d\tau = dt(1-u^2/c^2)^{\frac{1}{2}}. \tag{8.1}$$

Let the same clock C also be observed from another inertial frame K', which moves with constant velocity relative to K. Relative to this frame we shall have again

$$d\tau = dt'(1-u'^2/c^2)^{\frac{1}{2}}, \tag{8.2}$$

where u' is the velocity of the clock relative to K'. From eqns. (8.1) and (8.2) we conclude that $dt(1-u^2/c^2)^{\frac{1}{2}}$ has an invariant value. It always yields the time element $d\tau$ as indicated by a given clock C. No matter in which frame of reference we observe the corresponding dt, provided this is multiplied by $(1-u^2/c^2)^{\frac{1}{2}}$ we always get the same value $d\tau$. This quantity is called the proper time of the clock. Since the proper time $d\tau$ as indicated by the two positions of the hands of a clock cannot depend on some system of coordinates which moves relative to the clock, $d\tau$ must indeed have an invariant value. If the clock is moved for some time with a variable velocity u, at each time element we can apply (8.1). Adding up we thus get

$$\int_{\tau_1}^{\tau_2} d\tau = \int_{t_1}^{t_2} dt(1-u^2/c^2)^{\frac{1}{2}}, \tag{8.3}$$

or

$$\tau_2-\tau_1 = \int_{t_1}^{t_2} dt(1-u^2/c^2)^{\frac{1}{2}}. \tag{8.4}$$

The integral on the right-hand side again has an invariant value. In this relation u need not be constant either in direction or value and it represents the magnitude of the velocity.

The invariance of $dt(1-u^2/c^2)^{\frac{1}{2}}$ can also be confirmed by direct transformation from K to K'. Suppose K observes the clock (or any particle) moving with a velocity u between x and $x+dx$. Let K' observe the same clock (or particle) in the corresponding times and positions. We have to prove that from the Lorentz transformation it follows that

$$dt(1-u^2/c^2)^{\frac{1}{2}} = dt'(1-u'^2/c^2)^{\frac{1}{2}}. \tag{8.5}$$

The time element dt' is connected with dt according to the formula

$$dt' = \gamma\, dt+(\gamma v/c^2)\, dx = \gamma\{1+(v/c^2)(dx/dt)\}\, dt = \gamma(1+vu/c^2)\, dt. \tag{8.6}$$

In this equation v is the relative velocity between K and K'. (Note that we are not using a formula $dt' = \gamma\, dt$ which is applicable only if the clock or particle is at rest relative to K, $dx = dy = dz = 0$.) On the other hand, the transformation formula for a velocity along the x-axis is given by

$$u' = \frac{u+v}{1+uv/c^2}. \tag{8.7}$$

Inserting the values of dt' and u' according to (8.6) and (8.7) in the right-hand side of (8.5) it is found that it is identically equal to the left-hand side. In this proof we had to assume that the clock moves along the x-axis. If this is not the case, K and K' can use auxiliary frames which are properly inclined relative to the original frames. Any time observation in such auxiliary frames cannot differ from those of the original frames because the inclination is fixed.

A more direct, though formal, proof of (8.5) is obtained as follows. Since

$$c^2\, dt^2-(dx^2+dy^2+dz^2) = c^2\, dt'^2-(dx'^2+dy'^2+dz'^2), \tag{8.8}$$

we get

$$c^2\,dt^2\left(1-\frac{dx^2+dy^2+dz^2}{c^2\,dt^2}\right) = c^2\,dt'^2\left(1-\frac{dx'^2+dy'^2+dz'^2}{c^2\,dt'^2}\right). \tag{8.9}$$

Now $(dx^2+dy^2+dz^2)\,dt^{-2}$ is interpreted in K as a square of some velocity u from a point x, y, z to $x+dx$, $y+dy$, $z+dz$ in time dt. The same applies to K'. Consequently,

$$(1-u^2/c^2)\,dt^2 = (1-u'^2/c^2)\,dt'^2, \tag{8.10}$$

in agreement with (8.5). We now also see that

$$c^2\,d\tau^2 \equiv -ds^2. \tag{8.11}$$

The proper time is a useful concept in the formulation of relativistic dynamics as we shall see in Chapter 5.

9. A Geometrical Interpretation of the Lorentz Transformation

It is possible to illustrate the Lorentz transformation geometrically if we consider a four-dimensional space in which each point corresponds to an event x, y, z, t and vice versa. Since the transformation is linear and homogeneous the simplest geometrical interpretation is obtained if we look upon x, y, z, t as Cartesian coordinates. However, we have never assumed or laid down any condition which is equivalent to making the t-axis perpendicular to the other axes. In fact we shall see that this is in general impossible and that we must introduce a general Cartesian system.

Since we cannot draw a four-dimensional space over a two-dimensional plane, we shall ignore in the following the y- and z-axes, which at any rate do not change in the particular transformation (4.3), and assume the 'world' to be two-dimensional, its events being characterized by two coordinates x and t. In order to give the relativistic transformation a geometrical interpretation we then proceed as follows. We draw an arbitrary Cartesian system x, t and prescribe along the axes two gauges of geometrical lengths g_1 and g_2, one to represent a unit of distance, the other a unit of time (Fig. 4). Ignoring for a moment the actual relationship between the coordinates of K and K' we can construct for the latter, in a similar way as for K, a pair of axes, x', t' with two gauges g_1' and g_2'. Since we are interested primarily in homogeneous transformations we let K and K' have a common origin (see Fig. 5). In fact if we consider a continuous set of Cartesian systems with a common origin each with its own angle and pair of gauges, we obtain under certain conditions of continuity two gauge curves G_1 and

G_2 which are characteristic of some set of transformations. Given these curves with two parameters along them and the common origin, we can construct each system of axes by joining the points on G_1 and G_2 which belong to the same parameter value, to the origin O. We shall see in the following that the Lorentz transformation fixes two gauge curves in the two-dimensional world which enable us to determine the Cartesian (x, t) sets of all the possible frames of reference, and that in

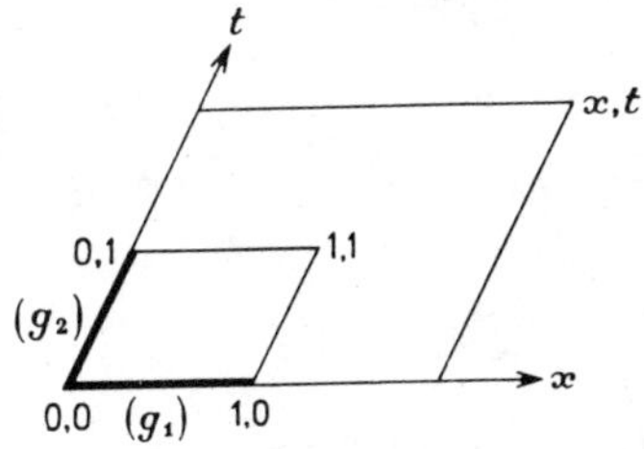

FIG. 4

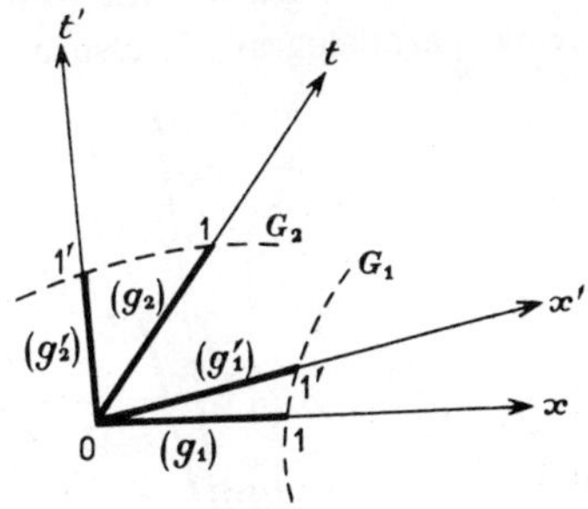

FIG. 5

this geometrical realization of the Lorentz transformation all its properties are reflected.†

In order to determine the Cartesian axes of K', once those of K have been arbitrarily selected, we proceed as follows. At first we look upon the transformation (4.3) not as a change from one set of axes into another but as a change from an arbitrary world point x, t into a second point x', t'. Each world point thus changes in a definite way into another world point. In this process a parallelogram changes into some other parallelogram (because the transformation is affine). Then we are going to look upon the parallelogram which is obtained from the fundamental parallelogram of K as the one which provides the new axes and new gauges of K'.

It will be convenient to introduce the variable $l = ct$. l has thus the dimension of a length and for simplicity we shall assume that the gauges of K along the x- and l-axes are equal. Eqns. (4.3) become:

$$x' = \gamma x + \gamma(v/c)l, \qquad l' = (\gamma v/c)x + \gamma l. \tag{9.1}$$

As the origin of the system of coordinates remains unchanged, we need only ask ourselves what happens to the points $P(1, 0)$ and $P(0, 1)$

† For rotations of orthogonal axes without change of gauge the gauge curves are two overlapping circles which merge into one.

which K' is going to call $P(1', 0')$ and $P(0', 1')$. From the transformation equations we get

$$x' = \gamma.1+\gamma(v/c).0 = \gamma,$$
$$l' = \gamma(v/c).1+\gamma.0 = \gamma v/c. \tag{9.2}$$

Hence $P(1', 0')$ has relative to x, t the coordinates γ and $\gamma v/c$. Similarly, we find that $P(0', 1')$ has the coordinates $\gamma v/c$ and γ. From this result we conclude that the points $P(1', 0')$ and $P(0', 1')$ are situated symmetrically with respect to the diagonal $OP(1, 1)$. Hence the new parallelogram is also a rhombus, the diagonal of which lies

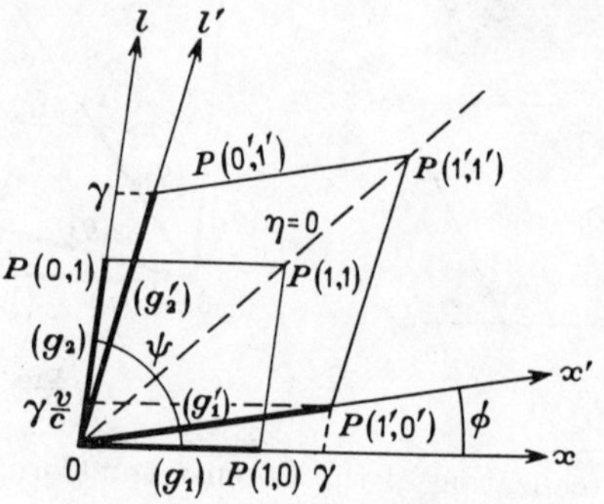

FIG. 6

on the same line as $OP(1, 1)$. This line is marked in Fig. 6 by $\eta = 0$, and it is common to all frames of reference. The gauges of any other observer, like those of K, will be equal in length along the x- and l-axes. As the geometrical length of the segment $O\gamma$ is γg, and the length of the segment $\gamma P(1', 0')$ is $(\gamma v/c)g$, and as we are going to look upon the length of $OP(1', 0')$ as equal to the gauge g' of K', we get from the figure the following relation between g and g':

$$\gamma^2 g^2 + \frac{\gamma^2 v^2}{c^2} g^2 + 2\gamma^2 \frac{v}{c} g^2 \cos\psi = g'^2 \tag{9.3}$$

or

$$\frac{g'^2}{g^2} = \gamma^2\left(1+\frac{2v}{c}\cos\psi+\frac{v^2}{c^2}\right). \tag{9.4}$$

Similarly, for the angle of contraction ϕ (see Fig. 6) we get

$$\frac{\sin\phi}{\sin(\psi-\phi)} = \frac{v}{c}. \tag{9.5}$$

Among the different possible observers there will exist one who is associated with an orthogonal pair of axes, and it will be convenient to refer to him as the original K. This in no way attributes to this particular observer any exceptional properties, but if we choose an orthogonal set for the first observer we simplify the calculation. For $\psi = 90°$, we thus get

$$\frac{g'^2}{g^2} = \gamma^2\left(1+\frac{v^2}{c^2}\right) = \frac{1+\beta^2}{1-\beta^2}, \tag{9.6}$$

$$\tan\phi = (v/c) = \beta. \tag{9.7}$$

Hence,
$$\frac{g'^2}{g^2} = \frac{1+\tan^2\phi}{1-\tan^2\phi} = \frac{1}{\cos 2\phi}. \tag{9.8}$$

From this it follows that the areas of the parallelograms are equal (Fig. 7). For if

$$2g'^2 \sin\tfrac{1}{2}\psi' \cos\tfrac{1}{2}\psi' = 2g^2 \sin\tfrac{1}{2}\psi \cos\tfrac{1}{2}\psi, \tag{9.9}$$

we get
$$\frac{g'^2}{g^2} = \frac{\sin\psi}{\sin\psi'}, \tag{9.10}$$

and in our case where $\psi = 90°$ and $\psi' = 90°-2\phi$, we confirm (9.8). The equality of the areas results from the determinant of the Lorentz transformation being equal to 1.

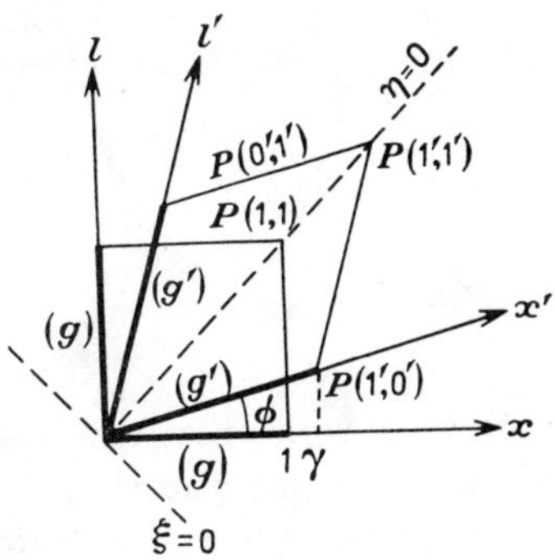

FIG. 7

Eqn. (9.8) is nothing but a parameter repesentation of the gauge curves. We shall now prove that the tip of g' regarded as a radius vector, moves along a hyperbola as we change $\phi = \tan^{-1}(v/c)$ (i.e. as we change the relative velocity between K' and K).

Let $g = 1$, so that the geometrical coordinates of $P(1', 0)$ relative to K are $x = \gamma$ and $l = \gamma(v/c) = \gamma\beta$. Then, using $\beta = l\gamma^{-1}$ and (9.8),

$$g'^2 = \frac{1+\beta^2}{1-\beta^2} = \frac{1+l^2/\gamma^2}{1-l^2/\gamma^2} = \frac{\gamma^2+l^2}{\gamma^2-l^2} = \frac{x^2+l^2}{x^2-l^2}. \tag{9.11}$$

On the other hand, from Fig. 7,

$$g'^2 = x^2+l^2. \tag{9.12}$$

From (9.11) and (9.12) we get

$$x^2-l^2 = 1. \tag{9.13}$$

This is the equation of a hyperbola. With

$$x-l = \eta, \qquad x+l = \xi, \tag{9.14}$$

it takes the form
$$\xi\eta = 1. \tag{9.15}$$

The lines $x = -l$ (i.e. $\xi = 0$) and $x = l$ (i.e. $\eta = 0$) represent the asymptotes of the hyperbola (Fig. 8). In a similar way the point $P(0', 1')$ lies on a hyperbola $x^2 - l^2 = -1$. Thus we find that the gauge lines consist of two hyperbolas, each having two branches. Since

$$x^2 - l^2 = x^2 - c^2t^2 = \pm 1,$$

the gauge lines represent all the events which belong to an interval of length ± 1 with respect to the origin $x = l = 0$. Because of the invariance of the interval s, any point on a gauge line moves along this

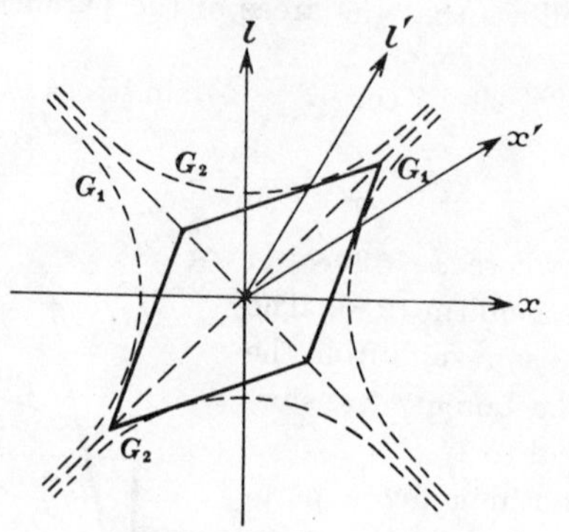

FIG. 8

curve when a Lorentz transformation is performed. This applies not only to the hyperbolas $x^2 - l^2 = \pm 1$ but also to any other hyperbola $x^2 - l^2 =$ const, which represents points of other value of s.† We also see that, through a Lorentz transformation, any straight line which passes through the origin changes into another straight line through the origin, and that pairs of lines which are symmetrical with respect to the line $\eta = 0$ change into lines which are again symmetrical in this respect.

If we look upon $ds^2 = dx^2 - dl^2$ as the square of a distance in the x, l plane, it defines a non-Euclidean metric, where in particular the distance from the origin is given by $x^2 - l^2$ and not by $x^2 + l^2$ as in the Euclidean case. We encounter positive, zero, and negative ds^2. From this point of view we can call the hyperbolas 'circles' because all the points on these curves have the same 'distance' from the origin. The Lorentz transformation can thus be looked upon as a rotation in the non-Euclidean plane in which the square of the distance is given by

† $y^2 + z^2$ is an invariant in our present case as we consider only a motion along the x-axes. Strictly speaking s^2 is equal to the invariant $x^2 + y^2 + z^2 - c^2t^2$, but to avoid using additional symbols we denote $x^2 - c^2t^2$, which is also an invariant in our present case, by s^2.

$ds^2 = dx^2 - dl^2$. The distance along each line which passes through the origin is equal to the geometrical distance when expressed in terms of the gauge along the considered line. The distance along the asymptotes is zero because along these lines the gauge is infinite (null lines). This follows also from $x^2 - l^2 = 0$. The geometrical interpretation of the Lorentz transformation as described above was first given by H. Minkowski. It is possible to introduce a Euclidean metric and make the Lorentz transformation look like an ordinary rotation, where a point moves on a circle (in a world x, y, t on the surface of a sphere, and in a world x, y, z, t on a hyper-spherical surface) by the following simple device. Instead of dealing with the variable $l = ct$, we use the variable $\lambda = ict$ (and $\lambda' = ict'$). The invariant interval is then given by

$$ds^2 = dx^2 + d\lambda^2 \tag{9.16}$$

and has thus the shape of a (distance)^2 in terms of Cartesian orthogonal coordinates in a Euclidean space. The invariance of this form (i.e. $dx^2 + d\lambda^2 = dx'^2 + d\lambda'^2$) then simply means that the axes are rotated when a Lorentz transformation is performed. In using the imaginary variable ict the geometrical interpretation becomes highly abstract and we have to rely more on analogies in algebraic relations. In terms of λ the Lorentz transformation takes the form

$$x' = \gamma x + (\gamma v/ic)\lambda,$$

$$\lambda' = -(\gamma v/ic)x + \gamma\lambda. \tag{9.17}$$

Let

$$\gamma = \cos\Phi, \tag{9.18}$$

$$\gamma v/ic = -\sin\Phi. \tag{9.19}$$

Then the relation $\cos^2\Phi + \sin^2\Phi = 1$ is satisfied:

$$\gamma^2 - \gamma^2 v^2/c^2 = \gamma^2(1 - v^2/c^2) = 1. \tag{9.20}$$

Hence, we can write the Lorentz transformation as

$$x' = \cos\Phi\, x - \sin\Phi\, \lambda, \qquad \lambda' = \sin\Phi\, x + \cos\Phi\, \lambda, \tag{9.21}$$

and it is directly evident that we are dealing here with a rotation by an 'angle' Φ. This angle is, however, always imaginary, because γ is real and always greater than 1.† From (9.18) and (9.19) we get

$$\tan\Phi = i(v/c). \tag{9.22}$$

Let

$$\Phi = i\Psi, \tag{9.23}$$

† That we are dealing here with an imaginary and not a complex angle can be seen from the expansion $\cos\Phi = 1 - \frac{1}{2}\Phi^2 + (1/4!)\Phi^4 - \dots$. As $\cos\Phi = \gamma$ is real the angle Φ can either be real or purely imaginary. Alternatively, this can be derived from

$$2\cos\Phi = \exp(i\Phi) + \exp(-i\Phi).$$

so that Ψ is a real angle. Then since

$$\cos\Phi = \cos i\Psi = \cosh\Psi, \qquad \sin\Phi = \sin i\Psi = i\sinh\Psi, \tag{9.24}$$

we get from (9.21) (using $l = -i\lambda$, $l' = -i\lambda'$)

$$x' = \cosh\Psi . x + \sinh\Psi . l, \qquad l' = \sinh\Psi . x + \cosh\Psi . l. \tag{9.25}$$

In this way we have regained the Lorentz transformation in real form which shows, however, more directly than earlier its nature as a hyperbolic rotation. The 'angle' Ψ can be shown to be equal to the area between the new x'-axis, the hyperbola G, and the old x-axis (Fig. 9).

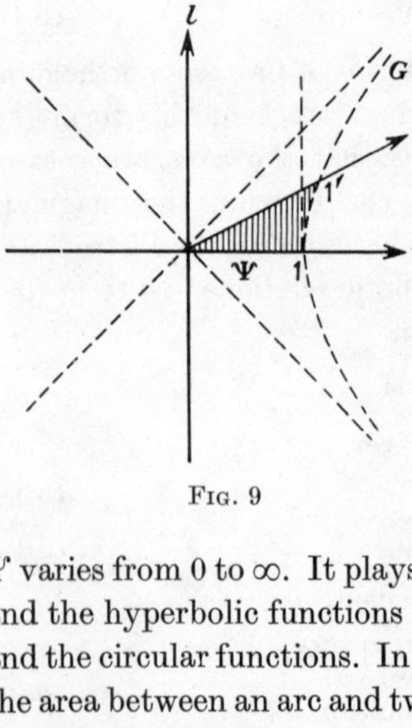

FIG. 9

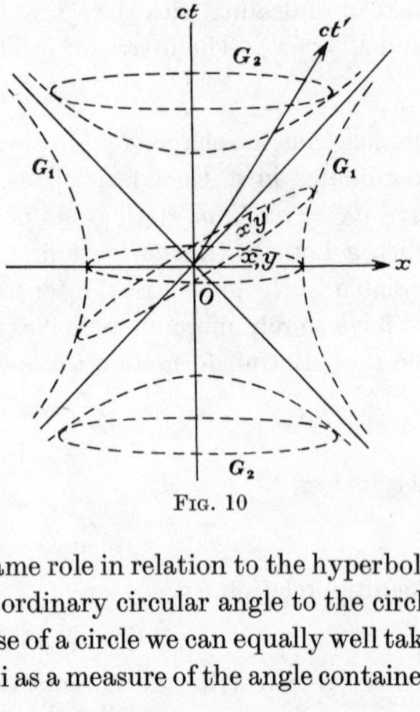

FIG. 10

Ψ varies from 0 to ∞. It plays the same role in relation to the hyperbola and the hyperbolic functions as the ordinary circular angle to the circle and the circular functions. In the case of a circle we can equally well take the area between an arc and two radii as a measure of the angle contained by them.

So far we have dealt with a one-dimensional space x and a two-dimensional world of events, x, t. It is also easy to visualize geometrically the Lorentz transformation when the space is two-dimensional (x, y) and the world of events three-dimensional. The hyperbola G_1 now becomes a connected hyperbolic gauge surface and the hyperbola G_2 a hyperboloid of two separated sheets. Instead of two light-lines we get two light-cones with a common apex at O (Fig. 10). Any plane through O which intersects the hyperbolic surface represents the x, y plane in some frame of reference. The radii of the ellipse of intersection represent the gauges for distance in the corresponding directions.

Hence, in each frame of reference the points on the periphery of the corresponding ellipse represent events which are simultaneous with the event at the origin, and consequently also among themselves, and of unit distance from O. In each frame the corresponding ellipse therefore appears as a circle which satisfies the equation $x^2+y^2 = 1$.

The time-axis ct points in such direction, that, together with the long axis of the ellipse, it forms an angle which is bisected by the light-cone. The relative velocity between K and K' depends on this angle only. If K' is changed into K^* by rotating only the space axes from x', y' to x^*, y^*, the t'-axis remains unchanged. x^*, y^* are, like x', y', a pair of conjugate axes of the ellipse.

Both in K and K' the respective ellipses yield the equations $x^2+y^2 = 1$ and $x'^2+y'^2 = 1$. This must not be taken to mean that x^2+y^2 is an invariant. These relations refer to different sets of events. The first of these relations refers to certain events in the x, y plane for which $t = 0$, while the second refers to events in the x', y' plane for which $t' = 0$. Only $x^2+y^2-c^2t^2$ is an invariant. If $t = 0$, we get for the interval of the events on the ellipse relative to the event at the origin $x^2+y^2 = 1$. Relative to K' the interval of these events is

$$x'^2+y'^2-c^2t'^2 = 1,$$

where t' varies from one point of the ellipse to the other. A similar statement applies to the events in K', which yield for $t' = 0$.

$$x'^2+y'^2 = 1.$$

When dealing with the events x, y, z, t we require a four-dimensional space and can then speak only in terms of analogies. The ellipses in each frame of reference which provide us with the gauges now become ellipsoids. Their equations are in each frame those of spherical surfaces $x^2+y^2+z^2 = 1$. The light-'cones' are three-dimensional spaces satisfying the equation $x^2+y^2+z^2-c^2t^2 = 0$, while the gauge surfaces are three-dimensional spaces satisfying the relation $x^2+y^2+z^2-c^2t^2 = \pm 1$.

Let us return to the two-dimensional case. With the help of the real x, ct plane we can confirm all the properties of the Lorentz transformation and give a convenient geometrical interpretation to physical events. Thus each material point will be associated with a world line which enables us to read the position of the point along the x-axis as a function of time. The different points on this line represent the different events associated with the material point, and as such have objective or absolute reality. Once we select a pair of axes x, l for one observer K, the world lines are all fixed in the plane and are entirely independent of the frame of reference. In other words we cannot associate with each frame a different world line, all belonging to the same particle. The position is rather such that the same world line with all its points is viewed from different frames of reference. A world point has absolute significance. Two events which coincide at x, y, z, t will do so no matter from which frame of reference this is viewed. On the other hand spatial coincidence or temporal coincidence (simultaneity) have only relative meaning and vary from one observer to another. Had an event not

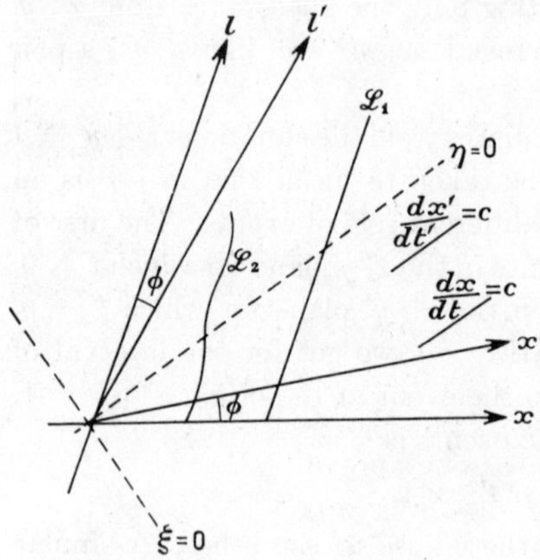

FIG. 11

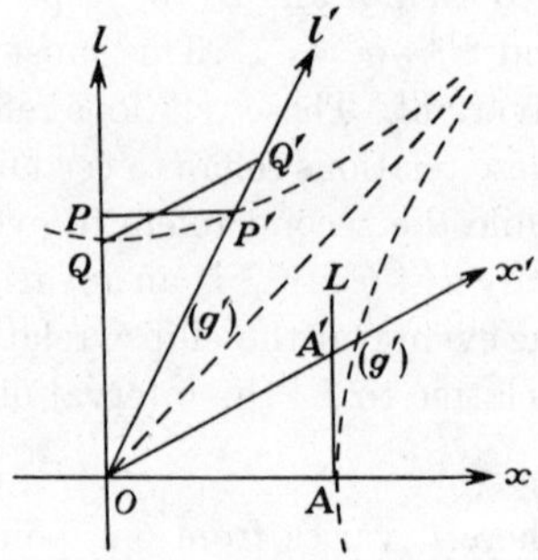

FIG. 12

possessed absolute significance there could be no question of transforming its coordinates from one frame to another.

As an example let us assume that it is known that relative to a system K a material point is at rest. Then the world line will be parallel to the l-axis (L_1 in Fig. 11). This world line with respect to K' will yield the result that the material point moves relative to K' with a velocity v. A point at rest relative to K' will be associated with a world line which is parallel to the l'-axis. The world line of a pulse of light is given by a line which is parallel to any one of the lines marked $\xi = 0$ and $\eta = 0$. These lines are common to all the frames of reference so that in each one of them the pulse is moving with a velocity c ($\xi = 0$ means $x = ct$ or $x' = ct'$, etc.), see (9.14). Since a material point cannot travel with a velocity which is greater than that of light it follows that all world lines of material points are directed at all points more steeply than a light-line.

To confirm the contraction of length when observed from a moving system, let a ruler $OA = 1$ rest in K. The world lines of the end points O and A are thus the lines Ol and L (see Fig. 12). The spatial distance relative to K' between any pair of simultaneous events which belong to these world lines is given by intersects which are parallel to the x'-axes, for instance OA'. As is seen from the figure, $OA' < g'$, so that the rod appears to K' to be shorter than a unit of length. In the same way let a clock rest at the origin of K'. The world line is thus Ol'. Two events separated by a unit of time ($3^{-1} \times 10^{-10}$ sec) in the history of this clock are given by the origin O and the point P'. The time distance between these events according to K is obtained by drawing parallels to the x-axis. These intersect the l-axis at O and P. As can be seen from the figure K maintains that more than one unit of time has elapsed so that the moving clock appears to him as going slow. Of course this will also appear to be the case for K' with regard to a clock which is at rest in K. The world line of this clock is Ol. The time distance of the different events which belong to this world line are obtained according to K' by drawing parallels to the x'-axis. If we take two events which are separated by a unit of time in K, say the events at O and Q, the parallels to the x'-axis will intersect the l'-axis at O and Q'. As is seen from Fig. 12, $OQ' > g'$, and K' maintains that the clock which rests in K goes slow. In the same way we can also prove that a rod at rest in K' appears shorter to an observer in K.

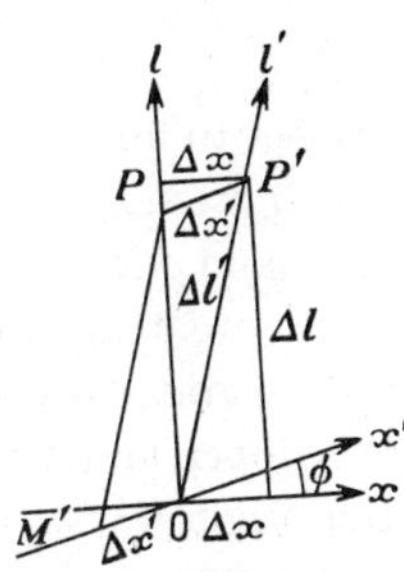

FIG. 13

We can also confirm that the relationship between the Cartesian axes of K and K' is such that they both obtain the same relative velocity v with respect to each other. For simplicity let K be associated with orthogonal axes x, l and let the axes of K' be inclined, say inwards, by an angle ϕ (Fig. 13). The world line of a point at the origin of K' and at rest relative to it proceeds along Ol'. Relative to K a change of position Δx has taken place in a time $\Delta l/c$. Hence the velocity is

$$v = \frac{\Delta x}{\Delta t} = c \tan \phi. \tag{9.26}$$

On the other hand, a point at rest relative to K, say at the origin, has a world line along Ol. Relative to K' this point appears to move in the negative direction of the x'-axis. If we consider $c^{-1}\Delta l'$ seconds where $\Delta l'$ is given by OP', the point will have moved a distance which is equal to OM'. Consequently

$$\frac{\Delta x}{\Delta x'} = \frac{\Delta l}{\Delta l'}, \tag{9.27}$$

and we find again
$$v' = \frac{\Delta x'}{\Delta t'} = c\tan\phi. \tag{9.28}$$

(In this calculation we do not use explicitly the gauges because they are equal for the x- and l-axes as well as the x'- and l'-axes respectively and cancel when calculating the ratios $\Delta x/\Delta l$ or $\Delta x'/\Delta l'$.)

In the geometrical interpretation of the Lorentz transformation we now obtain also a very convenient method of subdividing all the events into those which are time-like and those which are space-like, with

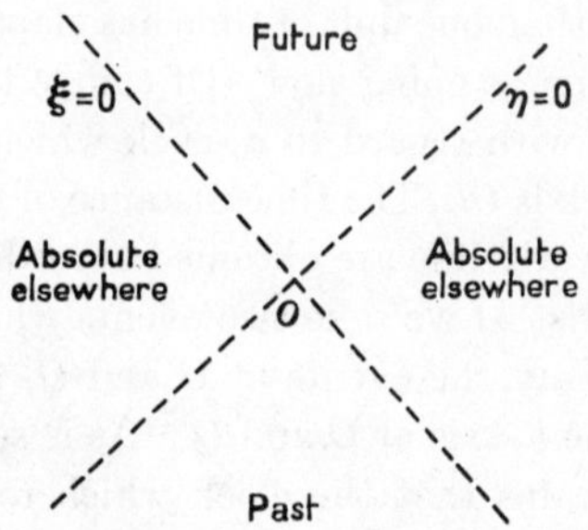

FIG. 14

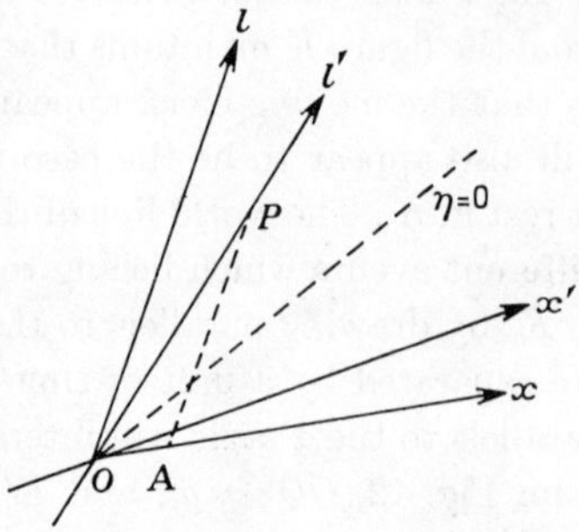

FIG. 15

respect to an event at the origin. We draw the lines $\xi = 0$ and $\eta = 0$ which are necessarily perpendicular to each other (Fig. 14). This fixes once and for all the light-lines (or light-'cones' in the general case) which are common to all observers. All the events along the light-lines are separated by a null-interval from the origin. Events lying inside the upper and lower cones are separated by time-like intervals from the origin ($c^2\,dt^2 > dx^2$). In the upper cone we have events which lie to the future of the event at the origin, the absolute future. The lower cone represents events which preceded the event at the origin, they are the absolute past. We are entitled to use the term absolute, because in all frames of reference the past with respect to the origin precedes the future. This follows from the fact that all l-axes can point only in the direction of the upper cone so that each l-axis is divided by the origin into positive and negative halves. It also follows that any event which is separated from the event at the origin by a time-like interval can be 'transformed to rest', i.e. there exists a system of coordinates in which the two events occur at the same place. We have only to join the two world points by a line and use it as an l-axis and draw the x-axis symmetrically with respect to the light-line $\eta = 0$ and the required frame of reference is obtained (Fig. 15).

The two side cones which form a connected single region represent all the events which are separated from the event at the origin by a space-like interval. Such events can never appear as occurring at the same place, but they can always be transformed to simultaneity. The system of coordinates in question is obtained by taking the x-axis along OP, and the l-axis symmetrically with respect to $\eta = 0$. The events at O and P will then appear simultaneous (Fig. 16).

For such events, past and future have no absolute significance. In

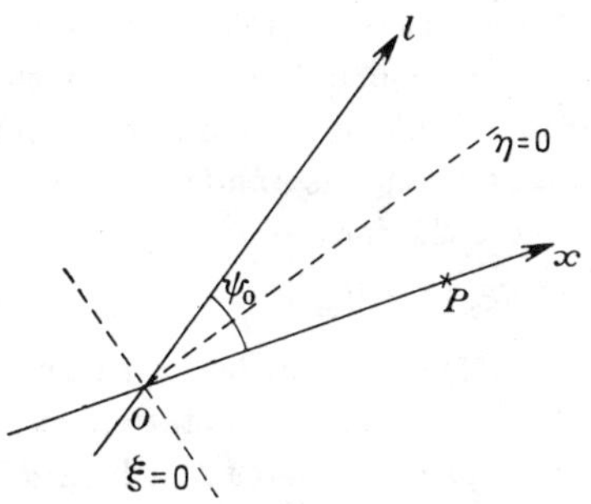

FIG. 16

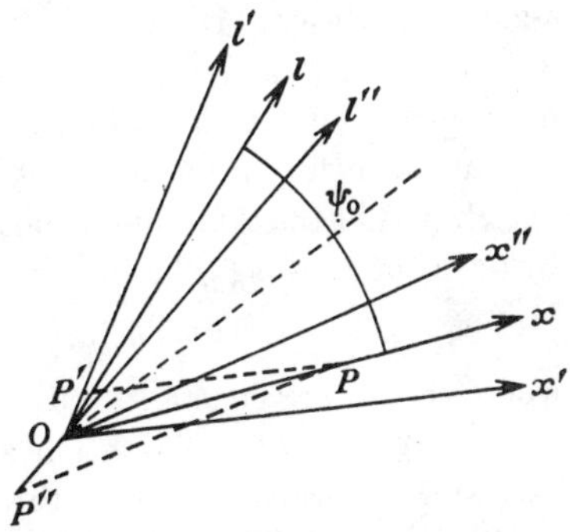

FIG. 17

all frames of reference which are more closely aligned to the light-lines, the event at O appears as following in time the event at P, whereas to all observers which are associated with x-axes of an angle ψ greater than ψ_0 (the one which corresponds with simultaneity) the event at P will appear as following in time the event at O.

Since the concepts of past and future have no absolute meaning for events which are separated by a space-like interval, we must consider the fate of the law of causality. For suppose that relative to K the event at P appears after the event at O and that it is assumed that the event at P was caused by the event at O (Fig. 17). Relative to another observer it may happen that the event at P will occur earlier than the event at O, so that the effect preceded the cause. In reality no difficulty arises for the law of causality. Events which are separated by a space-like interval are separated to such an extent in time that even a ray of light dispatched from O (or P) cannot reach P (or O) in time before the second event takes place. Now light travels with the greatest possible velocity, hence no action could be transmitted from O to cause the event at P. Events separated by a space-like interval cannot be interrelated through cause and effect and the difficulty for the law of causality does not arise.

We conclude this section with a remark concerning the orthogonality of the x-, λ-axes. As mentioned earlier, in deriving the Lorentz transformation we have made at no stage of the derivation an assumption which would fix the angle between the x- and ct-axes should these be given a geometrical interpretation. In fact we have seen that any angle ψ between 0 and 180° is taken up. It may therefore seem surprising that when we use the imaginary variable $\lambda = ict$ we find that the x-, λ-axes are always orthogonal. This is due to the very special way in which the imaginary variable is introduced. It is possible to introduce more general Cartesian axes, say ξ_1, ξ_2, provided the nature of the Lorentz transformation as a rotation of the plane is maintained, but ξ_1, ξ_2 have not the same simple physical significance as x and $t = -ic^{-1}\lambda$. In terms of ξ_1 and ξ_2 the interval will have the form

$$ds^2 = g_{11}\,d\xi_1^2+2g_{12}\,d\xi_1\,d\xi_2+g_{22}\,d\xi_2^2, \tag{9.29}$$

where the g_{ik} are constants with a value depending on the transformation $(x,\lambda) \rightarrow (\xi_1, \xi_2)$. To obtain a simple physical interpretation, it is necessary to transform back to a system of coordinates in which ds^2 has the normal form $dx^2+d\lambda^2$, where x is real and λ is imaginary, or dx^2-dl^2, when x and l are real. In the general theory of relativity the use of the general coordinates ξ_1, ξ_2 is essential and the g_{ik} are not constants but functions of ξ_1 and ξ_2 which depend on the gravitational field. There it is possible to regain the normal form only in a sufficiently small region of space-time where the gravitational field is eliminated by letting a frame of reference of x-, y-, z-axes and clocks fall freely in the gravitational field. In the special theory we can use the orthogonal x, λ for any finite region and the introduction of more general coordinates ξ_1, ξ_2 is not required; it would only make the formalism look more complicated without adding any advantage.

10. The Lorentz Group

Suppose we perform a transformation from K to K' where the relative velocity is v_1 and from K' to K'' where the relative velocity is v_2. The result of these transformations must be equivalent to a single transformation from K to K'' with a certain velocity v which is given by

$$v = \frac{v_1+v_2}{1+v_1\,v_2/c^2}. \tag{10.1}$$

Moreover, for each Lorentz transformation from K to K' there also exists the inverse transformation from K' to K. The identical transformation K to K is also a Lorentz transformation. Consequently we can say that all the mathematical conditions for the presence of a group of transformations are fulfilled. The Lorentz transformations for all possible relative velocities form a group.

It is interesting to go back to the transformation eqns. (2.35) and apply the group property to them, because in this way it is possible to give another proof of the relation $a(v) = (1-v^2/c^2)^{-\frac{1}{2}}$, and the law of composition of velocities. Let

$$\begin{aligned} x' &= a_1 x + a_1 v_1 t \\ y' &= y \\ z' &= z \\ t' &= v_1^{-1}(a_1 - a_1^{-1})x + a_1 t \end{aligned} \tag{10.2}$$

and

$$\begin{aligned} x'' &= a_2 x' + a_2 v_2 t' \\ y'' &= y' \\ z'' &= z' \\ t'' &= v_2^{-1}(a_2 - a_2^{-1})x' + a_2 t' \end{aligned} \tag{10.3}$$

where $a_1 = a(v_1)$ and $a_2 = a(v_2)$. These two transformations are equivalent to a single transformation from K to K'', which can be obtained by eliminating x', y', z', t' from (10.2) and (10.3). We get

$$\begin{aligned} x'' &= \{a_1 a_2 + a_2 v_2 v_1^{-1}(a_1 - a_1^{-1})\}x + a_1 a_2(v_1+v_2)t \\ y'' &= y \\ z'' &= z \\ t'' &= \{a_1 v_2^{-1}(a_2 - a_2^{-1}) + a_2 v_1^{-1}(a_1 - a_1^{-1})\}x + \{a_1 a_2 + a_1 v_1 v_2^{-1}(a_2 - a_2^{-1})\}t. \end{aligned} \tag{10.4}$$

Now from the principle of relativity the resulting transformation must have the same form as (10.2) or (10.3). In particular, the coefficient of x in the first of the equations (10.4) and the coefficient of t in the fourth of the equations must be equal. Consequently the following relation must hold:

$$a_1 a_2 + \frac{a_2 v_2}{v_1}(a_1 - a_1^{-1}) = a_1 a_2 + \frac{a_1 v_1}{v_2}(a_2 - a_2^{-1}) \tag{10.5}$$

or

$$v_1^{-2}(1 - a_1^{-2}) = v_2^{-2}(1 - a_2^{-2}). \tag{10.6}$$

In this equation we have on the left-hand side a function of v_1 and on the right-hand side a function of v_2, where v_1 and v_2 are independent variables. Therefore such an equation can hold only if both sides are equal to a universal constant. Denoting this constant by c^{-2} where c has the dimensions of a velocity, we get

$$a(v) = (1-v^2/c^2)^{-\frac{1}{2}}, \tag{10.7}$$

which confirms our previous result.

In terms of the Lorentz transformations the passage from K to K' and then to K'' is expressed as follows:

$$\begin{aligned} x' &= \gamma_1 x + \gamma_1 v_1 t \\ y' &= y \\ z' &= z \\ t' &= (\gamma_1 v_1/c^2)x + \gamma_1 t \end{aligned} \tag{10.8}$$

$$\begin{aligned} x'' &= \gamma_2 x' + \gamma_2 v_2 t' \\ y'' &= y' \\ z'' &= z' \\ t'' &= (\gamma_2 v_2/c^2)x' + \gamma_2 t'. \end{aligned} \tag{10.9}$$

The direct transformation from K to K'' which is obtained from (10.8) and (10.9) is given by

$$\begin{aligned} x'' &= \gamma_1\gamma_2(1+v_1 v_2/c^2)x + \gamma_1\gamma_2(v_1+v_2)t \\ y'' &= y \\ z'' &= z \\ t'' &= \gamma_1\gamma_2(v_1+v_2)c^{-2}x + \gamma_1\gamma_2(1+v_1 v_2/c^2)t. \end{aligned} \tag{10.10}$$

On the other hand, expressed in terms of the relative velocity v between K and K'', the transformation must have the form

$$\begin{aligned} x'' &= \gamma(v)x + \gamma(v)vt \\ y'' &= y \\ z'' &= z \\ t'' &= \gamma(v)vc^{-2}x + \gamma(v)t. \end{aligned} \tag{10.11}$$

On comparing coefficients between (10.10) and (10.11) we get

$$\gamma_1\gamma_2(1+v_1 v_2 c^{-2}) = \gamma(v), \tag{10.12}$$

$$\gamma_1\gamma_2(v_1+v_2) = \gamma(v)v. \tag{10.13}$$

Dividing (10.13) by (10.12) we get

$$v = \frac{v_1+v_2}{1+v_1 v_2/c^2}. \tag{10.14}$$

After some algebra it is also possible to confirm (10.7) $\{a(v) = \gamma(v)\}$.

The Lorentz group that we have encountered so far contains only one parameter, the relative velocity v. Had we allowed for any inclination between the space axes and for a velocity in an arbitrary direction we should have obtained a much larger group, the elements of which require six parameters for their specification. Three parameters, α, β, γ, may be said to arise from the rotation of the space axes of K' relative to those of K, and three other parameters from the components v_x, v_y, v_z which describe the motion of K' relative to K.

The sub-group which we have considered so far is commutative. If $L(v_1)$ denotes the matrix of one Lorentz transformation and $L(v_2)$ the matrix of a second Lorentz transformation, then

$$L(v_1)L(v_2) = L(v_2)L(v_1) = L\left[\frac{v_1+v_2}{1+v_1 v_2/c^2}\right]. \tag{10.15}$$

The full Lorentz group is not commutative, because rotations of axes are not commutative. If R_1 denotes the matrix of one rotation of space axes and R_2 the matrix of a second rotation, then in general

$$R_1 R_2 \neq R_2 R_1.$$

In other words, the order in which the two rotations are performed is relevant.

The result of two transformations is obtained by multiplying the matrices of each of the transformations with each other. For instance, for the sub-group which we have considered so far:

$$L(v_1)L(v_2) = \begin{bmatrix} \gamma(v_1) & 0 & 0 & \gamma(v_1)v_1 \\ 0 & 1 & 0 & 0 \\ 0 & 0 & 1 & 0 \\ c^{-2}\gamma(v_1)v_1 & 0 & 0 & \gamma(v_1) \end{bmatrix} \begin{bmatrix} \gamma(v_2) & 0 & 0 & \gamma(v_2)v_2 \\ 0 & 1 & 0 & 0 \\ 0 & 0 & 1 & 0 \\ c^{-2}\gamma(v_2)v_2 & 0 & 0 & \gamma(v_2) \end{bmatrix}$$

$$= \begin{bmatrix} \gamma(v_1)\gamma(v_2)(1+v_1 v_2/c_2) & 0 & 0 & \gamma(v_1)\gamma(v_2)(v_1+v_2) \\ 0 & 1 & 0 & 0 \\ 0 & 0 & 1 & 0 \\ c^{-2}\gamma(v_1)\gamma(v_2)(v_1+v_2) & 0 & 0 & \gamma(v_1)\gamma(v_2)(1+v_1 v_2/c^2) \end{bmatrix}. \tag{10.16}$$

Using (10.14) this yields $L(v)$. In the general case we shall write for $L(\mathbf{v}, \alpha, \beta, \gamma)$

$$L(v_x, v_y, v_z, \alpha, \beta, \gamma) = \begin{bmatrix} L_{11} & L_{12} & L_{13} & L_{14} \\ L_{21} & L_{22} & L_{23} & L_{24} \\ L_{31} & L_{32} & L_{33} & L_{34} \\ L_{41} & L_{42} & L_{43} & L_{44} \end{bmatrix}. \tag{10.17}$$

The characteristic of all Lorentz transformations is that they leave the interval $ds^2 = dx^2+dy^2+dz^2-c^2\,dt^2$ unchanged in value. In terms of x, y, z, ict they represent, to begin with, rotations of axes in the four-dimensional world of Minkowski. However, the mere invariance of ds^2 yields a group which is wider than the one we have had in mind so far and which is called the restricted group. Thus, the transformation $x' = x$, $y' = y$, $z' = z$, $t' = -t$ leaves ds^2 unchanged, but it interchanges past and future. If this is to be avoided we must insist upon $L_{44} > 0$ in all elements of the group. $L_{44} < 0$ will always involve time inversion. Another transformation which is not included in the restricted Lorentz group is the space inversion:

$$x' = -x, \quad y' = -y, \quad z' = -z, \quad t' = t.$$

This transformation takes place if K is a right-handed system and K' left-handed (or vice versa). To avoid such reflection we must add the condition that the determinant $\det L$ of the transformation matrix should be equal to $+1$. (The determinant of $x' = -x$, $y' = -y$, $z' = -z$, $t' = t$ is -1.) We have already proved when dealing with the special subgroup which refers to a motion along the x-axis that the determinant is $+1$ if the axes point in the same direction in both K and K'. (It would have turned out to be equal to -1 if the x- and x'-axes had pointed in opposite directions.) This property cannot change if the direction of motion between K and K' is changed or through rotation of K relative to K'. The proof that for any Lorentz transformation the determinant can be only ± 1 proceeds as follows. From

$$dx'_\kappa = \sum_{\mu=1}^{4} L_{\kappa\mu}\, dx_\mu = \sum_{\nu=1}^{4} L_{\kappa\nu}\, dx_\nu \quad (\kappa = 1, 2, 3, 4) \tag{10.18}$$

and

$$\sum_{\kappa=1}^{4} dx'^2_\kappa = \sum_{\mu=1}^{4} dx^2_\mu, \tag{10.19}$$

where $x_1 = x$, $x_2 = y$, $x_3 = z$, $x_4 = ict$, we get

$$\sum_\kappa \Big[\Big(\sum_\mu L_{\kappa\mu}\, dx_\mu\Big)\Big(\sum_\nu L_{\kappa\nu}\, dx_\nu\Big)\Big] = \sum dx^2_\mu. \tag{10.20}$$

Consequently,

$$\sum_\kappa L_{\kappa\mu} L_{\kappa\nu} = \delta_{\mu\nu}, \tag{10.21}$$

where $\delta_{\mu\nu} = 1$ for $\mu = \nu$ and 0 for $\mu \neq \nu$. Denoting the transposed matrix of L by $\tilde{L}$, so that $L_{\kappa\mu} = \tilde{L}_{\mu\kappa}$, (10.21) gives

$$\sum \tilde{L}_{\mu\kappa} L_{\kappa\nu} = \delta_{\mu\nu}. \tag{10.22}$$

Since the summation index κ is now consecutive, we may express (10.22) in matrix form as follows:

$$\tilde{L}L = 1. \tag{10.23}$$

The determinant of a matrix is not affected by the interchange of columns and rows. Consequently $\det \tilde{L} = \det L$, and (10.23) yields

$$\det L = \pm 1. \tag{10.24}$$

It follows further that any continuous set of Lorentz matrices L can have a determinant which is either $+1$ or -1, and since the identical transformation has a determinant $+1$ the restricted group can consist only of matrices with $\det L = +1$.

Lorentz transformations with a determinant -1 do not form a group, because the product of any two matrices yields a matrix with a determinant $+1$.

From (10.23) it also follows that a Lorentz transformation can be characterized by the relation

$$\tilde{L} = L^{-1}. \tag{10.25}$$

The relations (10.21) are known as the orthogonality relations. They are always satisfied when a rotation of axes takes place, so that one orthogonal set of axes changes into another such set.

When Lorentz transformations are taken to be rotations in the Minkowski world, in other words if we make use of the coordinate $x_4 = ict$, the coefficients L_{14}, L_{24}, L_{34} and L_{41}, L_{42}, L_{43} are purely imaginary, while all the other elements of the matrix $L_{\mu\nu}$ are real. (See (9.17); the argument which led to this statement is easily generalized to Lorentz transformations involving all four coordinates.) Hence, from (10.21) we get

$$L_{14}^2 + L_{24}^2 + L_{34}^2 + L_{44}^2 = 1, \tag{10.26}$$

$$L_{44}^2 = 1 - (L_{14}^2 + L_{24}^2 + L_{34}^2). \tag{10.27}$$

Since the expression in the bracket is negative it follows that $L_{44}^2 \geqslant 1$, or

$$|L_{44}| \geqslant 1. \tag{10.28}$$

This relation shows that between transformations with $L_{44} > 0$ and $L_{44} < 0$ there is a gap of at least 2 in the value of L_{44} which cannot be bridged by any continuous steps. It follows further that the full Lorentz group can be subdivided into four sets of transformations, each continuous within itself but not with respect to the others.

	I	*II*	*III*	*IV*
$\det L$	$+1$	-1	-1	$+1$
L_{44}	>0	>0	<0	<0

The first of these sets constitutes the restricted Lorentz group. The second set consists of Lorentz transformations with space inversions.

This set does not form a group. Sets I and II together form a group which is called the orthochronous Lorentz group. It consists of all the transformations which do not involve time reversal. Set III consists of Lorentz transformations which involve time inversion, they do not form a group, but I and III form a group. Set IV consists of Lorentz transformations combined with both space and time inversions. It is cutomary to call Lorentz transformations with $\det L = +1$, proper transformations, and those with $\det L = -1$, improper transformations. The proper transformations also form a group (I+IV).

The principle of relativity is based on the experimental evidence that it is not possible to distinguish between two systems of coordinates which move relative to each other along a straight line and to decide which moves and which is at rest. Similarly, a principle of relativity with respect to space inversion should rest on experimental evidence telling us that left-handedness and right-handedness of systems of coordinates can be experienced only in the relative sense. In other words, from physical behaviour it should not be possible to recognize a system of coordinates and decide whether it is left-handed or right-handed. Until recently it was believed that all physical laws are invariant against space inversion, which is another way of saying that left-handedness or right-handedness has only a relative meaning. In quantum theory this leads to the law of conservation of parity. Parity in quantum mechanics is an operator which is associated with space inversion. However, recent observations on electron-neutrino decay and some of the strange particles have led to the conclusion that parity is not conserved in these cases. Electron-neutrino decay of nuclei would therefore look different in left-handed and right-handed systems. Essentially it amounts to the following: a neutrino, so it turns out, can have its spin pointing only in one definite way along its direction of motion. Hence, two neutrinos moving in opposite directions have always opposite spins. Suppose, now, we have a system of coordinates and are uncertain whether it is left-handed or right-handed. We let a neutrino move along the positive direction of the z-axis. If its spinning motion would make the x-axis chase the y-axis the system is right-handed: in the opposite case it is left-handed. With the help of a neutrino we can therefore determine in each case whether a system is left-handed or right-handed. Before the discovery of the non-conservation of parity the opinion was held that a neutrino could have its spin both in a state of parallelism as well as anti-parallelism with respect to its direction of motion. Under such circumstances it would not have been possible to identify a system of coordinates as being left-handed or right-handed in an absolute manner, i.e. without reference to a second 'standard' system. Hence, the experimental fact that β-decay of nuclei leads to only one type of neutrino, which shows an asymmetry with respect to left and right, proves that the principle of relativity with respect to left-handed and right-handed systems does not hold. However, although the non-conservation of parity has dealt a blow to the principle of relativity with regard to space inversion, the consideration of this transformation is still of importance, because many physical laws happen to be invariant with respect to it.

The relativity with regard to time inversion is not in doubt at present.

11. The Thomas Precession

Let us consider as before two systems K and K' with parallel axes and moving relative to each other along the x- and x'-axes. In spite

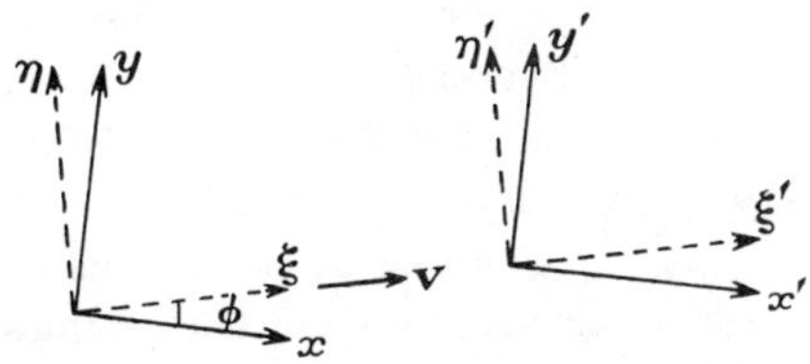

FIG. 18

of the Lorentz contraction the axes of K will appear as perpendicular to each other when viewed from K' and vice versa. Next let us rotate both frames of reference by an angle ϕ in the x, y plane. The following relations will then hold, if we denote the new coordinates again by x, y, z and x', y', z' respectively:

$$\begin{aligned} x'\cos\phi+y'\sin\phi &= \gamma(x\cos\phi+y\sin\phi+vt), \\ -x'\sin\phi+y'\cos\phi &= -x\sin\phi+y\cos\phi, \end{aligned} \tag{11.1}$$

the last relation corresponding to the old $y' = y$ (or $\eta' = \eta$ in Fig. 18).

Eliminating $\cos\phi$ and $\sin\phi$ from the left-hand sides we get, after adding the equations for the z and t coordinates,

$$\begin{aligned} x' &= x+[(\gamma-1)(x\cos\phi+y\sin\phi)]\cos\phi+\gamma v\cos\phi\, t, \\ y' &= y+[(\gamma-1)(x\cos\phi+y\sin\phi)]\sin\phi+\gamma v\sin\phi\, t, \\ z' &= z, \\ t' &= \gamma vc^{-2}(x\cos\phi+y\sin\phi)+\gamma t. \end{aligned} \tag{11.2}$$

In the rotated systems the relative velocity **v** has two components v_x and v_y and the following relations hold:

$$x\cos\phi+y\sin\phi = \frac{(\mathbf{xv})}{v}, \quad v_x = v\cos\phi, \quad v_y = v\sin\phi, \tag{11.3}$$

where **x** is a vector with the components x, y, z. Using (11.3) we may express (11.2) as follows:

$$\mathbf{x}' = \mathbf{x}+\frac{(\gamma-1)\mathbf{v}(\mathbf{xv})}{v^2}+\gamma\mathbf{v}t,$$

$$t' = \frac{\gamma(\mathbf{vx})}{c^2}+\gamma t. \tag{11.4}$$

In this form they are also valid when **v** is not necessarily a vector in the x, y plane.

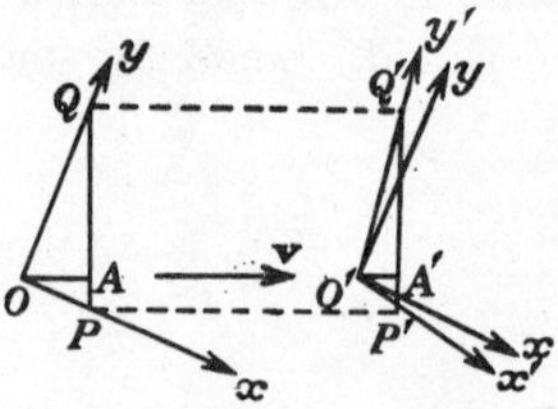

FIG. 19

Because of the Lorentz contraction, K will no longer appear as a set of perpendicular axes to an observer in K' and vice versa (see Fig. 19). Nevertheless, we shall regard K and K' as parallel simply because in such systems parallelism can always be restored by performing a rotation of equal magnitude in both frames.

Eqns. (11.4) do not represent the most general Lorentz transformation. This will be the case when the frames are not parallel. To obtain the relevant formula we need only consider a further transformation through a rotation R. In matrix notation we may write for the new space coordinates (say in K) $R\mathbf{x}$. The rotation of the axes in K will also affect the components of **v**, but the scalar product $(\mathbf{xv})$ will remain unchanged:

$$(\mathbf{xv}) = (R\mathbf{x}, R\mathbf{v}). \tag{11.5}$$

Consequently, instead of (11.4) we shall have in the most general case

$$\mathbf{x}' = R\mathbf{x}+(\gamma-1)\frac{(R\mathbf{v})(\mathbf{xv})}{v^2}+\gamma(R\mathbf{v})t,$$

$$t' = \frac{\gamma(\mathbf{vx})}{c^2}+\gamma t. \tag{11.6}$$

Let us now consider three frames of reference K, K', K'', where K' moves with a velocity v_1 relative to K along the x-axis and K'' moves relative to K' with a velocity v_2 along the y'-axis. We also assume that the axes of K' are parallel to those of K and the axes of K'' parallel to those of K'. We shall prove that the axes of K'' are not parallel to those of K, so that K and K'' cannot be brought to true parallelism by an equal rotation of axes.

The following relations hold:

$$\begin{aligned} x' &= \gamma_1 x+\gamma_1 v_1 t & x'' &= x' \\ y' &= y & y'' &= \gamma_2 y'+\gamma_2 v_2 t' \\ z' &= z & z'' &= z' \\ t' &= \gamma_1 v_1 c^{-2}x+\gamma_1 t & t'' &= \gamma_2 v_2 c^{-2}y'+\gamma_2 t', \end{aligned} \tag{11.7}$$

where
$$\gamma_1 = (1-v_1^2/c^2)^{-\frac{1}{2}}, \qquad \gamma_2 = (1-v_2^2/c^2)^{-\frac{1}{2}}. \tag{11.8}$$

Combining these transformations to obtain the relation between K and K'' we get

$$\begin{aligned} x'' &= \gamma_1 x+\gamma_1 v_1 t \\ y'' &= \gamma_2 y+\gamma_2 v_2\{(\gamma_1 v_1/c^2)x+\gamma_1 t\} \\ z'' &= z \\ t'' &= (\gamma_2 v_2/c^2)y+\gamma_2\{(\gamma_1 v_1/c^2)x+\gamma_1 t\}. \end{aligned} \tag{11.9}$$

Using (4.9) for the transformation of a velocity, we obtain for the velocity $\mathbf{v}$ of K'' relative to K the following result:

$$v_x = v_1, \quad v_y = v_2/\gamma_1, \quad v_z = 0. \tag{11.10}$$

We also have

$$v^2 = v_1^2+v_2^2/\gamma_1^2, \qquad \gamma = (1-v^2/c^2)^{-\frac{1}{2}}. \tag{11.11}$$

Had the axes of K'' been parallel to those of K, then with the relative velocity $\mathbf{v}$, (11.4) would have yielded for the x-axis

$$x'' = x+\frac{(\gamma-1)v_1}{v^2}\{xv_1+yv_2/\gamma_1\}+\gamma v_1 t. \tag{11.12}$$

The first equation in (11.9) cannot be made equivalent to this. Consequently the axes of K'' and K are not parallel to each other.

From the foregoing result we come to the conclusion that all the Lorentz transformations between parallel axes do not form a subgroup, as is, for instance, the case with all the transformations between parallel systems when the motion is along a given axis for all of them. The combination of two transformations ($K \to K'$ and $K' \to K''$), each of which is between parallel systems, does not in general yield a transformation of a type which holds between parallel systems. Since, however, it must be a Lorentz transformation, the combined transformation ($K \to K''$) can be represented as a Lorentz transformation between parallel systems together with a rotation.

Let us now assume that a system P is moved along a circular path parallel to itself and let us assume that an inertial system K is at rest relative to this circular path. From the previous result it follows that at each position of P its axes will appear rotated with respect to those of K, and that as a result of the continuous change of position and velocity of P, the rotation will also be a function of time and proceed with some angular velocity. Thus, from the theory of relativity it follows that the motion of P along a curved orbit, parallel to itself, appears in an inertial system as an orbital motion with a superimposed rotation of precession of the whole system. The first to call attention to this phenomenon was L. H. Thomas and this precession is known as the Thomas precession. Thomas derived the following formula for the angular velocity:

$$\boldsymbol{\omega}_T = -\frac{[\mathbf{v}, d\mathbf{v}/dt]}{v^2}(\gamma-1). \tag{11.13}$$

According to this formula, which will be proved further down, the angular velocity $\boldsymbol{\omega}_T$ is proportional to both v and the acceleration dv/dt, and the axis of rotation is perpendicular to the plane formed by these vectors. In a circular motion of radius r and uniform velocity v, the acceleration is $-v^2r^{-1}$, hence for the period τ_T of the Thomas precession as compared with the period T of the circular motion we get

$$\tau_T = \frac{T}{\gamma-1}. \tag{11.14}$$

Both T and τ_T are measured in an inertial frame of reference at rest relative to the circular orbit. For $v \ll c$ (11.13) and (11.14) become

$$\boldsymbol{\omega}_T = -\frac{[\mathbf{v}, d\mathbf{v}/dt]}{2c^2}, \tag{11.15}$$

$$\tau_T = 2Tc^2/v^2. \tag{11.16}$$

In this case τ_T is a much longer period than T; the precession is slow. For very high velocities τ_T becomes very small.

From the foregoing it follows that when a frame of reference P is carried along a curved path parallel to itself, on return to some original position O it is no longer parallel to its original orientation. The orientation is not integrable, since the deviation from the original orientation will depend on the shape of the orbit and the velocity along it. In this respect the orientation behaves similarly to the time indication of a clock which is taken round a closed path and is then compared with a clock which was left in the original position. When we speak of a change of orientation we mean, to be precise, the following. Let us assume that at a time t, as measured in K, the origin of P is at O and that at this moment it is at rest relative to an inertial frame K' which is parallel to K and is in motion relative to it. After a short time δt the origin of P is at some other point O', and it moves in a different direction with a velocity $\mathbf{v}+\delta\mathbf{v}$. Here we assume that P is parallel to an inertial frame K'', which is parallel to K' and is at rest relative to P, at this particular instant. In this way the orientation of P is kept parallel. At each stage it is parallel and at rest relative to some inertial frame K^n which is parallel to an inertial frame $K^{(n-1)}$ encountered a little earlier.

We shall now proceed with the proof of the Thomas formula. The simplest version of such a proof was given by S. Dancoff and D. R.

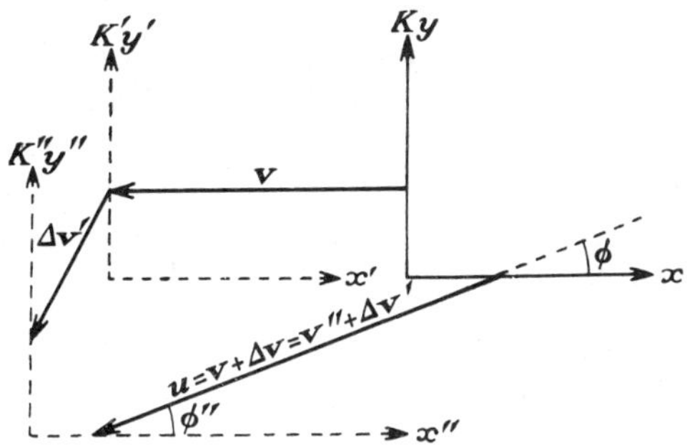

FIG. 20

Inglis. We consider an inertial frame of reference K and two auxiliary inertial frames K' and K'' which differ only slightly in their motion relative to K. We assume K' to be parallel to K and K'' to be parallel to K'. It will also be convenient to orient K and K' so that their relative motion is along the x-axes. Our object is to calculate the deviation from parallelism between K and K''.

In Fig. 20 the position of K' is shown relative to K one second after the coincidence of their origins (at $t = 0$), K'' is shown one second after the coincidence of its origin with that of K'. Consequently $\mathbf{v}$ is the relative velocity between K and K', $\Delta\mathbf{v}'$ is the relative velocity between K' and K''. This relative velocity appears in K as $\Delta\mathbf{v}$ so that the relative velocity between K and K'' is $\mathbf{u} = \mathbf{v}+\Delta\mathbf{v}$. On the other hand, the velocity $-\mathbf{v}$ between K' and K appears as a velocity $-\mathbf{v}''$ in K'' so that the relative velocity between K'' and K is given by $-\mathbf{u} = -\Delta\mathbf{v}'-\mathbf{v}''$. This is equal, apart from sign, to $\mathbf{v}+\Delta\mathbf{v}$. We have to prove that as a result of a Lorentz transformation the angles ϕ and ϕ'' which x and x'' respectively form with the relative velocity $\mathbf{u}$ between K and K'' are different and we wish to calculate this difference. In the calculation we shall make use of the infinitesimality of $\Delta\mathbf{v}'$ (or $\Delta\mathbf{v}$ as observed in K).

The velocity $\mathbf{u} = \mathbf{v}+\Delta\mathbf{v}$ of K'' relative to K is obtained by transforming $\Delta\mathbf{v}'$ from K' to K:

$$u_x = (\mathbf{v}+\Delta\mathbf{v})_x = \frac{\Delta v'_{x'}+v}{1+v\Delta v'_{x'}/c^2}, \tag{11.17}$$

$$u_y = (\mathbf{v}+\Delta\mathbf{v})_y = \frac{\Delta v'_{y'}}{\gamma_v(1+v\Delta v'_{x'}/c^2)}. \tag{11.18}$$

The angle between $\mathbf{u}$ and x is given by

$$\phi \cong \tan\phi = \frac{(\mathbf{v}+\Delta\mathbf{v})_y}{(\mathbf{v}+\Delta\mathbf{v})_x} = \frac{\Delta v'_{y'}}{\gamma_v(\Delta v'_{x'}+v)} \cong \frac{\Delta v'_{y'}}{\gamma_v v}. \tag{11.19}$$

In the same way the relative velocity $-\mathbf{u}$ between K'' and K is given by $-(\Delta\mathbf{v}'+\mathbf{v}'')$ and may be obtained by transforming $-\mathbf{v}$ from K' to K''. However, in this case the relative velocity between the inertial frames K' and K'' is infinitesimal. Consequently the Lorentz modification to the classical law of composition of velocity may be neglected, and we may use the classical law to calculate $\mathbf{u}$:

$$u_{x''} = (\mathbf{v}''+\Delta\mathbf{v}')_{x''} = v+\Delta v'_{x''},$$

$$u_{y''} = (\mathbf{v}''+\Delta\mathbf{v}')_{y''} = \Delta v'_{y''}. \tag{11.20}$$

Now, because of the assumed parallelism between K'' and K',

$$\Delta v'_{x''} = \Delta v'_{x'}, \qquad \Delta v'_{y''} = \Delta v'_{y'}. \tag{11.21}$$

Consequently, we get for the angle ϕ'' between x'' and $\mathbf{u}$

$$\phi'' \cong \tan\phi'' = \frac{u_{y''}}{u_{x''}} = \frac{\Delta v'_{y''}}{v+\Delta v'_{x''}} \cong \frac{\Delta v'_{y''}}{v} = \frac{\Delta v'_{y'}}{v}. \tag{11.22}$$

As will be seen ϕ'' is different from ϕ. For the difference we get†

$$\Delta\phi = \phi-\phi'' = -\frac{\Delta v'_{y'}}{v}(1-\gamma^{-1}). \tag{11.23}$$

This change in orientation may take place in a time element $\Delta t'$ as measured in K'. Hence

$$\omega_T = \frac{\Delta\phi}{\Delta t} = \frac{\gamma\Delta\phi}{\Delta t'} = -\frac{\gamma}{v}\frac{\Delta v'_{y'}}{\Delta t'}(1-\gamma^{-1}). \tag{11.24}$$

In the limit we are interested in the effect produced when $\Delta\mathbf{v}'$ is perpendicular to $\mathbf{v}$ ($\Delta v'_{x'} \to 0$). (11.18) then yields $\Delta v'_{y'} = \gamma\Delta v_y$. Also $\Delta t' = \gamma\Delta t$. Hence,

$$\omega_T = -\frac{\gamma}{v}\frac{dv_y}{dt}(1-\gamma^{-1}). \tag{11.25}$$

† We calculate an anti-clockwise rotation.

Introducing the acceleration $d\mathbf{v}/dt$, we have

$$\boldsymbol{\omega}_T = \frac{-[\mathbf{v}, d\mathbf{v}/dt]}{v^2}(\gamma-1). \tag{11.26}$$

We have described a mathematical method of ensuring that P is kept parallel to itself. Physically this may be realized by letting P carry a system which conserves a direction in space. Such a system is, for instance, a body with an angular momentum. Therefore, if such a body is carried in a curved orbit without applying any torque to it, it will appear to an inertial frame of reference as precessing with the Thomas precession.

An interesting application of the Thomas precession is encountered when dealing with the anomalous Zeeman effect and the fine structure of atomic levels. When S. Goudsmit and G. E. Uhlenbeck first introduced the electron spin, they remarked that the gyromagnetic ratio required to explain the anomalous Zeeman effect leads to an incorrect result when this ratio is used to explain the fine structure of the energy levels. A correct result can be obtained in the latter case if one assumes that the magnetic moment has half the value required in the case of the Zeeman effect. The gyromagnetic ratio g^{-1} of an electron (i.e. the ratio between angular momentum and magnetic moment) required to explain the anomalous Zeeman effect is given by

$$g_{\text{spin}} = e/mc, \tag{11.27}$$

where m is the mass and e the charge of the electron. The difference between energy levels of such an electron in a magnetic field $\mathbf{H}$ (Zeeman effect) is given by

$$\Delta E_{\text{Zeeman}} = \frac{h}{2\pi}\omega_L, \tag{11.28}$$

where h is Planck's constant and ω_L is the Larmor frequency of precession of the electron axis in the magnetic field $\mathbf{H}$:

$$\boldsymbol{\omega}_L = g\mathbf{H}. \tag{11.29}$$

Let us now consider an electron which moves in a circular orbit in the electric field $\mathbf{E}$ of a nucleus. According to a formula which is proved in section 13, the electron 'sees' a somewhat modified electric field $\mathbf{E}'$ and a magnetic field $\mathbf{H}'$ which is perpendicular to the orbit:

$$\mathbf{E}' = \frac{\mathbf{E}}{(1-v^2/c^2)^{\frac{1}{2}}} \cong \mathbf{E}, \tag{11.30}$$

$$\mathbf{H}' = \frac{[\mathbf{E}, \mathbf{v}]}{c(1-v^2/c^2)^{\frac{1}{2}}} \cong \frac{[\mathbf{E}, \mathbf{v}]}{c}, \tag{11.31}$$

where the approximation applies when $v/c \ll 1$. As a result of the field $\mathbf{H}'$, the axis of the electron precesses with a Larmor frequency which is given by

$$\boldsymbol{\omega}_L = g\mathbf{H}' = c^{-1}g[\mathbf{E}, \mathbf{v}]. \tag{11.32}$$

This precession causes the orbit to precess in such a way that angular momentum is conserved in direction and magnitude (spin and orbital angular momenta precess around their resultant). The contribution to the total energy which results from this interaction between spin and orbit can be shown to be given by

$$\Delta E_{\text{fine structure}} = \frac{\omega_L \mathbf{I}_s \mathbf{I}_l}{mvr}, \tag{11.33}$$

where $\mathbf{I}_s$ is the spin angular momentum and $\mathbf{I}_l$ is the orbital angular momentum of magnitude mvr. The different possible (quantized) angles between $\mathbf{I}_s$ and $\mathbf{I}_l$ are responsible for the fine structure of the energy levels, and the energy difference between these levels as it follows from (11.33) is proportional to ω_L, which is given by (11.32). However, as was remarked by Goudsmit and Uhlenbeck, the separation $\Delta E_{\text{f.st.}}$ between the levels is twice as large as the one which is actually observed. The discrepancy was removed by L. H. Thomas who called attention to the existence of the relativistic precession $\boldsymbol{\omega}_T$. The total precession of the electron axis is given by

$$\boldsymbol{\omega}'_L = \boldsymbol{\omega}_L + \boldsymbol{\omega}_T = c^{-1}g[\mathbf{E}, \mathbf{v}] - \frac{[\mathbf{v}, d\mathbf{v}/dt]}{2c^2}. \tag{11.34}$$

Now for $v \ll c$ we have†

$$m\, d\mathbf{v}/dt = -e\mathbf{E}. \tag{11.35}$$

Consequently,

$$\frac{1}{2c^2}\left[\mathbf{v}, \frac{d\mathbf{v}}{dt}\right] = -\frac{e}{2mc^2}[\mathbf{v}, \mathbf{E}] = \tfrac{1}{2}c^{-1}g[\mathbf{E}, \mathbf{v}], \tag{11.36}$$

and we get

$$\boldsymbol{\omega}'_L = \tfrac{1}{2}c^{-1}g[\mathbf{E}, \mathbf{v}]. \tag{11.37}$$

Thus $\boldsymbol{\omega}'_L = \frac{1}{2}\boldsymbol{\omega}_L$. Since $\boldsymbol{\omega}'_L$ and not $\boldsymbol{\omega}_L$ is responsible for the spin-orbit interaction we have to replace $\boldsymbol{\omega}_L$ by $\frac{1}{2}\boldsymbol{\omega}_L$ in (11.33) and this yields the correct formula for the fine structure.

The Thomas precession makes the magnetic field, which is seen by the electron, to be half as effective. Alternatively, we can say that the Thomas precession reduces the effective value of the magnetic spin moment by a factor of half. This effect, like the Thomas precession, exists only when the electron moves on a curved path.

The problem discussed above must not be confused with the one that

† (11.35) is a non-relativistic formula, but is sufficiently accurate in the present analysis.

arises from the fact that the gyromagnetic ratio of an orbital motion of the electron is twice as big as that of its gyromagnetic ratio of spin. Furthermore, as already mentioned, (11.35) is valid only approximately. In reality a further relativistic effect arises from the variation of mass with velocity.

12. The Invalidity of Classical Dynamics with respect to Lorentz Transformations

Classical dynamics was the stronghold of the principle of relativity before it was extended to the whole of physics. All dynamical observations pointed out that motion of inertial systems is relative and that no dynamical experiment can be devised which will enable us to tell whether the frame we are using is moving with great or small speed or whether it is at rest (relative to some hypothetical absolute frame of reference or relative to some other inertial frame without observing the latter). This principle of relativity also found its expression in the mathematical formulation of the laws of dynamics through an invariance with respect to a Galilei transformation. According to the most fundamental of these laws, if we consider a system of bodies or particles which are not subject to an external action, energy and momentum are conserved. For simplicity let us assume that the total energy consists of kinetic energy only and that the particles interact through elastic collisions. Let W be the kinetic energy, so that

$$W = \sum_{k=1}^{n} \tfrac{1}{2} m_k u_k^2, \tag{12.1}$$

where n is the number of particles, m_k their masses, and u_k their velocities. The momentum according to classical theory is given by

$$\mathbf{P} = \sum_{k=1}^{n} m_k \mathbf{u}_k. \tag{12.2}$$

According to the laws of conservation,

$$W = \sum \tfrac{1}{2} m_k u_k^2 = \text{const}, \tag{12.3}$$

$$\mathbf{P} = \sum m_k \mathbf{u}_k = \text{const}. \tag{12.4}$$

By this we mean that as time progresses and the particles change their position and velocities the above quantities remain unchanged. We shall now prove that these laws remain unaltered if we change the frame of reference and make use of a Galilei transformation. The masses of the particles remain unchanged, since they are defined in all frames of reference in the same way, namely by measuring the velocity which they acquire when a standard force is applied for a unit of time. The velocity transforms as follows:

$$u_{kx'} = u_{kx}+v, \qquad u_{ky'} = u_{ky}, \qquad u_{kz'} = u_{kz}. \tag{12.5}$$

Consequently,

$$u_k^2 = u_{kx}^2+u_{ky}^2+u_{kz}^2 = u_{kx'}^2+u_{ky'}^2+u_{kz'}^2-2u_{kx'}v+v^2 = u_k'^2-2u_{kx'}v+v^2, \tag{12.6}$$

and we get the following relation between the kinetic energies in K and K':

$$W' = W+\sum_{k=1}^{n}(m_k u_{kx})v+\Big(\sum_{k=1}^{n} m_k\Big)\tfrac{1}{2}v^2, \tag{12.7}$$

or

$$W' = W+P_x v+(\sum m_k)\tfrac{1}{2}v^2. \tag{12.8}$$

For the momentum we get

$$P_{x'} = \sum m_k u_{kx'} = \sum m_k(u_{kx}+v) = P_x+(\sum m_k)v, \tag{12.9}$$

$$P_{y'} = P_y, \qquad P_{z'} = P_z. \tag{12.10}$$

We can now see that, as v and $\sum m_k \frac{1}{2}v^2$ are constant, it follows that if W is constant this also applies to W'. $P_x v$ remains constant because P_x remains constant and v is a constant. From (12.9) we see that if P_x is constant this also applies to $P_{x'}$. The constancy of $P_{y'}$ and $P_{z'}$ follows from (12.10).

The classical laws of conservation of energy and momentum are, however, not invariant if a Lorentz transformation is performed. For instance, let us consider the law of conservation of momentum. If a Lorentz transformation is performed we get

$$u_{kx'} = \frac{u_{kx}+v}{1+u_{kx}v/c^2}, \tag{12.11}$$

$$\sum m_k u_{kx'} = \sum \frac{m_k u_{kx}+m_k v}{1+u_{kx}v/c^2}. \tag{12.12}$$

Thus, if $\sum m_k u_{kx'}$ is to remain constant, we find that not only

$$\sum m_k u_{kx} = \text{constant}, \tag{12.13}$$

but the following relation must also hold:

$$\sum \frac{m_k v}{1+u_{kx}v/c^2} = \text{constant}. \tag{12.14}$$

However, (12.14) is not necessarily a consequence of (12.13). Matters are not improved if we also consider the constancy of terms which follow from the conservation of energy. We cannot conclude that (12.14) necessarily holds if (12.13) is true.

We shall see later that according to the theory of relativity momentum and energy of non-interacting particles (except through elastic collisions) are defined by

$$P_x = \sum \frac{m_k u_{kx}}{(1-u_k^2/c^2)^{\frac{1}{2}}}, \qquad W = \sum \frac{m_k c^2}{(1-u_k^2/c^2)^{\frac{1}{2}}}. \tag{12.15}$$

Performing a Lorentz transformation we get

$$P_{x'} = \gamma\left[\sum\left(\frac{m_k u_{kx}}{(1-u_k^2/c^2)^{\frac{1}{2}}} + v\frac{m_k}{(1-u_k^2/c^2)^{\frac{1}{2}}}\right)\right] = \gamma\left(P_x + v\frac{W}{c^2}\right), \tag{12.16}$$

$$\frac{W'}{c} = \frac{\gamma v}{c}\sum\frac{m_k u_{kx}}{(1-u_k^2/c^2)^{\frac{1}{2}}} + \gamma\sum\frac{m_k c}{(1-u_k^2/c^2)^{\frac{1}{2}}} = \gamma\left(\frac{v}{c}P_x + \frac{W}{c}\right). \tag{12.17}$$

Consequently if P_x and W are conserved this also applied to $P_{x'}$ and W'. The conservation of energy and momentum as defined by the theory of relativity is Lorentz invariant.

As already mentioned, classical dynamics was the stronghold of the principle of relativity in the old sense. Both experimentally as well as from a theoretical point of view all inertial systems were equivalent. However, since optical observations force us to adopt the Lorentz transformation, it is necessary to modify classical dynamics. The fact that all inertial frames are equivalent from an experimental point of view remains unchanged but the laws must be altered so as to comply with the Lorentz transformation. In this connexion it may also be pointed out that the experimental test of the principle of relativity in the field of optics has been carried out to a much greater degree of accuracy than can be achieved with present means in the field of dynamics; hence the latter has to yield to optics, which points to a Lorentz and not a Galilei transformation. As may be expected, the modifications which the theory of relativity introduces in dynamics have only minute effects whenever we deal with velocities which are small compared with that of light. In its conception the new dynamics differs, however, a great deal from the old dynamics.

13. The Compatibility of Maxwell's Equations with Lorentz Transformations

Lorentz derived his relation between coordinates of moving observers by searching for a transformation which leaves Maxwell's equations of the electromagnetic field **E**, **H** unchanged. In this section we shall follow the reverse course, by proving that Maxwell's equations satisfy the principle of relativity in the Einstein–Lorentz sense. Since we are dealing with fields, i.e. with physical entities which are functions of space and time, the field components relative to K, i.e. $E_x(x,y,z,t)$, $E_y(x,y,z,t)$, $E_z(x,y,z,t)$,..., $H_z(x,y,z,t)$, can also be expressed in terms of x',y',z',t', or indeed any other set of variables which depend in a unique way on x,y,z,t. This does not change the significance of E_x,..., H_z as the components of the electromagnetic field along the x-, y-, z-axes as observed in K. By introducing new (auxiliary) variables only the functional form of E_x,..., H_z in its dependence on the various space-time points, in the new notation, is changed. In the same way the field equations take a new form.

The new functional dependence, say of E_x, is obtained from the relation

$$E_x(x,y,z,t) = E'_x(x',y',z',t'). \tag{13.1}$$

Both E_x and E'_x are the x-component of the electric field as observed by K, so that for any given space-time point

$$E_x = E'_x. \tag{13.2}$$

The difference lies only in E'_x being a different function of x', y', z', t' than E_x of x, y, z, t. The new function E'_x is obtained by substituting for x, y, z, t in $E_x(x,y,z,t)$ their values in terms of x', y', z', t', so that (13.1) is obtained. In a general way we can say that if a given observer substitutes for x, y, z, t some other variables ξ, η, ζ, τ this must not and cannot change the values of physical entities. They only become new functions of the new variables and behave in this respect formally like scalars.†

We shall denote the components of the electromagnetic field as they appear to K' by $E_{x'}$, $E_{y'}$, $E_{z'}$, $H_{x'}$, $H_{y'}$, $H_{z'}$. In the same way as for K, $E_{x'}$ is naturally a function of x', y', z', t', but can also be expressed as a function $E'_{x'}(x,y,z,t)$ if this is desired.

Again,
$$E_{x'}(x',y',z',t') = E'_{x'}(x,y,z,t). \tag{13.3}$$

Our object is to prove that from the fact that E_x,..., H_z satisfy Maxwell's equations in K, it is possible to determine certain functions,

$$E_{x'} = f_1(E_x, E_y, E_z, H_x, H_y, H_z),$$
$$\cdots\cdots\cdots\cdots$$
$$H_{z'} = f_6(E_x, E_y, E_z, H_x, H_y, H_z), \tag{13.4}$$

in such a way that the quantities $E_{x'}$,..., $H_{z'}$ satisfy similar equations of

† The concept of a scalar or any tensor will be defined in Chapter 2. The concept of a scalar arises only when a real change of frame takes place. In our present case, in K we retain the x, y, z, t system, so that the field components are throughout E_x, E_y, E_z, etc., but new auxiliary variables are used to specify the different space-time points. A transformation in the physical sense of the word (where scalars and tensors come into their own) takes place when we change from K to K'.

Maxwell relative to K'. In this way, for each electromagnetic field $E_x,\ldots, H_z$, K' will observe an electromagnetic field $E_{x'},\ldots, H_{z'}$ which satisfies relative to K' the same Maxwell equations as hold for $E_x,\ldots, H_z$ relative to K. We can then claim that an electromagnetic field **E**, **H** transforms according to (13.4) and that the transformed field, like the original field, satisfies Maxwell's equations. As a check for the full equivalence between K and K' we shall also have to satisfy ourselves that if, using (13.4), we express $E_x,\ldots, H_z$ in terms of $E_{x'},\ldots, H_{z'}$ we get the same functions $f_1,\ldots,f_6$ as before provided v is changed into $-v$. Moreover, the group property must hold, with the same multiplication law as for Lorentz transformations, because as we have already seen, the Lorentz transformation can be derived in its complete form from the group property.

Maxwell's equations in vacuum, for a region in which no charges or currents are present, have the following form:

$$
\begin{aligned}
-\frac{\partial E_y}{\partial z}+\frac{\partial E_z}{\partial y} &= -\mu\frac{\partial H_x}{\partial t}, & -\frac{\partial H_y}{\partial z}+\frac{\partial H_z}{\partial y} &= \epsilon\frac{\partial E_x}{\partial t},\\
-\frac{\partial E_z}{\partial x}+\frac{\partial E_x}{\partial z} &= -\mu\frac{\partial H_y}{\partial t}, & -\frac{\partial H_z}{\partial x}+\frac{\partial H_x}{\partial z} &= \epsilon\frac{\partial E_y}{\partial t},\\
-\frac{\partial E_x}{\partial y}+\frac{\partial E_y}{\partial x} &= -\mu\frac{\partial H_z}{\partial t}, & -\frac{\partial H_x}{\partial y}+\frac{\partial H_y}{\partial x} &= \epsilon\frac{\partial E_z}{\partial t},\\
\frac{\partial E_x}{\partial x}+\frac{\partial E_y}{\partial y}+\frac{\partial E_z}{\partial z} &= 0, & \frac{\partial H_x}{\partial x}+\frac{\partial H_y}{\partial y}+\frac{\partial H_z}{\partial z} &= 0.
\end{aligned} \tag{13.5}
$$

These equations apply in K where the cartesian coordinates x, y, z and a properly defined time are used. They can also be expressed in terms of x', y', z', t', which in K have not the significance of cartesian coordinates and time. However, as for each x, y, z, t there corresponds a certain x', y', z', t' and vice versa this is mathematically possible. In the new variables Maxwell's equations will take a new form. We can derive this form as follows. From $E'_x(x', y', z', t') = E_x(x, y, z, t)$ we get

$$
\begin{aligned}
\frac{\partial E_x}{\partial x} &= \frac{\partial E'_x}{\partial x'}\frac{\partial x'}{\partial x}+\frac{\partial E'_x}{\partial y'}\frac{\partial y'}{\partial x}+\frac{\partial E'_x}{\partial z'}\frac{\partial z'}{\partial x}+\frac{\partial E'_x}{\partial t'}\frac{\partial t'}{\partial x},\\
\frac{\partial E_x}{\partial y} &= \frac{\partial E'_x}{\partial x'}\frac{\partial x'}{\partial y}+\frac{\partial E'_x}{\partial y'}\frac{\partial y'}{\partial y}+\frac{\partial E'_x}{\partial z'}\frac{\partial z'}{\partial y}+\frac{\partial E'_x}{\partial t'}\frac{\partial t'}{\partial y},\\
\frac{\partial E_x}{\partial z} &= \frac{\partial E'_x}{\partial x'}\frac{\partial x'}{\partial z}+\frac{\partial E'_x}{\partial y'}\frac{\partial y'}{\partial z}+\frac{\partial E'_x}{\partial z'}\frac{\partial z'}{\partial z}+\frac{\partial E'_x}{\partial t'}\frac{\partial t'}{\partial z},\\
\frac{\partial E_x}{\partial t} &= \frac{\partial E'_x}{\partial x'}\frac{\partial x'}{\partial t}+\frac{\partial E'_x}{\partial y'}\frac{\partial y'}{\partial t}+\frac{\partial E'_x}{\partial z'}\frac{\partial z'}{\partial t}+\frac{\partial E'_x}{\partial t'}\frac{\partial t'}{\partial t}.
\end{aligned} \tag{13.6}
$$

We now assume that the new variables x', y', z', t' depend on the old variables x, y, z, t in accordance with a Lorentz transformation. Hence

$$\frac{\partial x'}{\partial x} = \gamma, \quad \frac{\partial y'}{\partial x} = 0, \quad \frac{\partial z'}{\partial x} = 0, \quad \frac{\partial t'}{\partial x} = \frac{\gamma v}{c^2},$$

$$\frac{\partial x'}{\partial y} = 0, \quad \frac{\partial y'}{\partial y} = 1, \quad \frac{\partial z'}{\partial y} = 0, \quad \frac{\partial t'}{\partial y} = 0,$$

$$\frac{\partial x'}{\partial z} = 0, \quad \frac{\partial y'}{\partial z} = 0, \quad \frac{\partial z'}{\partial z} = 1, \quad \frac{\partial t'}{\partial z} = 0,$$

$$\frac{\partial x'}{\partial t} = \gamma v, \quad \frac{\partial y'}{\partial t} = 0, \quad \frac{\partial z'}{\partial t} = 0, \quad \frac{\partial t'}{\partial t} = \gamma. \tag{13.7}$$

Consequently, we get for E_x, and in the same way for all other components, the following relations:

$$\frac{\partial E_x}{\partial x} = \gamma\frac{\partial E'_x}{\partial x'} + \gamma\frac{v}{c^2}\frac{\partial E'_x}{\partial t'}, \qquad \frac{\partial E_x}{\partial t} = \gamma v\frac{\partial E'_x}{\partial x'} + \gamma\frac{\partial E'_x}{\partial t'},$$

$$\frac{\partial E_y}{\partial x} = \gamma\frac{\partial E'_y}{\partial x'} + \gamma\frac{v}{c^2}\frac{\partial E'_y}{\partial t'}, \qquad \frac{\partial E_y}{\partial t} = \gamma v\frac{\partial E'_y}{\partial x'} + \gamma\frac{\partial E'_y}{\partial t'},$$

$$\frac{\partial E_z}{\partial x} = \gamma\frac{\partial E'_z}{\partial x'} + \gamma\frac{v}{c^2}\frac{\partial E'_z}{\partial t'}, \qquad \frac{\partial E_z}{\partial t} = \gamma v\frac{\partial E'_z}{\partial x'} + \gamma\frac{\partial E'_z}{\partial t'},$$

$$\frac{\partial E_x}{\partial y} = \frac{\partial E'_x}{\partial y'}, \qquad \frac{\partial E_y}{\partial y} = \frac{\partial E'_y}{\partial y'}, \qquad \frac{\partial E_z}{\partial y} = \frac{\partial E'_z}{\partial y'},$$

$$\frac{\partial E_x}{\partial z} = \frac{\partial E'_x}{\partial z'}, \qquad \frac{\partial E_y}{\partial z} = \frac{\partial E'_y}{\partial z'}, \qquad \frac{\partial E_z}{\partial z} = \frac{\partial E'_z}{\partial z'}. \tag{13.8}$$

Similar relations also hold for H_x, H_y, H_z. The general rule is that on substituting for x, y, z, t their equivalent in terms of x', y', z', t' the following operational relations hold:

$$\frac{\partial}{\partial x} = \gamma\frac{\partial}{\partial x'} + \gamma\frac{v}{c^2}\frac{\partial}{\partial t'}, \qquad \frac{\partial}{\partial y} = \frac{\partial}{\partial y'},$$

$$\frac{\partial}{\partial z} = \frac{\partial}{\partial z'}, \qquad \frac{\partial}{\partial t} = \gamma v\frac{\partial}{\partial x'} + \gamma\frac{\partial}{\partial t'}. \tag{13.9}$$

Let us now enter all the new expressions for $\partial E_x/\partial x, \ldots, \partial H_z/\partial z$ in Maxwell's equations. This will yield the new equations in terms of x', y', z', t' as they hold in K. From the first of Maxwell's equations

(13.5) and the divergence equation for H we get

$$\frac{\partial E'_z}{\partial y'}-\frac{\partial E'_y}{\partial z'} = -\mu\left(\gamma v\frac{\partial H'_x}{\partial x'}+\gamma\frac{\partial H'_x}{\partial t'}\right), \tag{13.10}$$

$$\gamma\frac{\partial H'_x}{\partial x'}+\frac{\gamma v}{c^2}\frac{\partial H'_x}{\partial t'}+\frac{\partial H'_y}{\partial y'}+\frac{\partial H'_z}{\partial z'} = 0. \tag{13.11}$$

If we eliminate $\partial H'_x/\partial x'$ we get

$$\gamma\frac{\partial E'_z}{\partial y'}-\mu v\gamma\frac{\partial H'_y}{\partial y'}-\gamma\frac{\partial E'_y}{\partial z'}-\mu v\gamma\frac{\partial H'_z}{\partial z'} = -\mu\frac{\partial H'_x}{\partial t'}, \tag{13.12}$$

or

$$\frac{\partial}{\partial y'}\gamma(E'_z-\mu vH'_y)-\frac{\partial}{\partial z'}\gamma(E'_y+\mu vH'_z) = -\mu\frac{\partial H'_x}{\partial t'}. \tag{13.13}$$

From this equation we learn that the first of Maxwell's equations also holds with respect to K' in which x', y', z', t' have the physical significance of space and time coordinates, whereby

$$E_{z'} = \gamma(E'_z-\mu vH'_y), \qquad E_{y'} = \gamma(E'_y+\mu vH'_z), \qquad H_{x'} = H'_x. \tag{13.14}$$

In these equations we can now replace E'_y, E'_z, H'_x, H'_y, H'_z by $E_y,\ldots, H_z$ (see 13.2). The change in notation is only required when carrying out the differentiations indicated in (13.9). We thus get

$$E_{z'} = \gamma(E_z-\mu vH_y), \qquad E_{y'} = \gamma(E_y+\mu vH_z), \qquad H_{x'} = H_x. \tag{13.15}$$

In the same way, by bringing to bear the first of the curl equations of H and the divergence equation of E, we find that they also hold in K', and that the remaining components of the electromagnetic field transform according to

$$H_{z'} = \gamma(H_z+\epsilon vE_y), \qquad H_{y'} = \gamma(H_y-\epsilon vE_z), \qquad E_{x'} = E_x. \tag{13.16}$$

Eqns. (13.15) and (13.16) express for each space-time point (called x, y, z, t by K and x', y', t', z' by K') the relation between the field components relative to K' in terms of the field components relative to K. However, on testing these transformation equations it will be found that not all the equations of Maxwell remain unchanged. For instance, from the divergence equation for E relative to K we get on using (13.15) or (13.14) and (13.9)

$$\gamma\left(\frac{\partial E_{x'}}{\partial x'}+\frac{\partial E_{y'}}{\partial y'}+\frac{\partial E_{z'}}{\partial z'}\right)-\mu v\gamma\left(\frac{\partial H_{z'}}{\partial y'}-\frac{\partial H_{y'}}{\partial z'}\right)+\frac{\gamma v}{c^2}\frac{\partial E_{x'}}{\partial t'} = 0. \tag{13.17}$$

This will yield

$$\frac{\partial E_{x'}}{\partial x'}+\frac{\partial E_{y'}}{\partial y'}+\frac{\partial E_{z'}}{\partial z'} = 0 \tag{13.18}$$

only if

$$\epsilon\mu = 1/c^2, \tag{13.19}$$

and by making use of the equation

$$\frac{\partial H_{z'}}{\partial y'}-\frac{\partial H_{y'}}{\partial z'}=\epsilon\frac{\partial E_{x'}}{\partial t'}, \tag{13.20}$$

which we already know to be valid in K'.

The required relation between ϵ and μ (13.19) can also be proved as follows. From the principle of relativity it follows that if we solve (13.15) and (13.10) and express E_x, E_y, E_z, H_x, H_y, H_z in terms of $E_{x'}$,..., $H_{z'}$ we should get the same equations provided v is changed into $-v$. This is a condition we have to lay down for every set of transformation equations. In actual fact, if we carry the calculation through we find

$$\begin{aligned} E_x &= E_{x'}, & H_x &= H_{x'}, \\ E_y &= \frac{E_{y'}-\mu v H_z}{\gamma(1-\epsilon\mu v^2)}, & H_y &= \frac{H_{y'}+\epsilon v E_{z'}}{\gamma(1-\epsilon\mu v^2)}, \\ E_z &= \frac{E_{z'}+\mu v H_y}{\gamma(1-\epsilon\mu v^2)}, & H_z &= \frac{H_{z'}-\epsilon v E_{y'}}{\gamma(1-\epsilon\mu v^2)}. \end{aligned} \tag{13.21}$$

These equations differ from (13.15) and (13.16) by more than the sign of v. They will, however, satisfy the principle of relativity if (13.19) holds.†

In general, when we have a set of equations in certain physical entities which remain unchanged under a Lorentz transformation, the transformation equations for the physical entities will automatically also satisfy the principle of relativity and have the same group properties as the Lorentz transformations, i.e. if v_1 and v_2 refer to two transformations, the parameter v of the combined transformation is equal to $(v_1+v_2)(1+v_1 v_2 c^{-2})^{-1}$.

It may be added that if Maxwell's equations are expressed in Gaussian units

$$c\,\mathrm{curl}\,\mathbf{E} = -\dot{\mathbf{H}}, \qquad c\,\mathrm{curl}\,\mathbf{H} = \dot{\mathbf{E}}, \tag{13.22}$$

which is equivalent to setting $\epsilon = \mu = c^{-1}$, the transformation equations for the field components become

$$\begin{aligned} E_{z'} &= \gamma\{E_z-(v/c)H_y\}, & H_{x'} &= H_x, \\ E_{y'} &= \gamma\{E_y+(v/c)H_z\}, & H_{y'} &= \gamma\{H_y-(v/c)E_z\}, \\ E_{x'} &= E_x, & H_{z'} &= \gamma\{H_z+(v/c)E_y\}. \end{aligned} \tag{13.23}$$

The invariance of Maxwell's equations with respect to the Lorentz transformation lends great support to the theory of relativity. A similar

† The relation $\epsilon\mu = c^{-2}$ is usually obtained by considering plane wave solutions of Maxwell's equations (c is the velocity of these waves). It is, however, more satisfactory to show that it is a necessary consequence of the invariance of Maxwell's equations, which imply that ϵ and μ are invariants.

calculation to that which was carried out above, but on the basis of a Galilei transformation, reveals that Maxwell's equations cannot remain unchanged. In other words, it is then impossible to determine a set of transformation equations for the electromagnetic field which will bring about the required invariance.

Maxwell's equations are also invariant with respect to space and time reflection (P and T) according to the transformation schemes in Table I, in which ρ is the charge density and **J** the current density.†

TABLE I

	P curl → −curl div → −div $\partial/\partial t \to \partial/\partial t$	T curl → curl div → div $\partial/\partial t \to -\partial/\partial t$	$PT = TP$ curl → −curl div → −div $\partial/\partial t \to -\partial/\partial t$
(**E**, **H**)	(−**E**, **H**) (**E**, −**H**) (−**E**, **H**) (**E**, −**H**)	(−**E**, **H**) (−**E**, **H**) (**E**, −**H**) (**E**, −**H**)	(**E**, **H**) −(**E**, **H**) −(**E**, **H**) (**E**, **H**)
(**J**, ρ)	(−**J**, ρ) (**J**, $-\rho$) (−**J**, ρ) (**J**, $-\rho$)	(**J**, $-\rho$) (**J**, $-\rho$) (−**J**, ρ) (−**J**, ρ)	−(**J**, ρ) (**J**, ρ) (**J**, ρ) −(**J**, ρ)

In Maxwell's equations ρ can be positive or negative. However, as long as we deal with a given charge, we cannot associate with the space inversion P the transformation $\rho \to -\rho$. Consequently only the first and third modes in the table apply. This makes **E** and **J** polar vectors (**E** → −**E**, **J** → −**J** under P) and **H** an axial vector (**H** → **H** under P). Moreover, there is no reason to assume that time reversal should change the sign of ρ, but as it changes the sign of a velocity ($v = dx/dt$) it changes the sign of a current density $J = \rho v$. Hence only the third mode of transformation can apply for the electromagnetic field.

In conclusion it may be pointed out that a transformation which leaves a set of equations invariant can also be used to obtain from a given solution of the equations a new solution. Thus, if $E_x, \ldots, H_z$ satisfy Maxwell's equations (relative to K), the linear combinations of these quantities as given in the right-hand sides of (13.15) and (13.16) (or 13.23) are also possible as a solution of the field equations.

† When charges and currents are present Maxwell's equations in Heaviside–Lorentz units are

$$\begin{aligned} c \operatorname{curl} \mathbf{H} &= \dot{\mathbf{E}} + \mathbf{J}, \\ \operatorname{div} \mathbf{H} &= 0; \\ c \operatorname{curl} \mathbf{E} &= -\dot{\mathbf{H}}, \\ \operatorname{div} \mathbf{E} &= \rho. \end{aligned} \qquad (13.24)$$

CHAPTER 2

THREE-DIMENSIONAL TENSORS

1. Introduction

In the previous chapter it was shown that the three space coordinates x, y, z combine together with the time t in the transformation which takes place between two inertial frames which move relative to each other. The transformation consists of four linear and homogeneous equations relating x', y', z', t' with x, y, z, t. For instance, for motion along the x-axes it is as follows:

$$\begin{bmatrix} x' \\ y' \\ z' \\ t' \end{bmatrix} = \begin{bmatrix} \gamma & 0 & 0 & \gamma v \\ 0 & 1 & 0 & 0 \\ 0 & 0 & 1 & 0 \\ \gamma v/c^2 & 0 & 0 & \gamma \end{bmatrix} \begin{bmatrix} x \\ y \\ z \\ t \end{bmatrix}. \tag{1.1}$$

We have seen that a Lorentz transformation is the only one which satisfies both the principle of relativity and the law of constancy of light velocity. We have also seen that Maxwell's equations satisfy the Einstein–Lorentz principle of relativity and as a consequence the components E_x, E_y, E_z, H_x, H_y, H_z combine together and transform in the above special case, according to the following scheme:

$$\begin{bmatrix} E_{x'} \\ E_{y'} \\ E_{z'} \\ H_{x'} \\ H_{y'} \\ H_{z'} \end{bmatrix} = \begin{bmatrix} 1 & 0 & 0 & 0 & 0 & 0 \\ 0 & \gamma & 0 & 0 & 0 & \gamma v/c \\ 0 & 0 & \gamma & 0 & -\gamma v/c & 0 \\ 0 & 0 & 0 & 1 & 0 & 0 \\ 0 & 0 & -\gamma v/c & 0 & \gamma & 0 \\ 0 & \gamma v/c & 0 & 0 & 0 & \gamma \end{bmatrix} \begin{bmatrix} E_x \\ E_y \\ E_z \\ H_x \\ H_y \\ H_z \end{bmatrix}. \tag{1.2}$$

This example leads us to expect that physical entities are in general characterized by a number of components which transform according to a linear and homogeneous transformation when a homogeneous Lorentz transformation takes place. Moreover, these transformations must depend in a definite way on the Lorentz transformations and exhibit the same group property. It will be convenient to call physical entities which behave in this way linear and homogeneous. The electromagnetic field is then such an entity with six components. Let a linear entity possess n components and let $F_1, \ldots, F_n$ be a set of possible values of this entity. Similarly, let $G_1, \ldots, G_n$ be another set of possible values, then $F_1+G_1,..., F_n+G_n$ is also a possible set of values since all three sets transform in the same way. We also note that if a linear and homogeneous entity vanishes relative to one frame of reference it vanishes relative to any other frame.

The following questions now arise: are there any limitations on the possible number of components and can a general rule be derived which will yield the transformation formula for any given and possible number of components? In the case of x, y, z, t we relied on the principle of relativity and the law of constancy of light velocity. In the case of $E_x,..., H_z$ we added Maxwell's equations to our stock of information. It so happens that these equations are invariant with respect to a Lorentz transformation and consequently this yielded (1.2) directly. In other cases we may be less fortunate as the exact (relativistic) laws which could have been used are not known.

To deal with the problem of determining in each case the linear transformations in a direct manner, as was done for x, y, z, t, would be a tedious task. In particular it would not be easy to establish the relationship between the coefficients of transformation and the relative velocity between the frames of reference. Only in the case of entities with four real components we might expect, with some degree of certainty, that they would have to transform like x, y, z, t because, as we have seen, there exists essentially only one transformation between four components which satisfies the principle of relativity and the law of constancy of light velocity. Fortunately, however, the problem with which we are now faced can be solved in a general manner if we make use of the requirement that all the transformations, for any possible number of components, must possess an inner link with each other—they must have the same group property.

Let $L(v_1)$ represent the matrix of a Lorentz transformation when the relative velocity between K and K' along the x-axes is v_1. Similarly, let $L(v_2)$ represent a transformation from K' to K'' when the relative velocity is v_2. Then a direct transformation $L(v_3)$ from K to K'' can also be obtained from $L(v_1)$ and $L(v_2)$ by multiplying these matrices:

$$L(v_1)L(v_2) = L(v_3), \tag{1.3}$$

and, as we have already seen in Chapter 1,

$$v_3 = \frac{v_1+v_2}{1+v_1 v_2/c^2}. \tag{1.4}$$

Eqns. (1.3) and (1.4) express the group property of the Lorentz transformation, or the law of multiplication of the elements of the group.

Let us now consider some other entity with a number of components which is not necessarily four. Let $S(v_1)$, $S(v_2)$, and $S(v_3)$ be the transformation matrices of this entity from K to K', K' to K'', and K to K'' respectively. These matrices must obey the same multiplication law as the matrices L:

$$S(v_1)S(v_2) = S(v_3), \tag{1.5}$$

where v_1, v_2 and v_3 are connected as before by (1.4). If this were not the case, we should get different laws of composition of velocities, depending on the entity observed. Since, however, only one law of composition of velocities can exist in nature, we necessarily come to the conclusion that all entities transform by means of matrices which exhibit the same group property as the Lorentz group.

Any set of matrices which satisfy a multiplication law which is identical with that of a given group is said to be a matrix representation of the group. More explicitly, we say a representation is present if we can associate with each element g_k of a group G a matrix M_k as follows:

$$\begin{matrix} g_1 & g_2 & g_3 & \cdots & g_k & \cdots & g_n, \\ M_1 & M_2 & M_3 & \cdots & M_k & \cdots & M_n, \end{matrix} \tag{1.6}$$

whereby, if

$$g_i g_k = g_l, \tag{1.7}$$

it also follows that

$$M_i M_k = M_l. \tag{1.8}$$

The case may arise when whole sub-sets of elements in the group are represented by the same matrix.

The transformation matrices of any linear and homogeneous physical entity must always form a representation of the Lorentz group, and from a mathematical point of view our task is thus to determine these representations. However, in this chapter we shall not deal as yet with the Lorentz group, but with one of its sub-groups. We shall assume that all the observers are at rest relative to each other ($v = 0$), and that the change in frame of reference merely consists in a rotation of axes. It will be convenient to deal at first with this special case and postpone the consideration of the Lorentz group to a later stage. The equivalence of all orthogonal Cartesian axes, no matter how inclined relative to each other, can be looked upon as resulting from a principle of relativity with respect to orientation, or from the isotropy of space.

2. Rotation of Axes

Let K and K' be two orthogonal Cartesian frames of reference which are inclined relative to each other and have the same origin. The transformation of coordinates which we are now considering is therefore

$$\begin{aligned} x' &= r_{1'1}x+r_{1'2}y+r_{1'3}z, \\ y' &= r_{2'1}x+r_{2'2}y+r_{2'3}z, \\ z' &= r_{3'1}x+r_{3'2}y+r_{3'3}z, \\ t' &= t, \end{aligned} \tag{2.1}$$

where the $r_{i'k}$ are the direction-cosines between the i'- and k-axes. The $r_{i'k}$ are not independent of each other. If (2.1) is to represent a rotation of orthogonal axes, the following orthogonality relations must hold:

$$\sum_{k'=1}^{3} r_{k'i}r_{k'l} = \delta_{il}, \tag{2.2}$$

where δ_{il} is equal to 1 for $i = l$ and to 0 for $i \neq l$. They can be proved in the same way as (10.21) in Chapter 1. In the present case we must use the invariance of $x^2+y^2+z^2$.

If we solve (2.1) and express x, y, z in terms of x', y', z', we get a new set of equations, the coefficients of which we denote by $r_{lm'}^{-1}$ and we know already from (10.25) in Chapter 1 that

$$r_{i'k} = r_{ki'}^{-1}. \tag{2.3}$$

Let us look upon x, y, z as a one-column matrix X and denote the matrices $(r_{i'k})$ and $(r_{lm'}^{-1})$ by R and R^{-1} respectively. Finally, let $\tilde{R}$ denote the transposed matrix in which corresponding columns and rows have been exchanged. Then we have in matrix notation:

$$X' = RX, \tag{2.4}$$

$$X = R^{-1}X', \tag{2.5}$$

$$R^{-1} = \tilde{R}. \tag{2.6}$$

Consider two rotations R_2 and R_1 performed one after the other. The result can also be achieved by a single rotation R_3. The following relation must hold:

$$R_1 R_2 = R_3. \tag{2.7}$$

For any set of square matrices, not necessarily orthogonal, the determinants obey the relation

$$\Delta_2\Delta_1 = \Delta_1\Delta_2 = \Delta_3. \tag{2.8}$$

However, in our present case of rotations, $\Delta_1 = \Delta_2 = \Delta_3 = 1$ and (2.8) is trivial.

Let us now assume that an observer in K measures an entity which is completely defined by just one value, i.e. one component. If this entity is to have a meaning which is independent of the frame of reference, then we must observe the same value in K' using the same units as in K. Hence, denoting the entity by F, its transformation law is given by

$$F' = F. \tag{2.9}$$

Physical entities of this type are called scalars. It often happens that the scalar has a value which is a function of x, y, z, t. Let $F(x,y,z,t)$ be the function expressing this value for each point x, y, z, t. Then the corresponding function $F'(x', y', z', t')$ in K' is obtained from the equation

$$F'(x', y', z', t) = F(x, y, z, t). \tag{2.10}$$

To obtain F' we have to replace in $F(x,y,z,t)$ the coordinates x, y, z by their values in terms of x', y', z' as obtainable from (2.1). The resulting function of x', y', z', t is the one required.

Next let us assume that relative to K a physical entity is observed which possesses two components F_1 and F_2. To get a case which goes beyond the one we have already considered, we must assume that neither F_1 nor F_2 are scalars. If both were scalars we should simply be faced with a case in which two scalar entities were observed at the same time, for instance, pressure and temperature. Let us observe the same entity also relative to K'. In this frame of reference we may find that the components are given by F'_1 and F'_2. Mathematically, the most general transformation or relation between F_1, F_2 and F'_1, F'_2 would have the form†

$$\begin{aligned} F'_1 &= f_1(F_1, F_2), \\ F'_2 &= f_2(F_1, F_2), \end{aligned} \tag{2.11}$$

where f_1 and f_2 are some functions which contain as parameters the coefficients of the rotation. Now in our present case, as already mentioned, when $v = 0$, the principle of relativity simplifies into a principle of relativity with respect to orientation or isotropy of space. Hence, if we solve (2.11) and express F_1, F_2 in terms of F'_1 and F'_2 we should get the same functions again provided r_{ik} is replaced by $r_{ik}^{-1} = r_{ki}$.

If the entity (F_1, F_2) is linear and homogeneous the transformation (2.11) must be linear and homogeneous, say

$$\begin{aligned} F'_1 &= T_{11} F_1 + T_{12} F_2, \\ F'_2 &= T_{21} F_1 + T_{22} F_2, \end{aligned} \tag{2.12}$$

† A more correct notation for the transformed entity is $F_{1'}$, $F_{2'}$. This emphasizes that we are dealing with the same entity viewed from two different frames of reference. However, it is not customary to use primed indices. We shall mostly use primed indices, as was done for instance in (2.1), but occasionally, when more convenient, unprimed indices will be used and the prime will be transferred to the letter which defines the tensor. Instead of $r_{i'k}$ we then use r_{ik}. It is clear that the first index refers to the new axes and the second to the old axes.

where the T_{ik} depend only on the relative inclination of the axes x, y, z and x', y', z'.

The transformation (2.12) for all possible inclinations between axes of K and K' must form a representation of the group of rotations. Let R_1, R_2, and R_3 be three different rotations of axes, whereby

$$R_1 R_2 = R_3. \tag{2.13}$$

Let $T(R_1)$, $T(R_2)$, and $T(R_3)$ be the transformation matrices of (F_1, F_2) for R_1, R_2, and R_3 respectively. Then these transformations must satisfy the condition

$$T(R_1)T(R_2) = T(R_3). \tag{2.14}$$

We have already enlarged on this aspect in section 1 of this chapter. At present when $v = 0$ we cannot use the argument we have advanced before with regard to the law of composition of velocities. However, the group property of the transformations T as expressed in (2.13) and (2.14) results from the following argument which is more general than the one we applied previously. Suppose $T(R_1)T(R_2) \neq T(R_3)$. Then, by transforming (F_1, F_2) directly from K_1 to K_3 we get a result which is different from the one obtained if the rotation is achieved in two steps, R_1 and R_2. Since this is not permissible (2.14) must hold.

From the foregoing it follows that to obtain the transformation matrices of entities with two components when the axes are rotated, we must seek two-dimensional representations of the group of rotations, i.e. a representation which consists of matrices with two columns and two rows.

In cases when the components F_1 and F_2 are functions of x, y, z, t, (2.12) takes the form

$$\begin{aligned} F_1'(x', y', z', t) &= T_{11} F_1(x, y, z, t) + T_{12} F_2(x, y, z, t), \\ F_2'(x', y', z', t) &= T_{21} F_1(x, y, z, t) + T_{22} F_2(x, y, z, t), \end{aligned} \tag{2.15}$$

where the T_{ab} $(a, b = 1, 2)$ are independent of x, y, z, t. In this case, to obtain the component $F_1'(x', y', z', t)$ we have to take the combination $T_{11} F_1(x, y, z, t) + T_{12} F_2(x, y, z, t)$ and express x, y, z in terms of x', y', z' as obtainable from (2.1). A similar procedure yields $F_2'(x', y', z', t)$.

If (F_1, F_2) is to represent an entity which does not consist simply of two scalars, the two-dimensional representation must be irreducible. In other words, the matrices must not have the form

$$\begin{pmatrix} T_{11} & 0 \\ 0 & T_{22} \end{pmatrix} \tag{2.16}$$

for all possible rotations. For if this were the case we could prove that the diagonal elements T_{11} and T_{22} are necessarily equal to 1. The only one-dimensional matrices which can represent the group of rotations consist of 'ones':

$$R_1 R_2 = R_3.$$
$$1.1 = 1, \tag{2.17}$$

Hence, in the reduced case, we necessarily have

$$\begin{bmatrix} T_{11} & 0 \\ 0 & T_{22} \end{bmatrix} = \begin{bmatrix} 1 & 0 \\ 0 & 1 \end{bmatrix}, \tag{2.18}$$

and for the components F_1, F_2 we get

$$F'_1 = F_1,$$
$$F'_2 = F_2, \tag{2.19}$$

so that each of the components represents a scalar.

In the more general case, if the entity has n components, the transformation law will have the form

$$F'_a = \sum_{b=1}^{n} T_{ab} F_b, \tag{2.20}$$

where the T_{ab} are constants depending only on the parameters of rotation r_{kl} and forming an n-dimensional representation of the group of rotations.

The determination of all the possible linear transformations of real variables which represent the group of rotations is essentially an algebraic problem and it is outside the scope of this book to deal with the matter systematically. There exists an infinite number of possible representations and those which are irreducible have the dimensions 1, 3, 5, 7,.... Hence irreducible linear and homogeneous entities with an even number of components do not exist.

The simplest possible representation of the group of rotations is the one which results from the way a scalar transforms. However, this representation is trivial. To begin with, the most general transformation which we might expect for an entity with a single component is

$$F'(x', y', z', t) = f_R\{F(x, y, z, t)\}, \tag{2.21}$$

or

$$F' = f_R(F), \tag{2.22}$$

where f_R is an arbitrary function, depending only on the considered rotation R, and yielding some representation of the group of rotations. However, if the entity F is linear and homogeneous, f_R must be a linear and homogeneous function of F, say

$$F' = T(R)F, \tag{2.23}$$

where $T(R)$ is a number independent of F and depending only on the rotation R. Now the transformation (2.23) must also be a representation of the group of rotations:

$$T(R_1)T(R_2) = T(R_1 R_2). \tag{2.24}$$

The only possible way of satisfying this relation for any pair of rotations is achieved by the following solution:

$$T(R_1) = T(R_2) = T(R_1 R_2) = 1. \tag{2.25}$$

In this way we obtain again the transformation of a scalar, which we have already introduced in a more direct way:

$$F' = 1.F = F. \tag{2.26}$$

The representation which is connected with scalars is the simplest possible. It represents each rotation by 1. The group property is satisfied and corresponds with $1.1 = 1$.

The next irreducible representation is three-dimensional. In this case the matrices $T_{ik}(R)$ may be taken simply as equal to the matrices R of the rotations themselves. Physical entities transforming according to this representation are called vectors. Vectors and coordinates transform in the same way. Familiar examples are forces, velocities, accelerations, etc.

The next irreducible representation is five-dimensional. Instead of considering this, it is more useful in physics to consider the transformations of entities which are obtainable from vectors. Let us form all the possible products of the components of two vectors **A** and **B**. Their number is nine: $A_x B_x$, $A_x B_y$,..., $A_z B_z$. These products are said to represent the direct product $\mathbf{A}\times\mathbf{B}$. The observer K' will also observe nine components $A_{x'} B_{x'}$,..., $A_{z'} B_{z'}$. It is evident that these quantities depend linearly on the corresponding products in K:

$$A_{k'} B_{l'} = \sum_{m,n} T_{k'l',mn} A_m B_n, \tag{2.27}$$

where k', l' are x', y', z' and m, n are x, y, z. The matrix in (2.27) contains 81 elements which are obtainable from the $r_{i'k}$ as follows:

$$T_{k'l',mn} = r_{k'm} r_{l'n}. \tag{2.28}$$

Thus the transformation matrix is obtained by multiplying each element of R by any other element of R (including the product of each element with itself). We may denote the transformation by $(R\times R)$, since T can be looked upon as a direct product of R with itself.

Let us consider the rotations R_1, R_2, and R_3 where $R_1 R_2 = R_3$. Then

$$(R_1\times R_1)(R_2\times R_2) = \{(R_1 R_2)\times(R_1 R_2)\} = (R_3\times R_3). \tag{2.29}$$

This follows directly from the fact that all the members of this equation represent matrices with 81 elements obtained from R_1 and R_2 in the same way.

Eqn. (2.29) demonstrates that the direct product $(R\times R)$ is also a representation of the group of rotations. This rule can be generalized in two ways. Firstly, from a representation R we can get others by forming the direct products $(R\times R)$, $\{(R\times R)\times R\} = (R\times R\times R)$, etc. Secondly, if M_k and N_k are sets of matrices which represent the group of rotations R, then $(M_k\times N_k)$ is also a representation of the same group. M_k and N_k need not have the same dimensions.

From the foregoing we conclude that the direct product of two vectors forms a linear entity with nine components. It transforms according to a nine-dimensional representation $(R\times R)$ of the group of rotations. An entity of the type $(\mathbf{A}\times\mathbf{B})$ is called a tensor of second rank. In the same way $(\mathbf{A}\times\mathbf{B}\times\mathbf{C})$ is a tensor of third rank transforming according to $(R\times R\times R)$. It will be convenient to look upon a vector as a tensor of first rank and a scalar as a tensor of zero rank. From now on a vector will be denoted by a symbol with one index, say F_k. In this notation we emphasize that it has three components according to three possible values of k: x, y, z (or 1, 2, 3 if x_1, x_2, x_3 is used instead of x, y, z). In the same way we shall denote a tensor of second rank by F_{kl}. This entity has two indices each running through x, y, z (or 1, 2, 3). The number of components is thus nine. A tensor of third rank F_{klm} has three indices and $3^3 = 27$ components, and so forth for higher tensors. Whenever possible we shall denote a scalar also by a capital letter, in this case without an index. However, to follow common usage, we shall have to deviate sometimes from this convention.

Let $(\mathbf{A}\times\mathbf{B})$ and $(\mathbf{C}\times\mathbf{D})$ be two tensors of second rank. Then any linear combination $m(\mathbf{A}\times\mathbf{B})+n(\mathbf{C}\times\mathbf{D})$ is also a tensor of second rank which transforms according to $(R\times R)$. This indicates that the most general tensor of second rank is not a direct product of two vectors, but can be obtained through suitable linear combinations of such direct products.

A tensor of second rank F_{kl} is not a primitive entity because the representation $(R\times R)$ turns out to be reducible. In other words, in the nine-dimensional space in which each axis is associated with one component of F_{kl} it is possible to introduce a system of coordinates (or change the tensor frame) in such a way that $(R\times R)$ takes the form

$$(R\times R)' = \begin{bmatrix} [1] & 0 & 0 \\ 0 & [3] & 0 \\ 0 & 0 & [5] \end{bmatrix}. \tag{2.30}$$

The numbers in brackets indicate the dimensions of the matrix. That a reduction is possible is seen from the fact that any tensor of second rank can be split into a sum of two tensors of second rank, a symmetrical tensor $S_{kl} = S_{lk}$ and an anti-symmetrical tensor $A_{kl} = -A_{lk}$, according to:

$$F_{kl} = \tfrac{1}{2}(F_{kl}+F_{lk})+\tfrac{1}{2}(F_{kl}-F_{lk}). \tag{2.31}$$

That each part in (2.31) is a tensor is proved as follows for the special tensor:

$$F_{kl} = A_k B_l \equiv \tfrac{1}{2}(A_k B_l+B_k A_l+A_k B_l-B_k A_l). \tag{2.32}$$

Each of the terms on the right-hand side represents the k, l component of a tensor. Consequently, since the sum of two tensors is also a tensor, $A_k B_l+B_k A_l = S_{kl}$ is a tensor and we also have $S_{kl} = S_{lk}$. Similarly, since the difference of two tensors is a tensor, again

$$A_k B_l-B_k A_l = A_{kl}$$

is a tensor with the property $A_{kl} = -A_{lk}$. The proof can be extended to more general tensors than in (2.32).

A symmetrical tensor has essentially six different components. Consequently, the part of $(R\times R)$ concerned with such tensors has six dimensions. In the same way, the part of $(R\times R)$ concerned with the anti-symmetrical tensor has only three dimensions because an anti-symmetrical tensor has essentially only three non-vanishing numbers which characterize it: $A_{xy} = -A_{yx}$, $A_{yz} = -A_{zy}$, and $A_{xz} = -A_{zx}$. The splitting of any tensor into a symmetrical and an anti-symmetrical part is accompanied by a reduction of $(R\times R)$ according to the scheme:

$$(R\times R)' = \begin{bmatrix} [3] & 0 \\ 0 & [6] \end{bmatrix}. \tag{2.33}$$

The three-dimensional part of $(R\times R)'$ is identical with the three-dimensional matrices R. Hence, as far as rotations are concerned, anti-symmetrical tensors of second rank may be looked upon as vectors. In ordinary vector analysis they are called vector products and denoted by $[\mathbf{A}\times\mathbf{B}]$ or $\mathbf{A}\wedge\mathbf{B}$. Vector products of vectors are said to be axial vectors, as distinguished from the ordinary vectors which are called polar vectors. Since x, y, z are the components of a typical polar vector, a reflection of axes $x' = -x$, $y' = -y$, $z' = -z$ changes the sign of the components of a polar vector. On the other hand, $x_1 y_2-x_2 y_1$, $x_1 z_2-x_2 z_1$, $y_1 z_2-z_2 y_1$ are the components of a typical axial vector $[\mathbf{X}_1, \mathbf{X}_2]$. On reflecting the axes with respect to the origin ($x'_1 = -x_1$, $x'_2 = -x_2$, etc.) the components of an axial vector, unlike those of a polar vector, do not change sign. From the foregoing it follows that anti-symmetrical tensors of second rank are identical with vectors only as long as rotations are considered. A difference between them arises when a reflection of axes is considered.

The six-dimensional part of $(R\times R)$ in (2.33) is reducible into two irreducible representations of one and five dimensions respectively, see (2.30). This is connected with the invariance of the sum of all the diagonal elements $(F_{xx}+F_{yy}+F_{zz})$ with respect to rotations of axes. This sum is called the trace of the tensor and its invariance follows from the elementary fact that $A_x B_x+A_y B_y+A_z B_z$ is nothing but the scalar product of **A** and **B** and, as suggested by its name, it retains its value when the axes are rotated. We now see that the six quantities which characterize a symmetrical tensor can always be made to depend on one quantity which does not change its value when the axes are rotated (a scalar or invariant), and five others, which transform as the components of an entity with five components.

3. Tensor Algebra and Analysis

The following are the main algebraic rules which apply in tensor theory:

(1) If $F_{kl...}$ and $G_{kl...}$ are tensors of the same rank then $aF_{kl...}+bG_{kl...}$ is also a tensor of the same rank (a, b any numbers).

(2) A direct product of two tensors $F_{kl...}$ of rank α and $G_{mn...}$ of rank β is a tensor $H_{kl...mn...}$ of rank $\alpha+\beta$.

(3) From any tensor of rank α it is possible to obtain a tensor of rank $\alpha-2$ by an algebraic process which is known as contraction. This consists of adding up all the components which have equal indices in a given pair, for instance in k, m:

$$G_l = \sum_{k=m=1}^{3} F_{klm} = \sum_{k=1}^{3} F_{klk} = F_{1l1}+F_{2l2}+F_{3l3}. \tag{3.1}$$

Thus, in this example contraction is taking place with regard to the first and third index. The resulting tensor is of rank 1, i.e. it is a vector.

(4) Any tensor can be symmetrized or anti-symmetrized with respect to any pair of indices. For example, $(\mathbf{A}\times\mathbf{B}\times\mathbf{C})\pm(\mathbf{C}\times\mathbf{B}\times\mathbf{A})$ is symmetrical or anti-symmetrical with respect to the first and third index. In particular, if we form all the permutations of $(\mathbf{A}\times\mathbf{B}\times\mathbf{C})$ and sum them we get a tensor which is symmetrical in any pair of indices, and if we take all the even permutations with positive sign and the odd permutations with negative sign the sum will yield a tensor which is anti-symmetrical in any pair of indices.

These four algebraic operations of linear combination, direct multiplication, contraction, and symmetrization or anti-symmetrization have absolute significance, i.e. they apply both in K and K'. For instance, if F_{mkl} and $F_{m'k'l'}$ are the components of a tensor of third rank relative to K and K' respectively, and if G_k and $G_{k'}$ are corresponding components of a vector, then if (3.1) holds in K, a similar relation holds in K':

$$G_{l'} = \sum_{k'=1}^{3} F_{k'l'k'}. \tag{3.2}$$

The algebraic operations mentioned are the only ones which enable us to obtain from one set of tensors a new set of such entities. To give just one example of an operation which does not lead to a tensor, we may consider the squaring of each element of a tensor. The result is not a tensor. The relation between $F^2_{k'l'}$ and F^2_{kl} is not that of a tensor transformation and does not yield a representation of the group of rotations.

The contraction of a tensor of rank 2 leads to a scalar, because the result has only one component:

$$S = \sum_{l=1}^{3} F_{ll}. \tag{3.3}$$

In particular, if a tensor happens to be a direct product of two vectors, $F_{kl} = A_k B_l$, its contraction leads to what is called in vector analysis the scalar product of the vectors:

$$S = \sum_{l=1}^{3} A_l B_l. \tag{3.4}$$

If a tensor happens to be a function of x, y, z, we can also add differentiation to the algebraic operations. By differentiating each component with respect to x, y, z we get from a tensor of rank α a new tensor of rank $\alpha+1$. For instance,

$$\frac{\partial F_{kl}}{\partial x_m} = G_{klm}. \tag{3.5}$$

Let us prove this rule in the simplest possible case, when the original tensor is of rank 0, i.e. a scalar $F(x,y,z,t)$. The derivatives relative to a frame of axes K are $\partial F/\partial x$, $\partial F/\partial y$, $\partial F/\partial z$.† In K' the scalar is

$$F'(x',y',z',t),$$

where

$$F(x,y,z,t) = F'(x',y',z',t), \tag{3.6}$$

and the derivatives are $\partial F'/\partial x'$, $\partial F'/\partial y'$, $\partial F'/\partial z'$. We have to prove that

† Extension through simple differentiation applies only for constant coordinate transformations, where the coefficients of transformation are independent of x, y, z, t. Only such transformations are considered in this book.

these three components depend on the former through a vector transformation. From (3.6) we get

$$\frac{\partial F'}{\partial x'} = \frac{\partial F}{\partial x}\frac{\partial x}{\partial x'} + \frac{\partial F}{\partial y}\frac{\partial y}{\partial x'} + \frac{\partial F}{\partial z}\frac{\partial z}{\partial x'},$$

$$\frac{\partial F'}{\partial y'} = \frac{\partial F}{\partial x}\frac{\partial x}{\partial y'} + \frac{\partial F}{\partial y}\frac{\partial y}{\partial y'} + \frac{\partial F}{\partial z}\frac{\partial z}{\partial y'},$$

$$\frac{\partial F'}{\partial z'} = \frac{\partial F}{\partial x}\frac{\partial x}{\partial z'} + \frac{\partial F}{\partial y}\frac{\partial y}{\partial z'} + \frac{\partial F}{\partial z}\frac{\partial z}{\partial z'}. \tag{3.7}$$

Now, from (2.1),
$$\frac{\partial x_l}{\partial x_{k'}} = r_{lk'}^{-1} = r_{k'l}. \tag{3.8}$$

Hence
$$\frac{\partial F'}{\partial x_{k'}} = r_{k'l}\frac{\partial F}{\partial x_l}. \tag{3.9}$$

This equation proves that $\partial F/\partial x_k$ transform like the components of a vector or like x, y, z.

From (3.6) we find

$$\frac{\partial F'}{\partial x'}dx' + \frac{\partial F'}{\partial y'}dy' + \frac{\partial F'}{\partial z'}dz' = \frac{\partial F}{\partial x}dx + \frac{\partial F}{\partial y}dy + \frac{\partial F}{\partial z}dz. \tag{3.10}$$

Since dx, dy, dz transform like x, y, z, they form three components of a vector. We have also proved that $\partial F/\partial x_l$ are components of a vector. This suggests that whenever we have $A_l B_l = A_{l'} B_{l'}$ and we know that A_l is a vector, we may conclude that B_l is also a vector.

From a vector we get through extension a tensor of rank two:

$$G_{kl} = \frac{\partial F_k}{\partial x_l}. \tag{3.11}$$

(That this is an absolute relation is suggested by $G_{kl}\Delta x_l = \Delta F_k$ which is obtained through contraction.)

By contraction we can lower the rank of this tensor by two and get a scalar. In vector analysis this is called the divergence of F_k:

$$S = \sum_{l=1}^{3}\frac{\partial F_l}{\partial x_l} = \frac{\partial F_x}{\partial x} + \frac{\partial F_y}{\partial y} + \frac{\partial F_z}{\partial z}. \tag{3.12}$$

We may treat $\partial/\partial x$, $\partial/\partial y$, $\partial/\partial z$ as components of a vector which is denoted by ∂_k. The scalar product of this vector with any vector F_k yields its divergence which is a scalar. It follows further that we can look upon $\nabla^2 = \partial^2/\partial x^2 + \partial^2/\partial y^2 + \partial^2/\partial z^2$ as a scalar operator. If $F(x, y, z)$

is a scalar, $\nabla^2 F$ is also a scalar, because the product of two scalars is again a scalar. Hence

$$\frac{\partial^2 F}{\partial x^2}+\frac{\partial^2 F}{\partial y^2}+\frac{\partial^2 F}{\partial z^2}=\frac{\partial^2 F'}{\partial x'^2}+\frac{\partial^2 F'}{\partial y'^2}+\frac{\partial^2 F'}{\partial z'^2}. \tag{3.13}$$

If we multiply ∇^2 by each component of a vector F_k (more generally of any tensor) we get a new vector (or tensor of the same rank). The same vector can be obtained by extending in two steps the rank of F_k ($F_k \to \partial F_k/\partial x_l \to \partial^2 F_k/\partial x_l \partial x_m$) and by subsequent contraction of the resulting tensor of third rank with respect to the lower indices:

$$\sum_{l=1}^{3} \frac{\partial^2 F_k}{\partial x_l \partial x_l}=\frac{\partial^2 F_k}{\partial x^2}+\frac{\partial^2 F_k}{\partial y^2}+\frac{\partial^2 F_k}{\partial z^2}. \tag{3.14}$$

Extension of a tensor can also be obtained when the tensor happens to depend on another tensor and a differentiation is performed. For example, let a scalar S depend on a tensor F_{kl}, then

$$\frac{\partial S}{\partial F_{kl}}=G_{kl} \tag{3.15}$$

is also a tensor. In this case the rank of the resulting tensor is higher by two than that of the original tensor S.

4. Some Special Tensors

No vector exists which has the same components in different systems of coordinates. There exist, however, tensors of higher rank which possess the same components in all the frames of reference. Such tensors are said to have invariant components.

Let us consider the Kronecker symbol δ_{kl}. If we treat it as a tensor of second rank and transform it accordingly we get (see 2.28)

$$\sum_{m,n=1}^{3} r_{k'm} r_{l'n} \delta_{mn}=\delta_{k'l'}. \tag{4.1}$$

We have yet to prove that the right-hand side is in fact a Kronecker symbol. This is achieved by actually carrying out the summation on the left-hand side. It yields

$$\sum_m r_{k'm} r_{l'm}=\sum_m r_{k'm} r^{-1}_{ml'}=\delta_{k'l'} \tag{4.2}$$

and the invariance of the components is thus proved. From δ_{kl} we can form higher invariant tensors through direct products $\delta_{kl}\delta_{mn}$, etc.

Another tensor with invariant components is the one which is of third rank and anti-symmetrical in every pair of indices. Let ϵ_{klm} be such a tensor. Then, by definition,

$$\epsilon_{klm} = -\epsilon_{lkm} = \epsilon_{mkl} = -\epsilon_{kml} = \epsilon_{lmk} = -\epsilon_{mlk}. \tag{4.3}$$

It follows that whenever two indices are equal the component is zero. Hence of the 27 components of the tensor, only 6 are non-vanishing. They all have the same absolute value, 3 being positive and the remainder negative. A completely anti-symmetrical tensor remains such a tensor when multiplied by a scalar. ϵ_{klm} is normalized in such a way that

$$\epsilon_{123} \equiv \epsilon_{xyz} = 1. \tag{4.4}$$

It then follows that all the even permutations of 1, 2, 3 correspond to components with $+1$ and all the odd permutations correspond to -1. By carrying out the transformation

$$\sum_{l,n,t=1}^{3} r_{k'l} r_{m'n} r_{s't} \epsilon_{lnt} = \epsilon'_{k'm's'} = (\det R)\epsilon_{k'm's'}, \tag{4.5}$$

we can prove that

$$\epsilon'_{kms} = \epsilon_{kms}. \tag{4.6}$$

A few remarks must now be added with regard to the behaviour of ϵ_{klm} and δ_{kl} when reflections of axes are considered.

Ordinary rotations combined with a reflection form a wider group of transformations which is called the full group of rotations. It is obtained through a direct product of the ordinary group of rotations with the group formed by E, P where E is the identical transformation $x' = x, \ldots, z' = z$ and P is the reflection $x' = -x, \ldots, z' = -z$. This group of two elements has two one-dimensional representations according to the schemes:

$$\begin{matrix} E & P \\ 1 & 1 \end{matrix} \qquad \begin{matrix} E & P \\ 1 & -1. \end{matrix} \tag{4.7}$$

From these we can construct various three-dimensional representations such as:

$$\overset{E}{\begin{bmatrix} 1 & 0 & 0 \\ 0 & 1 & 0 \\ 0 & 0 & 1 \end{bmatrix}} \qquad \overset{P}{\begin{bmatrix} 1 & 0 & 0 \\ 0 & 1 & 0 \\ 0 & 0 & -1 \end{bmatrix}} \tag{4.8}$$

but only the following can be used:

$$\overset{E}{\begin{bmatrix} 1 & 0 & 0 \\ 0 & 1 & 0 \\ 0 & 0 & 1 \end{bmatrix}} \overset{P}{\begin{bmatrix} 1 & 0 & 0 \\ 0 & 1 & 0 \\ 0 & 0 & 1 \end{bmatrix}} \quad \text{or} \quad \overset{E}{\begin{bmatrix} 1 & 0 & 0 \\ 0 & 1 & 0 \\ 0 & 0 & 1 \end{bmatrix}} \overset{P}{\begin{bmatrix} -1 & 0 & 0 \\ 0 & -1 & 0 \\ 0 & 0 & -1 \end{bmatrix}}. \tag{4.9}$$

The reason for this is that both E and P commute with all the rotations. In other words, if we first rotate and then reflect the axes or perform these operations in opposite order we get the same result. It follows that the three-dimensional matrices which represent E and P must all commute with the rotational matrices R. In accordance with a theorem by I. Schur about irreducible sets of matrices such as the R's, this can only be the case if the matrices representing E and P are multiples of a unit matrix. This is the case with (4.9) but not with (4.8).

Both representations of (4.9) can occur. The first one allows us to define vectors with components which do not change sign on reflection, the second representation enables us to define vectors which change sign whenever a reflection takes place. When we speak of a vector we usually think of components which change sign on reflection (like x, y, z). But in accordance with (4.9) we can also introduce vectors which do not change sign. Such vectors are called pseudo-vectors. We have already seen that if A_k and B_l are vectors (of same type), $A_k B_l - A_l B_k$ behaves like a vector which does not change sign on reflection. Hence it is a pseudo-vector. In the same way, in accordance with (4.7) we can distinguish between scalars and pseudo-scalars. The former remain unchanged when a reflection takes place, while pseudo-scalars change sign.

The transformation law of a tensor of second rank is the same as that of $\mathbf{A}\times\mathbf{B}$. If both vectors are of the same type the components of the tensor will not change sign on reflection, if they are of different type all the components of the tensor will change sign. All tensors of odd rank should normally change sign on reflection because by a tensor we mean in the first place an entity which transforms like the direct product of polar vectors. Tensors of odd rank which do not change sign on reflection must be distinguished from the former and are called pseudo-tensors. With tensors of even rank matters are reversed. Normally they should not change sign on reflection. If they do they are pseudo-tensors. Any tensor becomes a pseudo-tensor by multiplication by a pseudo-scalar and any pseudo-tensor becomes a tensor through such multiplication. Algebraically we can express the difference between tensors and pseudo-tensors by the following formulae:

$$\text{tensors:} \qquad F_{k'l'\dots} = \sum_{m,n\dots} r_{k'm} r_{l'n} \dots F_{mn\dots},$$

$$\text{pseudo-tensors:} \qquad F_{k'l'\dots} = r \sum r_{k'm} r_{l'n} \dots F_{mn\dots},$$

where r is the determinant of the transformation which takes place.

As long as we deal with rotations proper $r = 1$ and there is no difference between tensors and pseudo-tensors, but when we deal with a reflection $r = -1$ and the components of pseudo-tensors transform with a factor -1 as compared with those of a tensor of the same rank.

It is natural to define the tensors δ_{kl} and ϵ_{klm} as having invariant components also if reflections take place. Hence δ_{kl} being of second rank, is a tensor. On the other hand, ϵ_{klm} is of third rank and, since by definition it is to remain unchanged under reflection, it is a pseudo-tensor. (See (4.5) where for reflection $\det R = -1$.)

The direct product of two tensors is a tensor, of a tensor and a pseudo-tensor is a pseudo-tensor, and finally the direct product of two pseudo-tensors is a tensor.

5. Volume, Surface, and Line Integrals

Let us consider three infinitesimal polar vectors dA_k, dB_l, dC_m. They determine a small parallelepiped, the volume of which is given by

$$dV = \sum_{k,l,m=1}^{3} \epsilon_{klm}\, dA_k\, dB_l\, dC_m. \tag{5.1}$$

It is, however, more conventional to express this volume in the following equivalent form:

$$dV = \begin{vmatrix} dA_x & dB_x & dC_x \\ dA_y & dB_y & dC_y \\ dA_z & dB_z & dC_z \end{vmatrix} \equiv [dA_x, dB_y, dC_z]. \tag{5.2}$$

From (5.1) or (5.2) it is directly evident that the volume element, and hence any finite volume, is an invariant with respect to ordinary rotations but changes sign on reflection. Hence it is a pseudo-scalar.

Yet another way of expressing the volume element is obtained by introducing the tensor

$$dV_{klm} = \begin{vmatrix} dA_k & dB_k & dC_k \\ dA_l & dB_l & dC_l \\ dA_m & dB_m & dC_m \end{vmatrix}. \tag{5.3}$$

This tensor is anti-symmetrical in any pair of indices and can therefore be only a multiple of ϵ_{klm}. dV_{klm} is a tensor and ϵ_{klm} is a pseudo-tensor, hence the product is a pseudo-scalar and is in fact equal to the volume element. The following relation holds:

$$dV = \frac{1}{3!} \sum_{k,l,m} \epsilon_{klm}\, dV_{klm}. \tag{5.4}$$

Two tensors which are associated with each other through ϵ_{klm} and appropriate numerical factors are said to be dual to each other. Thus dV is the dual of dV_{klm} and vice versa. For the square of the volume we get

$$dV^2 = \frac{1}{3!} \sum_{k,l,m} dV_{klm} dV_{klm}. \tag{5.5}$$

If we use three vectors which are parallel to the three axes respectively, $dA_x = dx$, $dA_y = dA_z = 0$; $dB_y = dy$, etc., the volume element has the value

$$dV = dxdydz. \tag{5.6}$$

If we transform to another system of coordinates we again get the more complicated expressions (5.1), (5.2), or (5.4), because the three vectors are no longer parallel to the axes. However, in the new system of coordinates we can again subdivide any volume into parallelepipeds which are parallel to the axes. From the invariance of the volume it then follows that

$$\iiint dxdydz = \iiint dx'dy'dz'. \tag{5.7}$$

It must, however, be realized that the new quantities are not to be obtained from a transformation of dx, dy, dz as if these were three components of a vector. These, like dx', dy', dz', refer to three different vectors.

When a tensor field $F_{pq...}$ is integrated over a volume we can consider two types of integrations. In one case we multiply $F_{pq...}$ by dV at each point and integrate over the whole volume. The result is a tensor of the same rank:

$$\iiint F_{pq...}\, dV = G_{pq...}. \tag{5.8}$$

In the second case we multiply $F_{pq...}$ at each point by the dual dV_{klm} of dV. The integration yields a tensor, of a rank which is higher by three than that of the original tensor:

$$\iiint F_{pq...}\, dV_{klm} = G_{pq...klm}. \tag{5.9}$$

Next, we consider surface elements and surface integrals. Two infinitesimal vectors dA_k, dB_l define an infinitesimal parallelogram which has three projections in the xy, yz, and xz planes. They behave like the components of an anti-symmetrical tensor of second rank, which we denote by dV_{kl} and which is given by

$$dV_{kl} = \begin{vmatrix} dA_k & dB_k \\ dA_l & dB_l \end{vmatrix} = dA_k dB_l - dB_k dA_l \equiv [dA_k, dB_l]. \tag{5.10}$$

This yields

$$\begin{aligned} dV_{xy} &= dA_x dB_y - dB_x dA_y, \\ dV_{yz} &= dA_y dB_z - dB_y dA_z, \\ dV_{zx} &= dA_z dB_x - dB_z dA_x. \end{aligned} \tag{5.11}$$

These are the areas of the projections of the considered parallelogram on the three planes.

With the help of the tensor ϵ_{klm} we can form a dual vector $d\sigma_m$ defined as follows:

$$d\sigma_m = \frac{1}{2!}\sum_{k,l} \epsilon_{klm}\, dV_{kl}. \tag{5.12}$$

$d\sigma_m$ is a vector which is perpendicular to the surface element and has a length which is equal to the area of the parallelogram. This follows from

$$\sum_{m=1}^{3} d\sigma_m^2 = \tfrac{1}{2}\sum_{k,l} dV_{kl}\, dV_{kl} = dV_{xy}^2 + dV_{yz}^2 + dV_{zx}^2, \tag{5.13}$$

and

$$\sum d\sigma_m\, dA_m = \tfrac{1}{2}\sum_{k,l,m} \epsilon_{klm} \begin{vmatrix} dA_k & dB_k \\ dA_l & dB_l \end{vmatrix} dA_m$$

$$= \tfrac{1}{2}\Big(\sum_{k,l,m} \epsilon_{klm}\, dA_k\, dA_m\, dB_l - \sum_{k,l,m} \epsilon_{klm}\, dB_k\, dA_l\, dA_m\Big) = 0, \tag{5.14}$$

since ϵ_{klm} is anti-symmetrical and $dA_k\, dA_m$ and $dA_l\, dA_m$ are symmetrical tensors. Similarly,

$$d\sigma_m\, dB_m = 0. \tag{5.15}$$

Of surface integrals the most important is that in which a vector F_m is multiplied scalarly by $d\sigma_m$ and the integration is carried over the surface

$$\sum_{m=1}^{3} \iint F_m\, d\sigma_m = S. \tag{5.16}$$

In our present notation, Gauss's theorem which converts a volume integral into a surface integral takes the form

$$\sum_{k=1}^{3} \iiint \frac{\partial F_k}{\partial x_k}\, dV = \sum_{m=1}^{3} \iint F_m\, d\sigma_m, \tag{5.17}$$

where the surface is the one enclosing the considered volume. Stokes's theorem takes the form

$$\frac{1}{2}\sum_{k,l=1}^{3} \iint \left(\frac{\partial F_k}{\partial x_l} - \frac{\partial F_l}{\partial x_k}\right) dV_{kl} = \sum_{m=1}^{3} \int F_m\, dL_m = \int (F_x\, dx + F_y\, dy + F_z\, dz), \tag{5.18}$$

where dL_m is a line element (a vector on the contour of the considered surface). Here we use the product of a tensor with dV_{kl} and integrate over the surface and not the product with $d\sigma_m$. The dual of dL_m is a tensor of second rank $d\sigma_{kl}$:

$$d\sigma_{kl} = \frac{1}{1!}\epsilon_{klm}\, dL_m. \tag{5.19}$$

It represents a surface element which is perpendicular to dL_m and has an area which is numerically equal to the length of dL_m. An integral over a surface of the form

$$\iint F_{pq\ldots}\, d\sigma_{kl} = G_{pq\ldots kl} \tag{5.20}$$

raises the rank of the integrated tensor by two, whereas a line integral of the form

$$\int F_{pq\ldots}\, dL_m = G_{pq\ldots m} \tag{5.21}$$

raises the rank by one.

6. Covariance of Physical Laws with respect to Rotation of Axes

No observation exists which would lead us to assume that the ordinary space is not isotropic (principle of relativity with regard to orientation). Therefore all the Cartesian systems which differ from each other only through the orientation of the axes are equivalent and the physical laws must take the same form in all of them. Moreover, since it is permissible to use any Cartesian system, the transformation from one such system to another must be accompanied in a consistent manner by a corresponding transformation of the different physical entities. Of the various linear entities the only ones which behave properly in this sense are the tensors, because their transformations form representations of the group of rotations.

Physical laws in their most natural form will therefore present relations between tensors. If these relations are then to have the same form in all the equivalent frames of reference they must be of absolute nature, i.e. of the types listed in section 3. Thus let us assume that a physical law expresses a relation between two tensors of the following form:

$$F_{kl} + G_{kl} = \delta_{kl}. \tag{6.1}$$

Then a similar relation will also hold in K':

$$F_{k'l'} + G_{k'l'} = \delta_{k'l'}, \tag{6.2}$$

where $F_{k'l'}$ and $G_{k'l'}$ are the transformed tensors. On the other hand, if we square the elements of each tensor and find that the following relation holds in K:

$$F^2_{kl} + G^2_{kl} = \delta_{kl}, \tag{6.3}$$

this cannot represent a general physical law. In K',

$$F^2_{k'l'} + G^2_{k'l'} \neq \delta_{k'l'}. \tag{6.4}$$

We must look upon (6.3) as representing a relation which happens to be valid only under special circumstances.

Tensors are said to be covariant entities and absolute relations between them, which do not change their form as the coordinates are transformed, are said to be covariant relations. All physical laws must be expressible as covariant relations between tensors. Relations which are not covariant cannot represent physical laws. Thus, if ϕ is a scalar, such relations as

$$\frac{\partial\phi}{\partial x}+\frac{\partial\phi}{\partial y}+\frac{\partial\phi}{\partial z}=0 \tag{6.5}$$

or

$$\frac{\partial^3\phi}{\partial x^3}+\frac{\partial^3\phi}{\partial y^3}+\frac{\partial^3\phi}{\partial z^3}=0 \tag{6.6}$$

cannot represent physical laws because they are not obtainable from the scalar ϕ and the vector operator ∂_k through any of the relations listed in section 3. On the other hand, the relation

$$\frac{\partial^2\phi}{\partial x^2}+\frac{\partial^2\phi}{\partial y^2}+\frac{\partial^2\phi}{\partial z^2}\equiv\nabla^2\phi=S, \tag{6.7}$$

where S is a scalar, might well represent a physical law, because, as we have already seen, ∇^2 is a scalar operator. If (6.5) and (6.6) are transformed by rotating the axes, it will be found that in the new axes the relations take much more complicated forms. On the other hand (6.7) will not change its form.

Expressed in a more general way, any physical law should be capable of taking the form

$$f_i(F, F_k, F_{kl},...)=0 \quad (i=1,2,...,s) \tag{6.8}$$

where the f_i express absolute or covariant relations between the tensors $F, F_k, F_{kl},...$. If (6.8) holds in K, then a corresponding relation will hold in K':

$$f_i(F', F'_k, F'_{kl},...)=0 \quad (i=1,2,...,s). \tag{6.9}$$

In this section we shall examine the covariance of some physical laws with respect to rotation of axes. Let us begin with the equations of motion of a particle:

$$m\frac{d^2x}{dt^2}=F_x, \qquad m\frac{d^2y}{dt^2}=F_y, \qquad m\frac{d^2z}{dt^2}=F_z, \tag{6.10}$$

where m is the mass of the particle and F_x, F_y, F_z are forces acting in the x, y, z directions respectively. These equations present no difficulty, from a mathematical point of view, in being interpreted as covariant relations, with respect to rotation of axes. We have only to look upon m and dt as scalars, and F_x, F_y, F_z as components of a vector, and the covariance is established.

From an experimental point of view we confirm the invariance of m and dt with respect to rotation of axes, and the behaviour of a force as a vector (law of composition of forces—forces are added like vectors). Eqns. (6.10) may be written in tensorial form as follows:

$$m\frac{d^2x_k}{dt^2}-F_k=0, \tag{6.11}$$

where x_k is a vector with components x, y, z.

Next let us consider Maxwell's equations and their behaviour with respect to rotation of axes. These equations, when currents and charges are present, have the following form:

$$\begin{aligned}
&\frac{\partial E_z}{\partial y}-\frac{\partial E_y}{\partial z}=-\frac{1}{c}\frac{\partial H_x}{\partial t}, &&\frac{\partial H_z}{\partial y}-\frac{\partial H_y}{\partial z}=\frac{1}{c}\frac{\partial E_x}{\partial t}+J_x,\\
&\frac{\partial E_x}{\partial z}-\frac{\partial E_z}{\partial x}=-\frac{1}{c}\frac{\partial H_y}{\partial t}, &&\frac{\partial H_x}{\partial z}-\frac{\partial H_z}{\partial x}=\frac{1}{c}\frac{\partial E_y}{\partial t}+J_y,\\
&\frac{\partial E_y}{\partial x}-\frac{\partial E_x}{\partial y}=-\frac{1}{c}\frac{\partial H_z}{\partial t}, &&\frac{\partial H_y}{\partial x}-\frac{\partial H_x}{\partial y}=\frac{1}{c}\frac{\partial E_z}{\partial t}+J_z,\\
&\frac{\partial E_x}{\partial x}+\frac{\partial E_y}{\partial y}+\frac{\partial E_z}{\partial z}=\rho, &&\frac{\partial H_x}{\partial x}+\frac{\partial H_y}{\partial y}+\frac{\partial H_z}{\partial z}=0,
\end{aligned} \tag{6.12}$$

where E_x, E_y, E_z are the components of the electric field intensity, H_x, H_y, H_z the components of the magnetic field intensity, ρ the charge density, and J_x, J_y, J_z the components of the current density.

Maxwell's equations are symmetrical with respect to x, y, z ,therefore they cannot be inherently in conflict with the isotropy of space. Nevertheless, they could still be non-covariant (just as $\partial/\partial x+\partial/\partial y+\partial/\partial z$ is a non-covariant operator in spite of its symmetry with respect to x, y, z). The covariance of Maxwell's equations with respect to rotations will be proved if we can interpret them (or write them) as a set of absolute relations between tensors. Suppose we start by assuming that E_x, E_y, E_z are components of a vector E_k. Then it is impossible for H_x, H_y, H_z or their time derivatives to represent a vector. We try the next entity, a tensor of second rank. Since the magnetic field has only three components, it is natural to assume that they are the three non-vanishing components of an anti-symmetrical tensor of second rank:

$$H_x = H_{23} = -H_{32}, \quad H_y = H_{31} = -H_{13}, \quad H_z = H_{12} = -H_{21}. \tag{6.13}$$

Now from E_x, E_y, E_z and $\partial/\partial x$, $\partial/\partial y$, $\partial/\partial z$ we can form a tensor of second rank

$$F_{kl} = \frac{\partial E_k}{\partial x_l} = \partial_l E_k. \tag{6.14}$$

This may be decomposed into a symmetrical and an anti-symmetrical part. The latter is given by

$$\tfrac{1}{2}(F_{kl}-F_{lk}) = \frac{1}{2}\left(\frac{\partial E_k}{\partial x_l}-\frac{\partial E_l}{\partial x_k}\right). \tag{6.15}$$

The first three of Maxwell's equations can now be expressed in tensor form as follows:

$$\frac{\partial E_k}{\partial x_l}-\frac{\partial E_l}{\partial x_k} = -\frac{1}{c}\frac{\partial H_{kl}}{\partial t}. \tag{6.16}$$

Next we have to consider the remaining equations and see whether they also can be expressed in tensor form without upsetting (6.16). As far as the divergence of E is concerned, we have only to look upon ρ as a scalar, when we get the covariant equation

$$\sum_{l=1}^{3} \partial_l E_l = \rho. \tag{6.17}$$

Turning now to the divergence of H, this cannot be written in the same way as (6.17) because we have already assumed H to be an anti-symmetrical tensor of second rank. We can, however, express the divergence equation as follows:

$$\frac{\partial H_{kl}}{\partial x_m}+\frac{\partial H_{lm}}{\partial x_k}+\frac{\partial H_{mk}}{\partial x_l} = 0. \tag{6.18}$$

The left-hand side of this equation is a tensor of third rank, anti-symmetrical in any pair of indices. It can therefore only be a multiple of ϵ_{klm}. According to (6.18) the factor is zero. Only $k \neq l \neq m$ yield non-vanishing terms in the left-hand side of (6.18) and we thus get six equations according to the six permutations of k, l, m. However, they only differ from each other by a factor ± 1, so essentially we have a single equation.

Finally, we can look upon the current density as a vector J_k and the curl equation for the magnetic field becomes

$$\sum_{l=1}^{3}\frac{\partial H_{kl}}{\partial x_l} = \frac{1}{c}\frac{\partial E_k}{\partial t}+J_k. \tag{6.19}$$

It is readily seen by using (6.13) that (6.19) is equivalent to the part of (6.12) which refers to H. Eqn. (6.19) is a covariant relation between the vectors $\partial E_k/\partial t$, J_k, and a tensor which is obtained from H_{kl} by extension $(\partial H_{kl}/\partial x_m)$ and subsequent contraction $(l = m)$.

CHAPTER 3

MAXWELL'S THEORY IN TENSOR FORMULATION

1. Four-tensors

From the general considerations outlined in the introduction to Chapter 2 it follows that all physical laws must be expressible as absolute relations between entities which transform according to some representation of the Lorentz group. The derivation of these representations and the corresponding entities which transform according to them is greatly simplified by the use of the variable $x_4 = ict$, because then all entities with covariant significance are tensors in a four-dimensional space. We have already dealt briefly with tensors of a three-dimensional space and seen how they are connected with orthogonal transformations of orthogonal Cartesian coordinates, i.e. with rotations of the latter. In the same way when we deal with orthogonal Cartesian coordinates in a four-dimensional space and consider all the transformations from one orthogonal set into another, we get what may be termed rotations in a four-dimensional space. These rotations form a group. Entities transforming according to the representations of this group are tensors in a four-dimensional space. To distinguish them from the tensors of a three-dimensional space they are often called four-tensors. All the rules which apply in the field of three-dimensional tensors can be taken over without change and used in the four-dimensional space, and indeed in any other number of dimensions. Formally, to distinguish between four-tensors and three-tensors, we shall use Greek indices in the case of the former.

Through the introduction of the coordinate $x_4 = ict$, the invariant interval ds^2 takes the form

$$ds^2 = dx^2+dy^2+dz^2-c^2\,dt^2 = \sum_{\mu=1}^{4} dx_\mu^2. \tag{1.1}$$

ds^2 is an invariant for rotations of the orthogonal axes x_μ and only for such transformations. If follows, therefore, that Lorentz transformations are rotations in the four-dimensional world (R_4). It must be realized, however, that Lorentz transformations are of a restricted type and that not every rotation in the four-dimensional space R_4 is a Lorentz transformation. In other words, not every rotation in R_4 can have physical significance. We have already seen that apart from the orthogonality relations

$$\sum_{\kappa=1}^{4} \tilde{L}_{\mu\kappa} L_{\kappa\nu} = \delta_{\mu\nu} \tag{1.2}$$

or

$$\tilde{L}L = 1, \tag{1.3}$$

Lorentz transformations must also satisfy the conditions that the following quantities are imaginary:

$$L_{14}, \quad L_{24}, \quad L_{34}, \quad L_{41}, \quad L_{42}, \quad L_{43},$$

while all the other elements of a Lorentz matrix are real. Hence Lorentz transformations constitute only a sub-group in the group of rotations in R_4. When we refer to four-tensors we shall have in mind Lorentz rotations only and not all rotations.

All that has been said in connexion with three-tensors can be extended without any change into the field of four-tensors. The only difference is that in summations we have to sum up from 1 to 4 instead of 1 to 3.

A tensor of rank n transforms like the direct product of n four-vectors, $A_{1\mu}\times A_{2\nu}\times\ldots\times A_{n\kappa}$. Hence the transformation matrix is given by

$$(L\times L\times\ldots\times L) = (L_{\mu\mu'}\, L_{\nu\nu'} \ldots L_{\kappa\kappa'}) = (T_{\mu\nu\ldots,\mu'\nu'\ldots}). \tag{1.4}$$

A four-dimensional tensor calculus for the theory of relativity can also be developed on the basis of the invariant form

$$ds^2 = dx^2+dy^2+dz^2-c^2\,dt^2, \tag{1.5}$$

without introducing the imaginary $icdt$.† For this purpose we let

$$ds^2 = \sum_{\mu=1}^{4} dx_\mu\, dx^\mu, \tag{1.6}$$

where

$$dx_1 = dx^1 = dx, \quad dx_2 = dx^2 = dy, \quad dx_3 = dx^3 = dz,$$
$$-dx_4 = dx^4 = c\,dt. \tag{1.7}$$

Introducing the matrices $(g_{\mu\nu})$ and $(g^{\mu\nu})$ defined by

$$(g_{\mu\nu}) = \begin{bmatrix} 1 & 0 & 0 & 0 \\ 0 & 1 & 0 & 0 \\ 0 & 0 & 1 & 0 \\ 0 & 0 & 0 & -1 \end{bmatrix} = (g^{\mu\nu}), \tag{1.8}$$

we get

$$dx_\mu = \sum_{\nu=1}^{4} g_{\mu\nu}\, dx^\nu, \qquad dx^\mu = \sum_{\nu=1}^{4} g^{\mu\nu}\, dx_\nu, \tag{1.9}$$

$$ds^2 = \sum_{\mu,\nu} g_{\mu\nu}\, dx^\mu dx^\nu = \sum_{\mu,\nu} g^{\mu\nu} dx_\mu\, dx_\nu. \tag{1.10}$$

† Many authors use $ds^2 = -dx^2-dy^2-dz^2+c^2\,dt^2$. This is only of formal significance.

We call dx^μ a contravariant (four-) vector and dx_μ a covariant vector, The matrices of the Lorentz transformations we denote by $L^\kappa{}_\lambda$ and $L_\kappa{}^\lambda$ respectively.† Thus we have

$$dx^{\mu'} = \sum_\nu L^{\mu'}{}_\nu \, dx^\nu, \tag{1.11}$$

$$dx_{\mu'} = \sum_\nu L_{\mu'}{}^\nu \, dx_\nu. \tag{1.12}$$

The matrices $L^{\mu'}{}_\nu$ and $L_{\mu'}{}^\nu$ are related to each other as follows (where for convenience we have dropped the primes from the indices):

$$L^k{}_l = L_k{}^l \quad (k, l = 1, 2, 3), \tag{1.13}$$

$$L^k{}_4 = -L_k{}^4, \quad L^4{}_k = -L_4{}^k, \tag{1.14}$$

$$L^4{}_4 = L_4{}^4. \tag{1.15}$$

The Lorentz group is represented both by the matrices $L^\mu{}_\nu$ and $L_\mu{}^\nu$ so that each of these set of matrices forms a group. From the invariance of ds^2 we have, using (1.10),

$$\sum_{\kappa\lambda} g_{\kappa\lambda} \, dx^\kappa dx^\lambda = \sum_{\mu'\nu'} g_{\mu'\nu'} \, dx^{\mu'} dx^{\nu'}, \tag{1.16}$$

$$\sum_{\kappa\lambda} g^{\kappa\lambda} \, dx_\kappa \, dx_\lambda = \sum_{\mu'\nu'} g^{\mu'\nu'} \, dx_{\mu'} \, dx_{\nu'}, \tag{1.17}$$

and this yields for the Lorentz matrices, using (1.11) and (1.12),

$$\sum_{\mu'\nu'} g_{\mu'\nu'} \, L^{\mu'}{}_\kappa \, L^{\nu'}{}_\lambda = g_{\kappa\lambda}, \tag{1.18}$$

$$\sum_{\mu'\nu'} g^{\mu'\nu'} L_{\mu'}{}^\kappa L_{\nu'}{}^\lambda = g^{\kappa\lambda}. \tag{1.19}$$

We can look upon these relations as replacing our former orthogonality relations (when using ict).

Expressing (1.18) in matrix form and adopting from now on the convention that whenever an index appears in any term twice we must sum from 1 to 4 without exhibiting the summation sign, we get

$$\tilde{L}^\kappa{}_\mu g_{\mu\nu} L^\nu{}_\lambda = g_{\kappa\lambda}, \tag{1.20}$$

or

$$\tilde{L}\diagdown g \, L\diagdown = g, \tag{1.21}$$

where $L\diagdown$ is the matrix $L^\mu{}_\nu$ and g is the matrix $g_{\mu\nu}$. Taking determinants we get, since $\det g = -1$,

$$\det \tilde{L}\diagdown . \det L\diagdown = 1, \tag{1.22}$$

therefore

$$\det \tilde{L}\diagdown = \det L\diagdown = \pm 1. \tag{1.23}$$

† When writing $L^\mu{}_\nu$ and $L_\mu{}^\nu$ as matrices, the index nearer to L, (μ), refers to rows and the index further from L, (ν), to columns.

A similar result is obtained from (1.19) for $L_\mu{}^\nu$ ($\equiv L\diagup$). If (1.18) and (1.19) are applied to $\kappa = \lambda = 4$, where $g_{44} = g^{44} = -1$, we get

$$(L^1{}_4)^2+(L^2{}_4)^2+(L^3{}_4)^2-(L^4{}_4)^2 = -1, \tag{1.24}$$

$$(L_1{}^4)^2+(L_2{}^4)^2+(L_3{}^4)^2-(L_4{}^4)^2 = -1. \tag{1.25}$$

Hence
$$|L^4{}_4| \geqslant 1, \qquad |L_4{}^4| \geqslant 1.$$

Thus we find as before (section 10, Chapter 1) that the Lorentz group divides into four sets according to whether the determinant is ± 1 and according to whether $L^4{}_4$ is positive or negative, the absolute value being always greater than or equal to 1. This is, of course, not surprising, because the essential properties of a group cannot depend on the particular mathematical device used.

We can also prove that

$$L^\mu{}_\nu = L^{-1}{}_\nu{}^\mu \qquad (\tilde{L}^{-1}_{\diagdown} = L\diagup), \tag{1.26}$$

so that the matrices which determine the transformations of contravariant vectors are contragredient to those which determine the transformations of covariant vectors and vice versa.† Contragredience involves taking the reciprocal and subsequent transposition or vice versa, these operations being commutative. If A, B, and C are any matrices such that $AB = C$, then we have $\tilde{B}\tilde{A} = \tilde{C}$, $B^{-1}A^{-1} = C^{-1}$, $\tilde{A}^{-1}\tilde{B}^{-1} = \tilde{C}^{-1}$. As a result of this property of contragredience the matrices $L_\mu{}^\nu$ form a Lorentz group if this is the case with $L^\mu{}_\nu$. We shall denote the contragredient of L by L^c.

As before, when using ict, we can get new representations of the Lorentz group by forming direct products, using both L's and L^c's. Thus, a tensor of second rank, contravariant in both indices, transforms according to

$$F^{\mu'\nu'} = L^{\mu'}{}_\kappa L^{\nu'}{}_\lambda F^{\kappa\lambda}. \tag{1.27}$$

A covariant tensor of second rank transforms according to

$$F_{\sigma'\tau'} = L_{\sigma'}{}^\eta L_{\tau'}{}^\rho F_{\eta\rho}, \tag{1.28}$$

and a mixed tensor of second rank has one contravariant and one covariant index:

$$F^{\mu'}_{\sigma'} = L^{\mu'}{}_\kappa L_{\sigma'}{}^\eta F^\kappa_\eta. \tag{1.29}$$

† This is proved as follows: from (1.11), (1.9), and (1.12) we have

$$dx^{\lambda'} = L^{\lambda'}{}_\mu g^{\mu\nu} dx_\nu = L^{\lambda'}{}_\mu g^{\mu\nu} L^{-1}{}_\nu{}^{\kappa'} dx_{\kappa'},$$

therefore
$$L^{\lambda'}{}_\mu g^{\mu\nu} L^{-1}{}_\nu{}^{\kappa'} \equiv g^{\lambda'\kappa'},$$

hence,
$$L\diagdown g L^{-1}_{\diagup} = g, \quad \text{and so} \quad L\diagdown = g L\diagup g^{-1} = g L\diagup g.$$

On the other hand, from (1.21),

$$\tilde{L}\diagdown g L\diagdown = g.$$

Consequently,
$$\tilde{L}\diagdown g g L\diagup = 1,$$

$$\tilde{L}\diagdown L\diagup = 1, \qquad \tilde{L}\diagdown = L^{-1}_{\diagup}.$$

In general we have

$$F^{\mu'\nu'\cdots}_{\sigma'\tau'\cdots} = L^{\mu'}{}_{\kappa}\, L^{\nu'}{}_{\lambda} \ldots L_{\sigma'}{}^{\eta}\, L_{\tau'}{}^{\rho} \ldots F^{\kappa\lambda\cdots}_{\eta\rho\cdots} \tag{1.30}$$

When we look upon the elements of a tensor as a column, and use the direct product matrices for the transformation, we necessarily arrange the elements of the tensor in some order, and consequently the order in which the Lorentz matrices are written matters. Thus

$$(L^c \times L) \neq (L \times L^c).$$

However, this difference only involves permutation of indices. It forces us to distinguish between the columns $(F^{\mu}{}_{\sigma})$ and $(F_{\mu}{}^{\sigma})$. Denoting these columns by $F^{\cdot}{}_{.}$ and $F_{.}{}^{\cdot}$, we have instead of (1.29)

$$F^{\cdot}{}_{.}{}' = (L \times L^c) F^{\cdot}{}_{.}, \qquad F_{.}{}^{\cdot}{}' = (L^c \times L) F_{.}{}^{\cdot} \tag{1.31}$$

where, to be more explicit (and for simplicity letting μ and σ run through 1, 2 only),

$$F^{\cdot}{}_{.} = \begin{matrix} F^1{}_1 \\ F^1{}_2 \\ F^2{}_1 \\ F^2{}_2 \end{matrix} \equiv \begin{matrix} F^1_1 \\ F^1_2 \\ F^2_1 \\ F^2_2 \end{matrix} \qquad F_{.}{}^{\cdot} = \begin{matrix} F_1{}^1 \\ F_1{}^2 \\ F_2{}^1 \\ F_2{}^2 \end{matrix} \equiv \begin{matrix} F^1_1 \\ F^2_1 \\ F^1_2 \\ F^2_2 \end{matrix} \tag{1.32}$$

and the difference in the order of the elements of the tensor F^{μ}_{σ} is shown. As long as we use the explicit transformation formulae (1.27) to (1.30) and no ordering of elements is involved, there is no need to distinguish between $F^{\mu}{}_{\sigma}$ and $F_{\mu}{}^{\sigma}$ and we can simply use F^{μ}_{σ}, because

$$L^{\mu'}{}_{\kappa}\, L_{\sigma'}{}^{\eta} \equiv L_{\sigma'}{}^{\eta}\, L^{\mu'}{}_{\kappa}. \tag{1.33}$$

Using columns and matrices, (1.27) and (1.28) become

$$F^{\cdot\,\cdot}{}' = (L \times L) F^{\cdot\,\cdot}, \tag{1.34}$$

$$F_{..}{}' = (L^c \times L^c) F_{..}, \tag{1.35}$$

and more generally we have, for instance,

$$F^{\cdot}{}_{.}{}^{\cdot}{}' = (L \times L^c \times L) F^{\cdot}{}_{.}{}^{\cdot} \tag{1.36}$$

We have introduced the $g_{\mu\nu}$ and $g^{\mu\nu}$ as numerical matrices. We can now see, however, that they form invariant tensors of second rank by virtue of the relations (1.18) and (1.19). With the help of the $g_{\mu\nu}$ and $g^{\mu\nu}$ we can lower and raise indices not only of a four-vector, as was done in (1.9), but in any tensor. For example,

$$g_{\mu\lambda} F^{\lambda}_{\kappa} = F_{\mu\kappa}, \tag{1.37}$$

$$g^{\mu\lambda} F^{\kappa}_{\lambda} = F^{\mu\kappa}. \tag{1.38}$$

In the present case, tensors can be added provided they are of the same kind. In other words, they must have the same number of covariant and contravariant indices respectively. The sum $A_\kappa + B^\kappa$ has no significance since it is neither a covariant nor a contravariant vector. When the tensors are written as columns we must also insist on the same ordering of indices. $F^{\cdot}{}_{\cdot} + G^{\cdot}{}_{\cdot}$ is a tensor sum, say $H^{\cdot}{}_{\cdot}$, but not $F^{\cdot}{}_{\cdot} + G_{\cdot}{}^{\cdot}$; in unordered components we shall have $F^\mu_\sigma + G^\mu_\sigma = H^\mu_\sigma$. Contraction can take place only with respect to a covariant and a contravariant index. In particular, the contraction of $dx_\mu dx^\nu$ yields the invariant ds^2 according to (1.6). On the other hand, the contraction of $dx_\mu dx_\nu$ is not covariant, since it yields $dx^2 + dy^2 + dz^2 + c^2 dt^2$ which is not an invariant. Symmetrization and anti-symmetrization of tensors which contain both covariant and contravariant indices can proceed only within each group of indices. Thus $A_\mu B^\nu \pm B_\mu A^\nu$ are mixed tensors of second rank, which are, however, neither symmetrical in μ, ν nor anti-symmetrical. On the other hand, $A_\mu B_\nu \pm B_\mu A_\nu$ are symmetrical and anti-symmetrical respectively. The same applies to $A^\mu B^\nu \pm B^\mu A^\nu$.

In a formalism which uses ict no distinction exists between covariant and contravariant indices since here the metrical tensor $g_{\mu\nu}$ is identical with $\delta_{\mu\nu}$. Consequently, $dx_\mu \equiv dx^\mu$ and we may use either a top index or a bottom index throughout. In this case $L = L^c$ (1.3).†

In spite of the apparent complication involved in using covariant and contravariant indices, it offers certain advantages over the method which uses a Euclidean metric ($g_{\mu\nu} = \delta_{\mu\nu}$). To begin with it uses real quantities, thus leaving the introduction of the imaginary i for other purposes. Furthermore, it lends itself in a more natural way to the general theory of relativity.

A tensor of rank n has n indices each running from 1 to 4, hence it has 4^n components. Of special interest are the invariant tensors $\delta_{\mu\nu}$, which is symmetrical in the two indices, and $\epsilon_{\kappa\lambda\mu\nu}$, which is anti-symmetrical in any pair of indices (using ict). On using $x^4 = -x_4 = ct$, we have $g_{\mu\kappa} g^{\kappa\nu} = g_\mu{}^\nu = \delta_\mu{}^\nu$. $\delta_\mu{}^\nu = 1$ for $\mu = \nu$ and $= 0$ for $\mu \neq \nu$.

2. Maxwell's Equations in Tensor Form

As we have already shown in Chapter 1, Maxwell's equations are invariant with respect to Lorentz transformations. Consequently, it should be possible to express them as absolute or covariant relations between four-tensors, in the same way as was done in the previous

† When $x^4 = -x_4 = ct$ is used, neither $L\diagdown$ nor $L\diagup$ is self-contragredient but they are contragredient to each other (1.26). These representations are equivalent since $L\diagdown = gL\diagup g^{-1}$ (see footnote, p. 92).

chapter with regard to the group of rotations. In this way the covariance of Maxwell's equations with respect to Lorentz transformation should be directly displayed. On the other hand, in the case of dynamics we know that the pre-relativistic theory is not Lorentz invariant. Consequently, it is not possible to express it in tensor form. It is for this reason that we begin by dealing in this chapter with Maxwell's equations and defer relativistic dynamics to Chapter 5.

Introducing the notation x_1, x_2, x_3, x_4 instead of x, y, z, ict, we get

$$\begin{aligned}
&+\frac{\partial H_z}{\partial x_2}-\frac{\partial H_y}{\partial x_3}-\frac{\partial(iE_x)}{\partial x_4}=0\\
-\frac{\partial H_z}{\partial x_1}&+\frac{\partial H_x}{\partial x_3}-\frac{\partial(iE_y)}{\partial x_4}=0\\
+\frac{\partial H_y}{\partial x_1}-\frac{\partial H_x}{\partial x_2}&-\frac{\partial(iE_z)}{\partial x_4}=0\\
+\frac{\partial(iE_x)}{\partial x_1}+\frac{\partial(iE_y)}{\partial x_2}+\frac{\partial(iE_z)}{\partial x_3}&=0
\end{aligned} \tag{2.1}$$

$$\begin{aligned}
&-\frac{\partial(iE_z)}{\partial x_2}+\frac{\partial(iE_y)}{\partial x_3}+\frac{\partial H_x}{\partial x_4}=0\\
+\frac{\partial(iE_z)}{\partial x_1}&-\frac{\partial(iE_x)}{\partial x_3}+\frac{\partial H_y}{\partial x_4}=0\\
-\frac{\partial(iE_y)}{\partial x_1}+\frac{\partial(iE_x)}{\partial x_2}&+\frac{\partial H_z}{\partial x_4}=0\\
\frac{\partial H_x}{\partial x_1}+\frac{\partial H_y}{\partial x_2}+\frac{\partial H_z}{\partial x_3}&=0.
\end{aligned} \tag{2.2}$$

If we examine (2.1) we find that the components of the electromagnetic field may be looked upon as forming an anti-symmetrical tensor of second rank according to the scheme:

$$F_{\mu\nu}=\begin{bmatrix} 0 & H_z & -H_y & -iE_x \\ -H_z & 0 & H_x & -iE_y \\ H_y & -H_x & 0 & -iE_z \\ iE_x & iE_y & iE_z & 0 \end{bmatrix}. \tag{2.3†}$$

Eqns. (2.1) can then be summed up in a covariant tensor form as follows:

$$\frac{\partial F_{\mu\nu}}{\partial x_\nu}=0 \quad (\mu=1,2,3,4). \tag{2.4}$$

† We order the elements of the tensor in matrix form simply for convenience. A tensor of second rank has only superficial similarity with a matrix.

Eqns. (2.2) can be expressed similarly in tensor form if we introduce a second anti-symmetrical tensor $G_{\mu\nu}$ defined as follows:

$$G_{\mu\nu} = \begin{bmatrix} 0 & -E_z & +E_y & -iH_x \\ E_z & 0 & -E_x & -iH_y \\ -E_y & E_x & 0 & -iH_z \\ iH_x & iH_y & iH_z & 0 \end{bmatrix}. \tag{2.5}$$

We then get from (2.2)

$$\frac{\partial G_{\mu\nu}}{\partial x_\nu} = 0 \quad (\mu = 1, 2, 3, 4). \tag{2.6}$$

The tensor $G_{\mu\nu}$ is not unrelated to the tensor $F_{\mu\nu}$. In fact

$$G_{\mu\nu} = \frac{1}{i2!}\epsilon_{\mu\nu\kappa\lambda}F_{\kappa\lambda}. \tag{2.7}$$

For instance,

$$G_{12} = -E_z = \frac{1}{2i}(\epsilon_{1234}F_{34}+\epsilon_{1243}F_{43})$$

$$= \frac{1}{2i}(1.F_{34}-1.F_{43}) = \frac{F_{34}}{i}. \tag{2.8}$$

$G_{\mu\nu}$ is thus a dual to $F_{\mu\nu}$.

The relation between $G_{\kappa\lambda}$ and $F_{\mu\nu}$ can also be expressed as follows:

$$G_{\kappa\lambda} = -iF_{\mu\nu}, \tag{2.9}$$

where κ, λ, μ, ν are an even permutation of 1, 2, 3, 4.

We are now in a position to express Maxwell's equations in terms of the tensor $F_{\kappa\lambda}$ only. Eqn. (2.6) for $\mu = 1$ becomes

$$\frac{\partial G_{12}}{\partial x_2}+\frac{\partial G_{13}}{\partial x_3}+\frac{\partial G_{14}}{\partial x_4} = -i\left(\frac{\partial F_{34}}{\partial x_2}+\frac{\partial F_{42}}{\partial x_3}+\frac{\partial F_{23}}{\partial x_4}\right) = 0 \tag{2.10}$$

and generally, for any μ, we get

$$\frac{\partial F_{\kappa\lambda}}{\partial x_\mu}+\frac{\partial F_{\lambda\mu}}{\partial x_\kappa}+\frac{\partial F_{\mu\kappa}}{\partial x_\lambda} = 0. \tag{2.11}$$

Eqns. (2.4) and (2.11) express the full set of Maxwell's equations in tensor form and in terms of the tensor $F_{\kappa\lambda}$. The covariance of Maxwell's equations with respect to Lorentz transformations is thus made evident. This covariance is, however, also self-evident if we use $F_{\kappa\lambda}$ and $G_{\kappa\lambda}$ simultaneously.

3. The Transformation Law of the Electromagnetic Field Components

In section 13 of Chapter 1 we have proved by direct transformation that Maxwell's equations are invariant with respect to Lorentz transformations, and that as a consequence the electric and magnetic field components transform according to the following laws:

$$\begin{aligned} E_{z'} &= \gamma\{E_z-(v/c)H_y\} & H_{z'} &= \gamma\{H_z+(v/c)E_y\} \\ E_{y'} &= \gamma\{E_y+(v/c)H_z\} & H_{y'} &= \gamma\{H_y-(v/c)E_z\} \\ E_{x'} &= E_x & H_{x'} &= H_x. \end{aligned} \tag{3.1}$$

In the preceding section we have seen that the electromagnetic field must have the properties of an anti-symmetrical tensor of second rank. Therefore, from the transformation properties of such a tensor we should again be able to derive (3.1). We shall restrict ourselves again to motion along the x-axes. The transformation matrix of a tensor of second rank is given by

$$L(v)\times L(v) = \begin{bmatrix} \gamma & 0 & 0 & -i\gamma v/c \\ 0 & 1 & 0 & 0 \\ 0 & 0 & 1 & 0 \\ i\gamma v/c & 0 & 0 & \gamma \end{bmatrix} \times \begin{bmatrix} \gamma & 0 & 0 & -i\gamma v/c \\ 0 & 1 & 0 & 0 \\ 0 & 0 & 1 & 0 \\ i\gamma v/c & 0 & 0 & \gamma \end{bmatrix}. \tag{3.2}$$

The direct product is a matrix with $16\times 16 = 256$ elements, the general expression of which is

$$T_{pq} = L_{\kappa'\mu}L_{\lambda'\nu} \quad (p = \kappa', \lambda';\ q = \mu, \nu). \tag{3.3}$$

The index p runs from 1 to 16 and corresponds to the 16 combinations 1, 1; 1, 2; ...; 2, 1; 2, 2; ...; 4, 4 of the indices κ', λ' each running from 1 to 4. In the same way the index q corresponds to the 16 combinations of the indices μ, ν.

The transformation

$$F_{\kappa'\lambda'} = L_{\kappa'\mu}L_{\lambda'\nu}F_{\mu\nu} \tag{3.4}$$

is greatly facilitated for tensors of second rank if use is made of matrix multiplication. This is achieved if we rewrite (3.4) as follows:

$$F_{\kappa'\lambda'} = L_{\kappa'\mu}F_{\mu\nu}\tilde{L}_{\nu\lambda'}. \tag{3.5}$$

In this form summation indices μ and ν appear twice as consecutive indices, as in matrix multiplication. Hence we get

$$(F_{\kappa'\lambda'})$$

$$= \begin{bmatrix} \gamma & 0 & 0 & -i\gamma v/c \\ 0 & 1 & 0 & 0 \\ 0 & 0 & 1 & 0 \\ i\gamma v/c & 0 & 0 & \gamma \end{bmatrix} \begin{bmatrix} 0 & H_z & -H_y & -iE_x \\ -H_z & 0 & H_x & -iE_y \\ H_y & -H_x & 0 & -iE_z \\ +iE_x & +iE_y & +iE_z & 0 \end{bmatrix} \begin{bmatrix} \gamma & 0 & 0 & i\gamma v/c \\ 0 & 1 & 0 & 0 \\ 0 & 0 & 1 & 0 \\ -i\gamma v/c & 0 & 0 & \gamma \end{bmatrix}$$

$$= \begin{bmatrix} 0 & \gamma\left(H_z+\frac{v}{c}E_y\right) & -\gamma\left(H_y-\frac{v}{c}E_z\right) & -iE_x \\ -\gamma\left(H_z+\frac{v}{c}E_y\right) & 0 & H_x & -i\gamma\left(E_y+\frac{v}{c}H_z\right) \\ \gamma\left(H_y-\frac{v}{c}E_z\right) & -H_x & 0 & -i\gamma\left(E_z-\frac{v}{c}H_y\right) \\ +iE_x & +i\gamma\left(E_y+\frac{v}{c}H_z\right) & +i\gamma\left(E_z-\frac{v}{c}H_y\right) & 0 \end{bmatrix}. \tag{3.6}$$

A comparison of this anti-symmetrical matrix with $F_{\mu\nu}$ according to (2.3) confirms all the transformation formulae of the electromagnetic field derived previously.

4. The Energy-Momentum Tensor

If we form the direct product of the electromagnetic tensor $F_{\kappa\lambda}$ with itself, we get a tensor of fourth rank $F_{\kappa\lambda}F_{\mu\nu}$. If this tensor is contracted with regard to both κ and μ as well as λ and ν we get a scalar. We denote it by $4S^{(1)}$, so that

$$S^{(1)} = \tfrac{1}{4}F_{\kappa\mu}F_{\kappa\mu} = \tfrac{1}{4}F^2_{\kappa\mu}. \tag{4.1}$$

If we insert the values of the components as given by (2.3) we get

$$S^{(1)} = \tfrac{1}{2}[H_x^2+H_y^2+H_z^2-(E_x^2+E_y^2+E_z^2)]. \tag{4.2}$$

In ordinary vector notation,

$$S^{(1)} = \tfrac{1}{2}(\mathbf{H}^2-\mathbf{E}^2). \tag{4.3}$$

This scalar represents the difference between the density of magnetic and electric field energies. The same result, apart from sign, is obtained if we use $G_{\kappa\lambda}$ instead of $F_{\mu\nu}$, where $G_{\kappa\lambda}$ is given by (2.5).

Next, let us form the direct product of $G_{\kappa\lambda}$ and $F_{\mu\nu}$ and contract with respect to both indices. We again get a scalar which we denote by $S^{(2)}$:

$$S^{(2)} = G_{\kappa\mu}F_{\kappa\mu} = -4(E_xH_x+E_yH_y+E_zH_z) = -4(\mathbf{EH}). \tag{4.4}$$

We thus find that the scalar product of the vectors **E** and **H** is a relativistic invariant. For a plane electromagnetic wave $\mathbf{EH} = 0$ and $\mathbf{E}^2-\mathbf{H}^2 = 0$. It follows, therefore, from the invariance of these quantities that a transverse electromagnetic wave will appear as such in every inertial frame of reference. In particular, a photon will always appear as a photon. If $\mathbf{E}^2-\mathbf{H}^2 \neq 0$, while $\mathbf{EH} = 0$, it is always possible to determine a frame of reference in which either **E** or **H** is zero and the electromagnetic field appears either as purely electric (when $\mathbf{E}^2 > \mathbf{H}^2$) or purely magnetic (when $\mathbf{E}^2 < \mathbf{H}^2$). If $\mathbf{EH} \neq 0$, it is possible to determine a frame of reference in which the vertical component of **E** (relative to **H**) or the vertical component of **H** (relative to **E**) is made to vanish and we are left with an electromagnetic field containing an **E** and an **H** which are parallel to each other. This follows from the possibility of expressing

$$\mathbf{EH} = \mathbf{EH}_{\parallel}+\mathbf{EH}_{\perp} = \mathbf{EH}_{\parallel} \quad \text{or} \quad \mathbf{EH} = \mathbf{HE}_{\parallel}+\mathbf{HE}_{\perp} = \mathbf{HE}_{\parallel} \tag{4.5}$$

and

$$\mathbf{E}^2-\mathbf{H}^2 = \mathbf{E}^2-(\mathbf{H}_{\parallel}+\mathbf{H}_{\perp})^2 = \mathbf{E}^2-\mathbf{H}_{\parallel}^2-\mathbf{H}_{\perp}^2, \tag{4.6}$$

$$\mathbf{E}^2-\mathbf{H}^2 = (\mathbf{E}_{\parallel}+\mathbf{E}_{\perp})^2-\mathbf{H}^2 = \mathbf{E}_{\parallel}^2+\mathbf{E}_{\perp}^2-\mathbf{H}^2. \tag{4.7}$$

It can be shown that $S^{(1)}$ and $S^{(2)}$ are the only independent invariants which can be formed from the electromagnetic field tensor $F_{\kappa\lambda}$. All the other invariants are functions of $S^{(1)}$ and $S^{(2)}$. It is to be noted that the energy density $\frac{1}{2}(\mathbf{E}^2+\mathbf{H}^2)$ is not an invariant and will have different values for observers who move relative to each other.

From (4.4) we find the following relations:

$$\frac{\partial S^{(2)}}{\partial F_{\kappa\mu}} = G_{\kappa\mu}, \tag{4.8}$$

$$\frac{\partial S^{(2)}}{\partial G_{\kappa\mu}} = F_{\kappa\mu}. \tag{4.9}$$

If we form the direct product of $F_{\kappa\lambda}$ with itself and contract only with respect to one pair of indices we get a symmetrical tensor of second rank,

$$H_{\kappa\lambda} = F_{\kappa\mu}F_{\lambda\mu} = H_{\lambda\kappa}. \tag{4.10}$$

Instead of using this tensor it is more justifiable from a physical point of view (as we shall soon see) to subtract from it the tensor

$$S^{(1)}\delta_{\kappa\lambda} = \tfrac{1}{4}F_{\sigma\tau}^2\delta_{\kappa\lambda}. \tag{4.11}$$

The result is called the energy-momentum tensor, denoted by $T_{\kappa\lambda}$. We thus define

$$T_{\kappa\lambda} = F_{\kappa\mu}F_{\lambda\mu}-S^{(1)}\delta_{\kappa\lambda}. \tag{4.12}$$

Expressed explicitly in terms of **E** and **H** we get

$$T_{\kappa\lambda} = \begin{bmatrix} W-(H_x^2+E_x^2) & -(H_x H_y+E_x E_y) & -(H_x H_z+E_x E_z) & i[\mathbf{E},\mathbf{H}]_x \\ -(H_x H_y+E_x E_y) & W-(H_y^2+E_y^2) & -(H_y H_z+E_y E_z) & i[\mathbf{E},\mathbf{H}]_y \\ -(H_x H_z+E_x E_z) & -(H_y H_z+E_y E_z) & W-(H_z^2+E_z^2) & i[\mathbf{E},\mathbf{H}]_z \\ i[\mathbf{E},\mathbf{H}]_x & i[\mathbf{E},\mathbf{H}]_y & i[\mathbf{E},\mathbf{H}]_z & -W \end{bmatrix}, \tag{4.13}$$

where

$$W = \tfrac{1}{2}(\mathbf{E}^2+\mathbf{H}^2) \tag{4.14}$$

is the density of electromagnetic energy in the field.

$$\mathbf{P} = c[\mathbf{E},\mathbf{H}] \tag{4.15}$$

is the Poynting vector of energy flow.

5. The Laws of Conservation of Energy and Momentum

From the tensor $T_{\kappa\lambda}$ we can form the tensor of third rank $\partial T_{\kappa\lambda}/\partial x_\mu$. Contracting with respect to the indices κ and μ we get a vector which is called the vector divergence of the tensor.† In our present case it follows from Maxwell's equations (2.4), that this vector vanishes. The proof is as follows. From (4.11) and (4.12) which define $T_{\kappa\lambda}$ we get

$$\frac{\partial T_{\kappa\lambda}}{\partial x_\kappa} = \frac{\partial F_{\kappa\mu}}{\partial x_\kappa} F_{\lambda\mu} + F_{\kappa\mu}\frac{\partial F_{\lambda\mu}}{\partial x_\kappa} - \tfrac{1}{4}\delta_{\kappa\lambda}\frac{\partial F_{\sigma\tau}^2}{\partial x_\kappa}. \tag{5.1}$$

Since $\delta_{\kappa\lambda} = 0$ for $\kappa \neq \lambda$, we get further

$$\frac{\partial T_{\kappa\lambda}}{\partial x_\kappa} = \frac{\partial F_{\kappa\mu}}{\partial x_\kappa} F_{\lambda\mu} + F_{\kappa\mu}\frac{\partial F_{\lambda\mu}}{\partial x_\kappa} - \frac{1}{2}\frac{\partial F_{\sigma\tau}}{\partial x_\lambda} F_{\sigma\tau}. \tag{5.2}$$

Because of Maxwell's equations (2.4) the first term on the right-hand side vanishes.

Using the second set of Maxwell's equations (2.11), the third term on the right-hand side of (5.2) yields, after changing σ, τ to κ, μ,

$$\frac{1}{2}\frac{\partial F_{\kappa\mu}}{\partial x_\lambda} F_{\kappa\mu} = \frac{1}{2}\left(\frac{\partial F_{\kappa\lambda}}{\partial x_\mu} + \frac{\partial F_{\lambda\mu}}{\partial x_\kappa}\right) F_{\kappa\mu}. \tag{5.3}$$

Altogether we now have

$$\frac{\partial T_{\kappa\lambda}}{\partial x_\kappa} = \frac{1}{2}\left(\frac{\partial F_{\lambda\mu}}{\partial x_\kappa} F_{\kappa\mu} - \frac{\partial F_{\kappa\lambda}}{\partial x_\mu} F_{\kappa\mu}\right). \tag{5.4}$$

In the last term we can exchange the indices κ and μ since the summation is over both of them:

$$\frac{\partial T_{\kappa\lambda}}{\partial x_\kappa} = \frac{1}{2}\left(\frac{\partial F_{\lambda\mu}}{\partial x_\kappa} F_{\kappa\mu} - \frac{\partial F_{\mu\lambda}}{\partial x_\kappa} F_{\mu\kappa}\right). \tag{5.5}$$

† The divergence of a vector yields a scalar divergence, that of a tensor of second rank a vector divergence, and so on to higher tensors.

Now, the electromagnetic field tensor is anti-symmetrical; consequently $F_{\kappa\mu} = -F_{\mu\kappa}$, $F_{\lambda\mu} = -F_{\mu\lambda}$, and the right-hand side of (5.5) vanishes. This yields

$$\frac{\partial T_{\kappa\lambda}}{\partial x_\kappa} = 0 \quad (\lambda = 1, 2, 3, 4). \tag{5.6}$$

Eqns. (5.6) justify the choice of $T_{\kappa\lambda}$. It expresses the laws of conservation of momentum and energy. Let us begin with the latter. For this purpose we must consider the fourth component ($\lambda = 4$) of (5.6):

$$\frac{\partial T_{\kappa 4}}{\partial x_\kappa} = \frac{i\partial[\mathbf{E},\mathbf{H}]_x}{\partial x} + \frac{i\partial[\mathbf{E},\mathbf{H}]_y}{\partial y} + \frac{i\partial[\mathbf{E},\mathbf{H}]_z}{\partial z} - \frac{\partial W}{ic\partial t} = 0. \tag{5.7}$$

In ordinary vector notation and symbols we thus get

$$c\,\mathrm{div}[\mathbf{E},\mathbf{H}] + \partial W/\partial t = 0; \quad \text{or} \quad \mathrm{div}\,\mathbf{P} + \partial W/\partial t = 0. \tag{5.8}$$

This equation expresses in differential form the law of conservation of energy. The change of energy density per unit of time is equal to the net inflow of energy current per unit of volume.

The other components in (5.6) have the form

$$\frac{\partial T_{xx}}{\partial x} + \frac{\partial T_{xy}}{\partial y} + \frac{\partial T_{xz}}{\partial z} + \frac{1}{c}\frac{\partial}{\partial t}[\mathbf{E},\mathbf{H}]_x = 0,$$

$$\frac{\partial T_{yx}}{\partial x} + \frac{\partial T_{yy}}{\partial y} + \frac{\partial T_{yz}}{\partial z} + \frac{1}{c}\frac{\partial}{\partial t}[\mathbf{E},\mathbf{H}]_y = 0,$$

$$\frac{\partial T_{zx}}{\partial x} + \frac{\partial T_{zy}}{\partial y} + \frac{\partial T_{zz}}{\partial z} + \frac{1}{c}\frac{\partial}{\partial t}[\mathbf{E},\mathbf{H}]_z = 0, \tag{5.9}$$

where the quantities $T_{xx}, \ldots, T_{zz}$ are identical with the corresponding elements in $T_{\kappa\lambda}$ (4.13). These equations are usually referred to as Maxwell's stress relations. However, when dealing with a field in vacuum it is more natural to look upon $\mathbf{T}_a = (T_{ax}, T_{ay}, T_{az})$ $(a = x, y, z)$ as defining a flow of an a-component of momentum density. Conservation of momentum then leads us to look upon

$$\mathbf{G} = (1/c)[\mathbf{E},\mathbf{H}] = \mathbf{P}/c^2 \tag{5.10}$$

as momentum density. Eqns. (5.9) would then have the same form as (5.8):

$$\mathrm{div}\,\mathbf{T}_a + (\partial/\partial t)G_a = 0 \quad (a = x, y, z). \tag{5.11}$$

It should be realized, however, that neither (5.8) nor (5.11) is in relativistic form, although they hold in each frame of reference. In other words ($[\mathbf{E},\mathbf{H}]$, W) do not form a four-vector, nor do ($\mathbf{T}_a$, G_a) form a set of three four-vectors. To obtain the equations (5.8) and (5.11) in a new frame we have to perform a tensor transformation of $T_{\kappa\lambda}$. The relativistic form of these laws of conservation in differential form is (5.6). We are not concerned here with the conservation of two independent entities, but of a four-vector of energy and momentum. In section 8 we shall return to these laws of conservation and express them in integral form. This will lead us to the four-vector mentioned.

From (4.15) and (5.10) we find that flow of electromagnetic energy and electromagnetic momentum are vectors which are parallel to each other, obeying the relation

$$\mathbf{G} = (1/c^2)\mathbf{P}. \tag{5.12}$$

This relation applies in any field which satisfies relativistic field equations. It shows that wherever there is a flow of energy there also exists momentum.

It may be added that from a law of conservation such as (5.6) the tensor $T_{\kappa\lambda}$ cannot be defined uniquely, because we can always add a divergence-free tensor to $T_{\kappa\lambda}$ without upsetting (5.6). The law of conservation of angular momentum requires $T_{\kappa\lambda}$ to be symmetrical. We shall return to this question in Chapter 4 on general field theory.

6. Maxwell's Equations in Integral Form

In this section it will be shown that, using the four-dimensional tensor formalism, it is possible to express Maxwell's equations in integral form, just as much as this is possible in ordinary three-dimensional vector analysis. In addition, this will provide us with an opportunity of discussing the different types of integrals (volume, surface, etc.) which occur in the four-dimensional theory of relativity.

In three-dimensional vectors the integral form of Maxwell's equations when no charges and currents are present is given by

$$\oint_{\Sigma} E_n \, d\sigma = 0, \qquad \oint_{\Sigma} H_n \, d\sigma = 0, \tag{6.1}$$

$$c \oint E_l \, dl = -\oint_{\Sigma} \dot{H}_n \, d\sigma, \qquad c \oint H_l \, dl = \oint_{\Sigma} \dot{E}_n \, d\sigma. \tag{6.2}$$

These relations express the following experimental results:

(*a*) The total flux of **E** or **H** measured over a closed surface vanishes.

(*b*) The closed line integrals of **E** and **H** are equal to the total flux of magnetic and electric displacement current respectively through any surface which is bounded by the considered loop. By means of Gauss's and Stokes's theorems the above integral relations can be changed into differential form. These theorems yield for any vector field the following relations:

$$\int \operatorname{div} \mathbf{A}\, dv = \oint_{\Sigma} A_n\, d\sigma, \tag{6.3}$$

$$\int \operatorname{curl}_n \mathbf{A}\, d\sigma = \oint A_l\, dl. \tag{6.4}$$

Applying them to Maxwell's integral equations we get

$$\operatorname{div} \mathbf{E} = 0, \qquad \operatorname{div} \mathbf{H} = 0, \tag{6.5}$$

$$c \operatorname{curl} \mathbf{E} = -\dot{\mathbf{H}}, \qquad c \operatorname{curl} \mathbf{H} = \dot{\mathbf{E}}. \tag{6.6}$$

Once this relationship, via Gauss's and Stokes's theorems, between the integral relations and the differential equations is realized, it is not difficult to read in the latter the original physical meaning as conveyed by the integral relations. We shall now derive the integral form of Maxwell's equations in four-dimensional tensor form.

In a four-dimensional space we have a greater variety of integrals as compared with the ordinary space. We encounter integrals over the four-dimensional volume, integrals over hyper-surfaces of three dimensions, integrals over two-dimensional surfaces, and finally integrals along lines.

As in three dimensions, a four-dimensional parallelepiped is associated with a number of vectors, four in the present case. From four vectors dA_κ, dB_λ, dC_μ, dD_ν we can form a tensor of fourth rank, which is anti-symmetrical in any pair of indices, so that essentially only one value is associated with the parallelepiped. Denoting this tensor by $dV_{\kappa\lambda\mu\nu}$, it is given by

$$dV_{\kappa\lambda\mu\nu} = \begin{vmatrix} dA_\kappa & dA_\lambda & dA_\mu & dA_\nu \\ dB_\kappa & dB_\lambda & dB_\mu & dB_\nu \\ dC_\kappa & dC_\lambda & dC_\mu & dC_\nu \\ dD_\kappa & dD_\lambda & dD_\mu & dD_\nu \end{vmatrix} \equiv [dA_\kappa, dB_\lambda, dC_\mu, dD_\nu]. \tag{6.7}$$

The tensor $dV_{\kappa\lambda\mu\nu}$ is anti-symmetrical in any pair of indices, as is directly evident from the determinant which is often denoted as indicated in (6.7). Because it is anti-symmetrical in any pair of indices it can only be a multiple of the tensor $\epsilon_{\kappa\lambda\mu\nu}$. It is convenient to include a factor i and define dV as follows:

$$dV_{\kappa\lambda\mu\nu} = i\, dV \epsilon_{\kappa\lambda\mu\nu}. \tag{6.8}$$

dV is a pseudo-scalar and by definition it is equal to the volume of the parallelepiped, a real quantity. Now from the definition of $\epsilon_{\kappa\lambda\mu\nu}$ we have†

$$\epsilon_{\kappa\lambda\mu\nu}\,\epsilon_{\kappa\lambda\mu\nu} = 4!. \tag{6.9}$$

Hence

$$dV = \frac{1}{i4!}\,\epsilon_{\kappa\lambda\mu\nu}\,dV_{\kappa\lambda\mu\nu}. \tag{6.10}$$

Other equivalent forms for expressing the volume dV are as follows:

$$dV = (1/i)\epsilon_{\kappa\lambda\mu\nu}\,dA_\kappa\,dB_\lambda\,dC_\mu\,dD_\nu, \tag{6.11}$$

$$dV = \frac{1}{i}\begin{vmatrix} dA_{x_1} & dA_{x_2} & dA_{x_3} & dA_{x_4} \\ dB_{x_1} & dB_{x_2} & dB_{x_3} & dB_{x_4} \\ dC_{x_1} & dC_{x_2} & dC_{x_3} & dC_{x_4} \\ dD_{x_1} & dD_{x_2} & dD_{x_3} & dD_{x_4} \end{vmatrix} \equiv \frac{1}{i}\,[dA_{x_1}, dB_{x_2}, dC_{x_3}, dD_{x_4}]. \tag{6.12}$$

Since dV is a pseudo-scalar, its square is a scalar and we have

$$dV^2 = -(1/4!)dV_{\kappa\lambda\mu\nu}\,dV_{\kappa\lambda\mu\nu}. \tag{6.13}$$

If the faces of the parallelepiped are parallel to the four axes we get $dV = i^{-1}\,dx_1\,dx_2\,dx_3\,dx_4 \equiv i^{-1}\,d^4x$. From the invariance of the volume it follows that if in each frame of reference we use volume elements with faces which are parallel to the axes, we shall have, as already mentioned earlier for three-dimensional space,

$$\iiiint dx_1\,dx_2\,dx_3\,dx_4 = \iiiint dx_1'\,dx_2'\,dx_3'\,dx_4' = ic\iiiint dxdydzdt. \tag{6.14}$$

Volume integrals over a four-dimensional space will have the form:

$$\iiiint F_{\sigma\tau\ldots}\,dV_{\kappa\lambda\mu\nu} = G_{\sigma\tau\ldots\kappa\lambda\mu\nu} \tag{6.15}$$

and

$$\iiiint F_{\sigma\tau\ldots}\,dV = G_{\sigma\tau\ldots}. \tag{6.16}$$

Next we consider threefold integrals which arise if we integrate over a three-dimensional hyper-surface. In the four-dimensional space a three-dimensional parallelepiped is determined by three vectors dA_κ, dB_λ, dC_μ. It yields four projections on the hyper-planes x_1, x_2, x_3; x_1, x_2, x_4; x_1, x_3, x_4; x_2, x_3, x_4. (In the ordinary space, a two-dimensional parallelepiped has three projections: on the xy, yz, and xz planes.) From the three vectors we can form through direct product and permu-

† There are $1.2.3.4 = 4!$ permutations of 1, 2, 3, 4. Since any of the components of $\epsilon_{\kappa\lambda\mu\nu} = \pm 1$ we get the result which is expressed in (6.9).

tation a tensor of third rank which is anti-symmetrical in all its indices. We denote it by $dV_{\kappa\lambda\mu}$. It is given by

$$dV_{\kappa\lambda\mu} = \begin{vmatrix} dA_\kappa & dA_\lambda & dA_\mu \\ dB_\kappa & dB_\lambda & dB_\mu \\ dC_\kappa & dC_\lambda & dC_\mu \end{vmatrix} \equiv [dA_\kappa, dB_\lambda, dC_\mu]. \tag{6.17}$$

A tensor of third rank has $4^3 = 64$ components, but when it is anti-symmetrical in all its indices only 24 components are non-vanishing. However, these depend essentially only on four different numbers, say a, b, c, and d. (1, 2, 3; 2, 1, 3; etc., yield $\pm d$, 124, 214,... yield $\pm c$, 134, 314,... yield $\pm b$, and finally 234, 324,... yield $\pm a$.) Of these numbers d is real while a, b, c are imaginary.

With the help of the tensor $\epsilon_{\kappa\lambda\mu\nu}$ we can form a dual pseudo-tensor, which in this case is a pseudo-four-vector. To follow more usual custom we denote it by $d\sigma_\nu$ instead of dV_ν:

$$i^{-1}d\sigma_\nu = (1/3!)\epsilon_{\kappa\lambda\mu\nu}dV_{\kappa\lambda\mu}. \tag{6.18}$$

The division by 3! makes the values of $i^{-1}d\sigma_1$, $i^{-1}d\sigma_2$, $i^{-1}d\sigma_3$, $i^{-1}d\sigma_4$ equal to the four numbers a, b, c, d of the tensor $dV_{\kappa\lambda\mu}$. (In (6.18) all six permutations of κ, λ, μ occur, each contributing to the sum with the same value. Thus, dV_{123} appears together with dV_{213}, dV_{231}, etc. The same applies to dV_{124}, dV_{234}, and dV_{134}.)

We look upon $d\sigma_\nu$ as characterizing the hyper-surface element which is determined by dA_κ, dB_λ, dC_μ. $d\sigma_\nu$ is perpendicular to every one of these vectors:

$$d\sigma_\nu dA_\nu = d\sigma_\nu dB_\nu = d\sigma_\nu dC_\nu = 0. \tag{6.19}$$

The length of $d\sigma_\nu$ is equal to the volume of the parallelepiped multiplied by i. By definition this volume is given by

$$d\sigma^2 = (1/3!)\, dV_{\kappa\lambda\mu}\, dV_{\kappa\lambda\mu}, \tag{6.20}$$

and the following relation holds:

$$d\sigma^2 = -d\sigma_\nu^2. \tag{6.21}$$

Integrals over the three-dimensional hyper-surface are of the type:

$$\iiint F_{\sigma\tau\ldots} dV_{\kappa\lambda\mu} = G_{\sigma\tau\ldots\kappa\lambda\mu}, \tag{6.22}$$

$$\iiint F_{\sigma\tau\ldots} d\sigma_\kappa = G_{\sigma\tau\ldots\kappa}. \tag{6.23}$$

As an example let $F_{\sigma\tau\ldots}$ be a vector F_σ; by contracting $\iiint F_\sigma d\sigma_\kappa$ we get a scalar S:

$$S = \iiint F_\kappa d\sigma_\kappa. \tag{6.24}$$

By subdividing the hyper-surface into parallelepipeds with faces which are parallel to the hyper-planes $(x_1 x_2 x_3)$, $(x_1 x_2 x_4)$, $(x_1 x_3 x_4)$, and $(x_2 x_3 x_4)$, we get

$$S = \frac{1}{i}\Big[\iiint F_1\,dx_2\,dx_3\,dx_4 + \iiint F_2\,dx_1\,dx_3\,dx_4 + \iiint F_3\,dx_1\,dx_2\,dx_4 + \\ + \iiint F_4\,dx_1\,dx_2\,dx_3\Big]. \qquad (6.25)$$

In a second system of coordinates K', we get a similar expression for S if we divide the hyper-planes again as was done in K.

The next type of integrals relates to surface elements of two dimensions. From two vectors dA_κ, dB_λ we can form the anti-symmetrical tensor of second rank:

$$dV_{\kappa\lambda} = \begin{vmatrix} dA_\kappa & dA_\lambda \\ dB_\kappa & dB_\lambda \end{vmatrix} \equiv [dA_\kappa, dB_\lambda]. \qquad (6.26)$$

A tensor of second rank has $4^2 = 16$ components. However, when it is anti-symmetrical only six numbers characterize it, in accordance with the six projections of the two-dimensional parallelogram on the planes (x_1, x_2), (x_2, x_3), (x_1, x_3), (x_1, x_4), (x_2, x_4), and (x_3, x_4).

To follow the same procedure as before, we form the dual tensor which in this case is equally of second rank,

$$d\sigma_{\mu\nu} = (1/2i)\epsilon_{\kappa\lambda\mu\nu}\,dV_{\kappa\lambda}. \qquad (6.27)$$

$d\sigma_{\mu\nu}$ is orthogonal to the vectors dA_κ, dB_κ and to the tensor $dV_{\kappa\lambda}$ obtained from them:

$$\sigma_{\kappa\lambda}\,dA_\lambda = \sigma_{\kappa\lambda}\,dB_\lambda = \sigma_{\kappa\lambda}\,dV_{\kappa\lambda} = 0. \qquad (6.28)$$

Integrals over two-dimensional surfaces will be of the form

$$\iint F_{\sigma\tau\ldots}\,dV_{\kappa\lambda} = G_{\sigma\tau\ldots\kappa\lambda}, \qquad (6.29)$$

$$\iint F_{\sigma\tau\ldots}\,d\sigma_{\mu\nu} = H_{\sigma\tau\ldots\mu\nu}. \qquad (6.30)$$

Using (6.30) and integrating a tensor of second rank F we get after contracting with respect to the two indices:

$$S = \iint F_{\mu\nu}\,d\sigma_{\mu\nu}. \qquad (6.31)$$

Since $d\sigma_{\mu\nu}$ is an anti-symmetrical tensor, the symmetrical part $S_{\mu\nu}$ of $F_{\mu\nu}$ gives no contribution to this integral (calling it I we have

$$2dI \equiv S_{\mu\nu}\,d\sigma_{\mu\nu} + S_{\nu\mu}\,d\sigma_{\nu\mu} = S_{\mu\nu}\,d\sigma_{\mu\nu} - S_{\mu\nu}\,d\sigma_{\mu\nu} = 0).$$

Hence,

$$S = \iint F_{\mu\nu}\,d\sigma_{\mu\nu} = \tfrac{1}{2}\iint (F_{\mu\nu} - F_{\nu\mu})\,d\sigma_{\mu\nu}. \qquad (6.32)$$

By dividing the surface of integration through planes which are parallel to (x_1, x_2), (x_1, x_3), etc., we get

$$S = i^{-1} \iint [F_{12}\,dx_3\,dx_4 + F_{13}\,dx_2\,dx_4 + \ldots + F_{34}\,dx_1\,dx_2]. \tag{6.33}$$

Finally, we have to consider line integrals of the form

$$I_\lambda = \int S\,dL_\lambda, \tag{6.34}$$

$$I_{\kappa\lambda} = \int F_\kappa\,dL_\lambda, \tag{6.35}$$

and so on to higher tensors. Here again, we introduce the dual tensor

$$d\sigma_{\lambda\mu\nu} = (1/i)\epsilon_{\kappa\lambda\mu\nu}\,dL_\kappa. \tag{6.36}$$

The theory of all these integrals is governed by a general theorem, which applies independently of the dimensions of the space. In the four-dimensional space the theorem may be expressed by the following three equations:

$$\sum_{\kappa=1}^{4} \iiiint \frac{\partial F_{\sigma\tau\ldots}}{\partial x_\kappa}[dA_\kappa, dB_\lambda, dC_\mu, dD_\nu] = \iiint F_{\sigma\tau\ldots}[dK_\lambda, dL_\mu, dM_\nu], \tag{6.37}$$

$$\sum_{\kappa=1}^{4} \iiint \frac{\partial F_{\sigma\tau\ldots}}{\partial x_\kappa}[dA_\kappa, dB_\lambda, dC_\mu] = \iint F_{\sigma\tau\ldots}[dK_\lambda, dL_\mu], \tag{6.38}$$

$$\sum_{\kappa=1}^{4} \iint \frac{\partial F_{\sigma\tau\ldots}}{\partial x_\kappa}[dA_\kappa, dB_\lambda] = \int F_{\sigma\tau\ldots}\,dK_\lambda. \tag{6.39}$$

In each of these relations the integral on the left-hand side is the extension of a given tensor $F_{\sigma\tau\ldots}$, and in each case the contraction with respect to one index κ leads from an integral over a 'volume' to one over its 'surface'.

The corresponding relations in a three-dimensional space are:

$$\sum_{k=1}^{3} \iiint \frac{\partial F_{pq\ldots}}{\partial x_k}[dA_k, dB_l, dC_m] = \iint F_{pq\ldots}[dK_l, dL_m], \tag{6.40}$$

$$\sum_{k=1}^{3} \iint \frac{\partial F_{pq\ldots}}{\partial x_k}[dA_k, dB_l] = \int F_{pq\ldots}\,dK_l. \tag{6.41}$$

These relations contain, among others, the theorems of Gauss and Stokes as special cases. In the same way, the corresponding four-dimensional theorems of Gauss and Stokes can be derived from (6.37) to (6.39).

Let us derive Gauss's theorem from (6.40). This is achieved if we let $F_{pq...}$ be a vector F_s. Multiplying the equation by ϵ_{slm} and contracting with respect to s, l, m, we get

$$\sum_{k,l,m,s=1}^{3} \iiint \frac{\partial F_s}{\partial x_k} \epsilon_{slm}[dA_k, dB_l, dC_m] = \sum_{l,m,s=1}^{3} \iint F_s \epsilon_{slm}[dK_l, dL_m]. \tag{6.42}$$

In the left-hand side only the completely anti-symmetrical part of the integrand $(\partial F_s/\partial x_k)\epsilon_{slm}$ can contribute to the integral. This part can only be a scalar multiplied by ϵ_{klm}. The only scalar in question here is $\frac{1}{3}(\partial F_s/\partial x_s)$ because the integrand is already anti-symmetrical in l, m. Had this not been the case we should have had to divide by 3!. Hence we rewrite the left-hand side as follows:

$$\frac{1}{3}\sum_{k,l,m,s=1}^{3} \iiint \frac{\partial F_s}{\partial x_s} \epsilon_{klm}[dA_k, dB_l, dC_m] = \frac{3!}{3}\sum \iiint \frac{\partial F_s}{\partial x_s}\, dV, \tag{6.43}$$

where dV is the volume element of the three-dimensional space (see Chapter 2, eqn. (5.4)). In the same way, introducing the dual vector $d\sigma_s$ of $[dK_l, dL_m]$ in the right-hand side of (6.42), we get

$$\sum_{l,m,s=1}^{3} \iint F_s \epsilon_{slm}[dK_l, dL_m] = 2! \sum \iint F_s\, d\sigma_s. \tag{6.44}$$

Combining (6.43) and (6.44) we get Gauss's theorem

$$\sum_{s=1}^{3} \iiint \frac{\partial F_s}{\partial x_s}\, dV = \sum_{s=1}^{3} \iint F_s\, d\sigma_s. \tag{6.45}$$

In more conventional form it is expressed as follows:

$$\iiint \left(\frac{\partial F_x}{\partial x}+\frac{\partial F_y}{\partial y}+\frac{\partial F_z}{\partial z}\right) dxdydz = \iint (F_x\, dydz + F_y\, dxdz + F_z\, dxdy). \tag{6.46}$$

Similarly, it is possible to derive Stokes's theorem from (6.41). Let $F_{pq...}$ be a vector F_l. Then on contraction with respect to l we get for the left-hand side

$$\sum_{k,l=1}^{3} \iint \frac{\partial F_l}{\partial x_k}[dA_k, dB_l] \equiv \frac{1}{2}\sum_{k,l} \int \left\{\left(\frac{\partial F_l}{\partial x_k}+\frac{\partial F_k}{\partial x_l}\right)+\left(\frac{\partial F_l}{\partial x_k}-\frac{\partial F_k}{\partial x_l}\right)\right\}[dA_k\, dB_l]. \tag{6.47}$$

Again, as $[dA_k, dB_l]$ is an anti-symmetrical tensor, only the anti-symmetrical part $(\partial F_l/\partial x_k)-(\partial F_k/\partial x_l)$ contributes to this integral. Hence (6.41) yields

$$\frac{1}{2}\sum_{k,l} \iint \left(\frac{\partial F_l}{\partial x_k}-\frac{\partial F_k}{\partial x_l}\right)[dA_k, dB_l] = \sum_{l=1}^{3} \int F_l\, dK_l. \tag{6.48}$$

This is Stokes's theorem in tensor form. Let k, $l = x, y$. Then the left-hand side yields for the part which relates to the xy-plane

$$\frac{1}{2}\int\int\left(\frac{\partial F_y}{\partial x}-\frac{\partial F_x}{\partial y}\right)[dA_x, dB_y]+\frac{1}{2}\int\int\left(\frac{\partial F_x}{\partial y}-\frac{\partial F_y}{\partial x}\right)[dA_y, dB_x]. \tag{6.49}$$

Let $dA_x = dx$, $dA_y = 0$, $dB_x = 0$, $dB_y = dy$. (6.49) then yields

$$\int\int\left(\frac{\partial F_y}{\partial x}-\frac{\partial F_x}{\partial y}\right)dxdy. \tag{6.50}$$

Dealing with the other terms which relate to the (y,z) and (x,z) planes in the same way, from (6.48) we get Stokes's theorem in its more conventional form:

$$\int\int\left(\frac{\partial F_y}{\partial x}-\frac{\partial F_x}{\partial y}\right)dxdy+\int\int\left(\frac{\partial F_z}{\partial y}-\frac{\partial F_y}{\partial z}\right)dydz+\int\int\left(\frac{\partial F_x}{\partial z}-\frac{\partial F_z}{\partial x}\right)dxdz$$
$$= \int (F_x dx+F_y dy+F_z dz). \tag{6.51}$$

Gauss's theorem, as it applies to the four-dimensional space, is expressed in the following relation, which is obtained from (6.37) if $F_{\sigma\tau\ldots}$ is assumed to be a vector F_σ:

$$\int\int\int\int\frac{\partial F_\kappa}{\partial x_\kappa}dV = \int\int\int F_\kappa d\sigma_\kappa. \tag{6.52}$$

The following form is sometimes used:

$$\int\int\int\int \text{DIV}\,\mathbf{F}\,dV = \int\int\int \mathbf{F}\,d\boldsymbol{\sigma}, \tag{6.53}$$

where

$$\text{DIV}\,\mathbf{F} = \frac{\partial F_\sigma}{\partial x_\sigma} \tag{6.54}$$

and $d\boldsymbol{\sigma}$ is a four-dimensional vector with components $d\sigma_\kappa$.

Stokes's theorem is derivable from (6.39). It takes the form

$$\frac{1}{2}\int\int\left(\frac{\partial F_\kappa}{\partial x_\lambda}-\frac{\partial F_\lambda}{\partial x_\kappa}\right)[dA_\kappa, dB_\lambda] = \int F_\sigma dK_\sigma. \tag{6.55}$$

Often the following notation is used:

$$\text{CURL}_{\kappa,\lambda}\mathbf{F} = \frac{\partial F_\kappa}{\partial x_\lambda}-\frac{\partial F_\lambda}{\partial x_\kappa}. \tag{6.56}$$

From (6.38) we can also obtain the following relation for an antisymmetrical tensor $A_{\kappa\lambda}$:

$$\int\int\int\frac{\partial A_{\kappa\lambda}}{\partial x_\kappa}d\sigma_\lambda = \int\int A_{\kappa\lambda}\,d\sigma_{\kappa\lambda}. \tag{6.57}$$

Let us now identify A with the electromagnetic tensor F as defined in (2.3). Using the relation (2.4),

$$\frac{\partial F_{\kappa\lambda}}{\partial x_\kappa} = 0, \tag{6.58}$$

we find

$$\iint F_{\kappa\lambda}\, d\sigma_{\kappa\lambda} = 0. \tag{6.59}$$

In the same way, for the dual tensor $G_{\kappa\lambda}$ we get

$$\iint G_{\kappa\lambda}\, d\sigma_{\kappa\lambda} = 0. \tag{6.60}$$

Eqns. (6.59) and (6.60) express Maxwell's equations in integral form. They are the tensor equivalent of (6.1) and (6.2). In (6.59) and (6.60) the differentiations with respect to x, y, z as well as those with respect to t have been eliminated. This is in step with the characteristic feature of the theory of treating x, y, z and t in a similar way.

7. The Presence of Sources

So far we have mainly considered an electromagnetic field without including the sources which produce it. We shall now assume that these are given as charge densities $\rho(x, y, z, t)$ and current densities $\mathbf{J}(x, y, z, t)$. They are not independent of each other, since they have to satisfy the equation of continuity:

$$\frac{\partial \rho}{\partial t} + \operatorname{div} \mathbf{J} = 0. \tag{7.1}$$

This equation suggests that J_x, J_y, J_z, $ic\rho$ form a four-vector. We shall denote it by J_κ so that we have

$$J_1 = J_x, \quad J_2 = J_y, \quad J_3 = J_z, \quad J_4 = ic\rho, \tag{7.2}$$

and assume that the vector is time-like.

In terms of this vector the equation of continuity takes the covariant form

$$\frac{\partial J_\kappa}{\partial x_\kappa} = 0. \tag{7.3}$$

From the equation of continuity alone it is not possible to conclude that $\mathbf{J}$, $ic\rho$ necessarily represent a four-vector. They could transform, for instance, so that in K' we should get an equation which differs from (7.1) by some factor. However, the acceptance of J_κ as a four-vector is further enhanced through the possibility of expressing Maxwell's equations with sources in covariant form:

$$\frac{\partial F_{\lambda\kappa}}{\partial x_\kappa} = c^{-1}J_\lambda, \tag{7.4}$$

$$\frac{\partial F_{\kappa\lambda}}{\partial x_\mu}+\frac{\partial F_{\lambda\mu}}{\partial x_\kappa}+\frac{\partial F_{\mu\kappa}}{\partial x_\lambda} = 0. \tag{7.5}$$

We must now consider the interaction between the field and the sources. We may expect the force with which the field reacts on the charges and currents to depend on the one hand on the field and on the other hand on the four-vector of current and charge J_κ. From these entities we can form a four-vector

$$K_\lambda = c^{-1}F_{\lambda\kappa}J_\kappa. \tag{7.6}$$

K_λ yields for the three space coordinates the Lorentz force

$$\mathbf{K} = \rho\mathbf{E}+c^{-1}[\mathbf{J},\mathbf{H}]. \tag{7.7}$$

This is easily verified if we substitute the values of $F_{\kappa\lambda}$ in terms of $\mathbf{E}$ and $\mathbf{H}$ (see 2.3).

The time component in (7.6) ($\lambda = 4$) yields

$$K_4 = c^{-1}(F_{41}J_1+F_{42}J_2+F_{43}J_3) = ic^{-1}(\mathbf{E}\mathbf{J}). \tag{7.8}$$

Hence this component, apart from the factor i, yields the power per unit of volume connected with the current $\mathbf{J}$.

The presence of charges and currents requires a reconsideration of the energy-momentum tensor. In deriving (5.6) we applied Maxwell's equation without charges and currents. In our present case we have

$$\frac{\partial F_{\kappa\mu}}{\partial x_\kappa}F_{\lambda\mu} = -c^{-1}J_\mu F_{\lambda\mu} = -K_\lambda \tag{7.9}$$

and this is the only change in the calculation which followed (5.2). Hence

$$\frac{\partial T_{\kappa\lambda}}{\partial x_\kappa} = -K_\lambda. \tag{7.10}$$

Instead of (5.8) we now get

$$c\,\mathrm{div}[\mathbf{E},\mathbf{H}]+\frac{\partial W}{\partial t} = -\mathbf{E}\mathbf{J}. \tag{7.11}$$

Similarly, we now have

$$\mathrm{div}\,\mathbf{T}_a = -\frac{1}{c}\frac{\partial}{\partial t}[\mathbf{E},\mathbf{H}]_a-K_a = \frac{\partial}{\partial t}\left(-\frac{1}{c}[\mathbf{E},\mathbf{H}]_a-\int\limits_{-\infty}^{t}K_a\,dt\right). \tag{7.12}$$

From (7.11) we see that the law of conservation of energy holds if we take into account a term ($-\mathbf{E}\mathbf{J}$) which represents either energy delivered to the field or absorbed from it depending on the angle between

E and **J**. In the same way we can say that the law of conservation of momentum holds if we take into account the terms $\int K_x\,dt$, $\int K_y\,dt$, $\int K_z\,dt$ of momentum acquired by the charges through the action of the field.

8. The Energy-Momentum Vector and its Conservation

The quantity $c\mathbf{G} = [\mathbf{E}, \mathbf{H}]$ and the energy W are both components of the energy-momentum tensor. Now, from the equation of conservation of energy one might feel tempted to assume that $c\mathbf{G}$ and iW form a four-vector because

$$\operatorname{div}[\mathbf{E},\mathbf{H}] + \frac{1}{c}\frac{\partial W}{\partial t} = \frac{\partial(cG_x)}{\partial x} + \frac{\partial(cG_y)}{\partial y} + \frac{\partial(cG_z)}{\partial z} + \frac{1}{c}\frac{\partial W}{\partial t} = 0. \tag{8.1}$$

However, from such an equation we cannot conclude that $(c\mathbf{G}, iW)$ or, since c is an invariant, $(\mathbf{G}, iW/c)$ is a four-vector. This would have been the case only if the whole expression (8.1) were equal to a non-vanishing invariant. If we transform $(\mathbf{G}, iW/c)$ like a vector into $(\mathbf{G}', iW'/c)$, it differs from the $(\mathbf{G}', iW'/c)$ obtained by transforming the energy-momentum tensor, although in each case the divergence remains zero.

Matters are different when we consider integrals over space regions. In this section we shall prove that for fields covering finite regions and containing no charges and currents the following four integrals form a four-vector P_λ:

$$P_\lambda = \frac{1}{ic}\iiint T_{4\lambda}\,dxdydz. \tag{8.2}$$

By this we mean more specifically the following: suppose a second observer K' who moves relative to the first is also forming the integrals (8.2). If the first observer considers the field at a given time t, as is implied in (8.2), this field will appear to the second observer as with a time which varies from one field point to the other. However, like the first observer, K' takes a definite t'-time and considers the field at his $t' =$ constant, and forms the integrals

$$P'_\lambda = \frac{1}{ic}\iiint T'_{4\lambda}\,dx'dy'dz'. \tag{8.3}$$

What is now claimed is that as a result of (8.1), which is the law of conservation of energy and momentum in differential form, P'_λ is obtainable from P_λ by a four-vector transformation.

Without the law of conservation P'_λ in (8.3) and P_λ in (8.2) could have no transformational relation to each other, since both observers refer to different times.

Instead of considering the integral (8.2) let us consider a space-like 'surface' and form the four-vector

$$P_\lambda(\sigma) = -\frac{1}{c}\iiint\limits_\sigma T_{\kappa\lambda}(\sigma)\,d\sigma_\kappa. \tag{8.4}$$

By a space-like surface is meant one in which any two points can be joined by a space-like vector only. $d\sigma_\kappa$ is a four-vector which is perpendicular to the hyper-surface element of magnitude $d\sigma$ (Fig. 21).

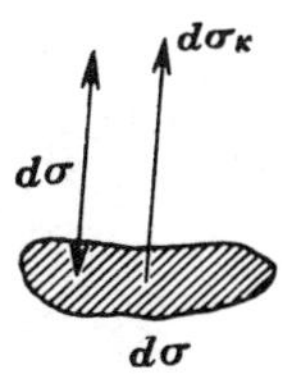

Fig. 21

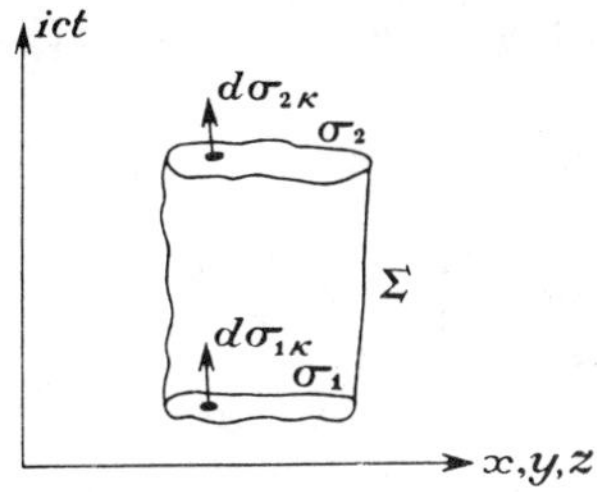

Fig. 22

In (8.4) $P_\lambda(\sigma)$ is a four-vector if both observers refer to the same space-like surface, as will be assumed now.

Let us now consider a four-dimensional region which is bounded by two space-like surfaces σ_1 and σ_2 and a cylindrical hyper-surface Σ over which the field may be assumed to be sufficiently small to be neglected (Fig. 22). In the whole region we have then

$$\frac{\partial T_{\kappa\lambda}}{\partial x_\kappa} = 0, \tag{8.5}$$

$$T_{\kappa\lambda} = 0 \quad (\text{over } \Sigma). \tag{8.6}$$

By Gauss's theorem we get

$$\iiiint \frac{\partial T_{\kappa\lambda}}{\partial x_\kappa}\,d^4x = \iiint\limits_{\Sigma,\sigma_1,\sigma_2} T_{\kappa\lambda}\,d\sigma_\kappa = \iiint\limits_{\sigma_2} T_{\kappa\lambda}\,d\sigma_{2\kappa} - \iiint\limits_{\sigma_1} T_{\kappa\lambda}\,d\sigma_{1\kappa}, \tag{8.7}$$

and because of (8.5),

$$\iiint\limits_{\sigma_2} T_{\kappa\lambda}\,d\sigma_{2\kappa} = \iiint\limits_{\sigma_1} T_{\kappa\lambda}\,d\sigma_{1\kappa}; \tag{8.8}$$

therefore
$$P_\lambda(\sigma_2) = P_\lambda(\sigma_1). \tag{8.9}$$

Eqn. (8.9) expresses the law of conservation of momentum and energy in integral form, for as we shall see P_1, P_2, P_3 represent the total momentum in the field and icP_4 represents the total energy. According to (8.9), P_λ does not depend on the particular space-like surface σ which we may consider. In particular, if σ_1 is $t_1 =$ constant, σ_2 is $t_2 =$ constant, we have $P_\lambda(t_2) = P_\lambda(t_1) = P_\lambda$ which is called the energy-momentum vector.

Relative to the second observer (in K') we have again

$$P'_\lambda(\sigma'_2) = P'_\lambda(\sigma'_1). \tag{8.10}$$

Both observers are now allowed to change the space-like surfaces at will and the restriction introduced earlier according to which both observers have to use the same space-like surface σ if P_λ is to be a four-vector can be dropped. In particular K and K' can use the flat surfaces $t =$ constant and $t' =$ constant, and it follows that P'_λ in (8.3) is obtainable from P_λ in (8.2) by vector transformation. Without the law of conservation (8.5) in differential form (or (8.9) in integral form) this result would not have been achieved and $P_\lambda(\sigma)$ in (8.4) would behave as a four-vector only if both observers referred to the same surface σ.

P_λ as given by (8.2) (or 8.3) has the following components:

$$P_a = \iiint G_a \, dxdydz, \qquad P_4 = (i/c) \iiint W \, dxdydz, \tag{8.11}$$

so that the identification of P_λ as the energy-momentum vector is justified.

The property of the four-vector P_λ as defined in (8.4) to remain a four-vector even when various observers refer to different space-like surfaces, holds only if no charges and currents are present. When these are present P_λ ($t =$ const.) and P'_λ ($t' =$ const.) are not obtainable from each other by a four-vector transformation. This fact has led until recently to confusions and wrong conclusions about the dynamical properties of electrons, in particular with regard to the origin of their mass and internal stability. The situation has been clarified independently by Kwal and Rohrlich, who firstly reiterated the fact that in the general case when $\partial T_{\kappa\lambda}/\partial x_\lambda \neq 0$, a four-vector can be obtained only if the surface is specified, and secondly pointed out that when dealing with a moving electron the surface must bear an invariant relation to the world line of the electron. We shall deal with this matter in Chapter 5.

The fact that a pulse of light of any shape with finite dimensions has a total momentum and a total energy (both purely electromagnetic) which transform like a four-vector, make it possible to attribute to it particle properties, as is done in quantum theory.

It may be added that the law of conservation of charge in differential form,

$$\frac{\partial J_\kappa}{\partial x_\kappa} = 0, \tag{8.12}$$

leads in a similar way to a law of conservation of total charge. Mathematically all we have to do is to replace $T_{\kappa\lambda}$ by J_κ. The tensor $T_{\kappa\lambda}$ leads to a four-vector of total energy and momentum. In a similar way, the vector J_κ leads to a scalar which we denote by e and call the total charge. Let

$$e(\sigma) = -\frac{1}{c}\iiint J_\kappa \, d\sigma_\kappa. \tag{8.13}$$

Then, as in the case of the tensor $T_{\kappa\lambda}$, we find that for each of the observers

$$e(\sigma_1) = e(\sigma_2). \tag{8.14}$$

This relation expresses the law of conservation of charge in integral form. It follows further that in (8.13) the invariance of $e(\sigma)$ will also be maintained if both observers do not refer to the same surface σ. In particular, if they use t = constant and t' = constant, we get, since $d\sigma_1 = d\sigma_2 = d\sigma_3 = 0$, $d\sigma_4 = i\,dxdydz$, and $J_4 = ic\rho$,

$$e(\sigma) = e = -\frac{1}{c}\iiint J_4 d\sigma_4 = \iiint \rho \, dxdydz, \tag{8.15}$$

and

$$e = \iiint \rho \, dxdydz = \iiint \rho' \, dx'dy'dz'. \tag{8.16}$$

This relation justifies us in regarding e as the total charge. We have proved both its invariance and its conservation by using the law of conservation in differential form. In (8.16) neither ρ nor $dxdydz$ are scalars, but their product is a scalar.

The use of space-like surfaces, and not the more special surfaces t = constant which lose this property when we change the frame of reference, has proved particularly useful in the quantum theory of fields.

9. The Retarded and Advanced Potentials

In electromagnetic theory it is found convenient to introduce auxiliary potentials, $\mathbf{A}$ and ϕ, which satisfy the equations

$$\frac{1}{c^2}\frac{\partial^2 \mathbf{A}}{\partial t^2} - \nabla^2 \mathbf{A} = \frac{\mathbf{J}}{c}, \tag{9.1}$$

$$\frac{1}{c^2}\frac{\partial^2 \phi}{\partial t^2} - \nabla^2 \phi = \rho, \tag{9.2}$$

$$\operatorname{div} \mathbf{A} + \frac{1}{c}\frac{\partial \phi}{\partial t} = 0. \tag{9.3}$$

These three equations form a substitute for Maxwell's equations (13.24) on p. 65. Eqn. (9.3) which links $\mathbf{A}$ and ϕ is known as the

Lorentz condition. From **A** and ϕ the electromagnetic field can be obtained through the relations

$$\mathbf{H} = \operatorname{curl} \mathbf{A}, \tag{9.4}$$

$$\mathbf{E} = -\left(\operatorname{grad} \phi + \frac{1}{c}\frac{\partial \mathbf{A}}{\partial t}\right). \tag{9.5}$$

Let **A** and ϕ be a set of potentials satisfying (9.1)–(9.3) and let χ be an arbitrary scalar function. Then the transformation,

$$\mathbf{A}^* = \mathbf{A} - \operatorname{grad} \chi, \tag{9.6}$$

$$\phi^* = \phi + \frac{1}{c}\frac{\partial \chi}{\partial t}, \tag{9.7}$$

will lead to a set of potentials $\mathbf{A}^*$, ϕ^* which yield the same electromag-

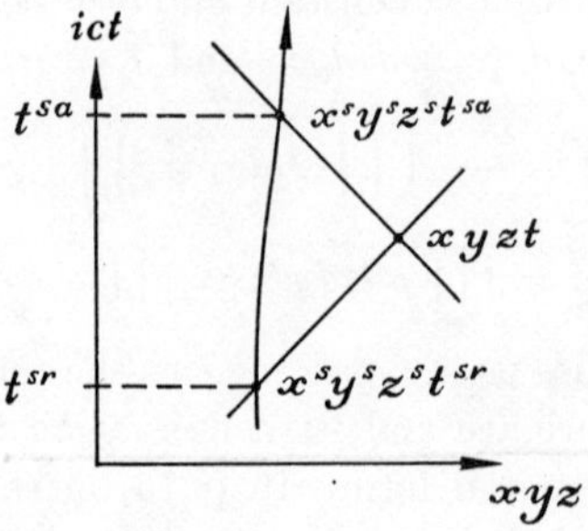

FIG. 23

netic field as **A** and ϕ. The transformation (9.6) and (9.7) is called a gauge transformation (of second kind). Although $\mathbf{A}^*$ and ϕ^* yield the same electromagnetic field as **A** and ϕ, they need not satisfy (9.1)–(9.3). This will be the case only if χ satisfies the wave equation

$$\frac{1}{c^2}\frac{\partial^2 \chi}{\partial t^2} - \nabla^2 \chi = 0. \tag{9.8}$$

It can be shown that a particular solution of (9.1) and (9.2) is given by

$$\mathbf{A}(x, y, z, t) = \frac{1}{4\pi c} \iiint \frac{\mathbf{J}(x^s, y^s, z^s, t^{sr})\, dx^s dy^s dz^s}{r}, \tag{9.9}$$

$$\phi(x, y, z, t) = \frac{1}{4\pi} \iiint \frac{\rho(x^s, y^s, z^s, t^{sr})\, dx^s dy^s dz^s}{r}, \tag{9.10}$$

where x, y, z, t are a field space-time point, x^s, y^s, z^s, t^s are to be taken over the current and charge region, and t^{sr} is the retarded time, i.e. the time obtained from the intersection of the light cone, with an apex at the field point, with the world lines of the charge elements (see Fig. 23). r is the spatial distance between the field point and a source point at the retarded time

$$r = [(x-x^s)^2+(y-y^s)^2+(z-z^s)^2]^{\frac{1}{2}}. \tag{9.11}$$

From the equation of the light cone

$$-c^2(t-t^{sr})^2+(x-x^s)^2+(y-y^s)^2+(z-z^s)^2 = 0 \tag{9.12}$$

follows

$$t^{sr} = t-\frac{r}{c}. \tag{9.13}$$

Any solution of eqns. (9.1) and (9.2) can be obtained by adding to the particular integral an appropriate solution of the homogeneous equations, obtained when **J** and ρ vanish. It should always be possible to get solutions which satisfy the Lorentz condition, for otherwise **A** and ϕ cannot represent the required potentials. Moreover, it can be shown† that the solutions (9.9) and (9.10) already satisfy the Lorentz condition as a result of the equation of continuity which links **J** and ρ:

$$\operatorname{div} \mathbf{J}+\frac{\partial \rho}{\partial t} = 0. \tag{9.14}$$

From a relativistic point of view this is not surprising, because $(\mathbf{J}, ic\rho)$ is a four-vector and as we shall see, this holds also for **A** and $i\phi$. As a result of the integrals (9.9) and (9.10), the relation (9.14) between **J** and ρ is reflected in a similar relation between **A** and ϕ which is the Lorentz condition.

It is possible to introduce potentials which do not satisfy the Lorentz condition. In particular one may choose the Coulomb gauge which is given by

$$\operatorname{div} \mathbf{A} = 0. \tag{9.15}$$

In this case ϕ satisfies a Poisson equation $\nabla^2\phi = \rho$, where ρ is not only a function of x, y, z but also of t, so that the equation is quasi-static. As to **A** it satisfies a more complicated equation than (9.1). The Coulomb gauge is particularly useful when a separation between longitudinal and transverse components of the field is desirable. But the price paid is loss of relativistic covariance. In other words, the mentioned separation of the field components is not Lorentz invariant and holds only in the particular Lorentz frame in which the separation has been carried out.

The next object is to express the introduction of the potentials in

† See, for instance, J. A. Stratton, *Electromagnetic Theory*, pp. 428–30.

covariant form. If **A** and $i\phi$ are assumed to form a four-vector, (9.1)–(9.3) may be summed up as follows:

$$\partial_\kappa^2 A_\lambda = -\frac{1}{c} J_\lambda, \tag{9.16}$$

$$\partial_\kappa A_\kappa = 0, \tag{9.17}$$

where
$$\partial_\kappa^2 = \frac{\partial^2}{\partial x_1^2}+\frac{\partial^2}{\partial x_2^2}+\frac{\partial^2}{\partial x_3^2}+\frac{\partial^2}{\partial x_4^2} \equiv \Box, \tag{9.18}$$

and
$$J_\lambda = (\mathbf{J}, ic\rho) \tag{9.19}$$

is the current density four-vector. The operator ∂_κ^2 is called the d'Alembertian. The relationship between **E**, **H**, and **A**, ϕ (see (9.4) and (9.5) becomes

$$F_{\kappa\lambda} = \partial_\kappa A_\lambda - \partial_\lambda A_\kappa. \tag{9.20}$$

It is easily verified that if A_λ satisfies (9.16) and (9.17) $F_{\kappa\lambda}$ satisfies Maxwell's equations:

$$\frac{\partial F_{\kappa\lambda}}{\partial x_\kappa} = \frac{\partial}{\partial x_\kappa}(\partial_\kappa A_\lambda - \partial_\lambda A_\kappa) = \partial_\kappa^2 A_\lambda - \frac{\partial}{\partial x_\lambda}(\partial_\kappa A_\kappa), \tag{9.21}$$

so that
$$\frac{\partial F_{\kappa\lambda}}{\partial x_\kappa} = -\frac{1}{c} J_\lambda. \tag{9.22}$$

Also,

$$\frac{\partial F_{\kappa\lambda}}{\partial x_\mu}+\frac{\partial F_{\lambda\mu}}{\partial x_\kappa}+\frac{\partial F_{\mu\kappa}}{\partial x_\lambda} = \frac{\partial}{\partial x_\mu}(\partial_\kappa A_\lambda - \partial_\lambda A_\kappa)+\frac{\partial}{\partial x_\kappa}(\partial_\lambda A_\mu - \partial_\mu A_\lambda)+$$
$$+\frac{\partial}{\partial x_\lambda}(\partial_\mu A_\kappa - \partial_\kappa A_\mu) = 0. \tag{9.23}$$

As to the Gauge transformation, it takes the form

$$A_\lambda^* = A_\lambda - \partial_\lambda \chi \tag{9.24}$$

and it is evident that if A_λ^* is to satisfy the Lorentz condition, the scalar χ must satisfy the homogeneous wave equation

$$\partial_\lambda A_\lambda^* = \partial_\lambda A_\lambda - \partial_\kappa^2 \chi = -\partial_\lambda^2 \chi, \tag{9.25}$$

so that if the left-hand side is to vanish this must also apply to $\partial_\lambda^2 \chi$. It also follows that A^* satisfies the inhomogeneous wave equation

$$\partial_\kappa^2 A_\lambda^* = \partial_\kappa^2 A_\lambda - \partial_\kappa^2 \partial_\lambda \chi = -\frac{1}{c} J_\lambda - \partial_\lambda(\partial_\kappa^2 \chi) = -\frac{1}{c} J_\lambda. \tag{9.26}$$

Eqns. (9.16) and (9.17) are invariant against space and time inversions. In this connexion **A** and ϕ transform as **J** and ρ (see p. 65, for instance):

$$P\colon \mathbf{A} \to -\mathbf{A}, \quad \phi \to \phi$$
$$T\colon \mathbf{A} \to -\mathbf{A}, \quad \phi \to \phi. \tag{9.27}$$

From the invariance of (9.16) and (9.17) with regard to these transformations it follows that on performing them on the solution (9.9) and (9.10), we may get new solutions. Firstly, let us express the solution in the form

$$A^r_\mu(x,y,z,t) = \frac{1}{4\pi c}\iiint \frac{J_\mu\{x^s,y^s,z^s,t-(r/c)\}}{r}\,dx^s dy^s dz^s. \tag{9.28}$$

A space reflection has no effect on this integral, since r is an invariant with regard to this transformation. Consequently a space reflection does not lead to a new solution. Next, let us consider time inversion. The condition

$$t^s = t-\frac{r}{c} \tag{9.29}$$

goes over into $-t^s = -t-\frac{r}{c}$ or $t^s = t+\frac{r}{c}$ and this condition leads to the new solution

$$A^a_\mu(x,y,z,t) = \frac{1}{4\pi c}\iiint \frac{J_\mu(x^s,y^s,z^s,t^{sa})}{r}\,dx^s dy^s dz^s, \tag{9.30}$$

where

$$t^{sa} = t+\frac{r}{c}. \tag{9.31}$$

In (9.31), r is the distance between the field point and a charge element at the advanced time (see Fig. 23).

Any solution of (9.16) and (9.17) can be expressed both in terms of the retarded potential as well as in terms of the advanced particular solution. The solution of the homogeneous equation required (under any additional boundary conditions) will depend on which particular integral is used. Let these be denoted by A^{in}_μ and A^{out}_μ, then

$$\begin{aligned} A_\mu(x_\kappa) &= A^{\text{in}}_\mu(x_\kappa)+A^r_\mu(x_\kappa) \\ &= A^{\text{out}}_\mu(x_\kappa)+A^a_\mu(x_\kappa). \end{aligned} \tag{9.32}$$

The notation here is adapted to a condition in which the wave represents an outgoing spherical wave at a great distance from the currents. The retarded integral represents all the requirements with regard to the condition at a great distance, and so the solution of the homogeneous equation must be an incoming wave. On the other hand, in the second form of the solution, the particular integral represents an incoming wave and at infinity this must be compensated by an outgoing wave.

Any linear combination of the particular integrals A^r and A^a with coefficients which add up to 1, is also a particular solution of the inhomogeneous equation and can be used with an appropriate solution of the homogeneous equation. Here it is important to realize that whenever the fundamental equations are invariant against time reversal, no preferred direction in time is involved (no time arrow). This is the case with Maxwell's equations. In practice we are, however, interested in solutions of Maxwell's equations which are not symmetric in time, but this is not connected, as might be thought, with the law of causality. More specifically, in practice we are interested in solutions which represent, at great distance from the currents, a spherical wave which travels outwards. A time non-symmetric solution of the inhomogeneous equation representing a spherical wave converging from outer regions and arriving just in time to coincide with a current is not observed, although it would be in accord with Maxwell's equations and cannot be ruled out on grounds of causality. The reason for the occurrence in practice of a solution which is asymmetric in time must be sought outside the content of Maxwell's equations. The corresponding motion of an electron involves loss of energy and constitutes an irreversible process. Wheeler and Feynman have put forward a theory which, as far as Maxwell's equations are concerned, is completely symmetrical in time but links the irreversibility with an absorber which they regard as essential in understanding the behaviour of an electron. The absorber consists of all the other charged particles in the world. If only a single electron existed it would not radiate, even if accelerated. Hogarth, using the time-symmetrical feature of the Wheeler–Feynman theory, deviates from them with regard to the time arrow. According to Hogarth, this is linked with the expansion of the universe, which makes the absorbing universe of the future different from that of the past. Hoyle and Narlikar have extended and somewhat modified the theory of Hogarth. We shall return to this matter later.

A time-symmetric solution of the inhomogeneous equation is given by

$$A_\mu^+(x_\kappa) = \tfrac{1}{2}A_\mu^r(x_\kappa) + \tfrac{1}{2}A_\mu^a(x_\kappa) = f(t^r) + f(t^a). \tag{9.33}$$

Since time reversal exchanges the roles of t^r and t^a, we see that the solution A_μ^+ retains its form under time reversal. It also satisfies the condition that the coefficients must add up to 1.

Another linear combination of retarded and advanced potentials which we shall require when dealing with the equation of motion of an electron is the antisymmetrical

$$A_\mu^-(x_\kappa) = \tfrac{1}{2}A_\mu^r(x_\kappa) - \tfrac{1}{2}A_\mu^a(x_\kappa). \tag{9.34}$$

On time reversal this function gets multiplied by -1. The main thing about the coefficients here is that they add up to 0. (The choice of $\frac{1}{2}$ and $-\frac{1}{2}$ is only for later convenience.) Consequently, A_μ^- is not a solution of the inhomogeneous equation, but since it corresponds to a current which vanishes everywhere, it is a solution of the homogeneous equation. Nevertheless, it bears a certain correlation to the current, but even for a moving point charge it is free of singularities along the world line.

In the expressions (9.28) and (9.30) for the retarded and advanced potentials respectively, it is not immediately evident that the factor of J_μ under the integral is an invariant as it should be if A_μ is to be a four-vector. Since all the world points on the world lines of the charge elements affecting the field point x_κ must lie on the light cone through x_κ (on the backward cone in the retarded case and on the forward cone in the advanced case), it amounts to proving that in integrating J_μ over the light cone, where (9.13) and (9.31) hold,

$$\frac{dx^s dy^s dz^s}{c(t-t^{sr})} \quad \text{and} \quad \frac{dx^s dy^s dz^s}{c(t^{sa}-t)} \tag{9.35}$$

are invariants. To prove this, consider a simple Lorentz transformation involving motion along the x-axis. Without loss of generality it may be assumed that $t = 0$ at the field point. Since for motion along x-axis, dy^s and dz^s are invariants, it remains to prove that

$$\frac{dx_1}{x_4} \quad \text{is an invariant} \tag{9.36}$$

if

$$x_1^2+x_4^2 = 0. \tag{9.37}$$

Let the Lorentz transformation be given by

$$\begin{aligned} x_1' &= \cosh\psi . x_1+\sinh\psi . i^{-1}x_4, \\ i^{-1}x_4' &= \sinh\psi . x_1+\cosh\psi . i^{-1}x_4. \end{aligned} \tag{9.38}$$

Hence,

$$\frac{i\,dx_1'}{x_4'} = \frac{\cosh\psi . dx_1+\sinh\psi . i^{-1}dx_4}{\sinh\psi . x_1+\cosh\psi . i^{-1}x_4}. \tag{9.39}$$

Now, from (9.37),

$$dx_4 = \frac{-x_1\,dx_1}{x_4} \tag{9.40}$$

and it now follows that

$$\frac{dx_1'}{x_4'} = \frac{dx_1}{x_1}, \tag{9.41}$$

or,

$$\frac{dx'}{ct'^r} = \frac{dx}{ct^r}, \qquad \frac{dx'}{ct'^a} = \frac{dx}{ct^a}. \tag{9.42}$$

This situation is illustrated in Fig. 24. Here the Lorentz transformation is described by a shift of the point x to x'. As illustrated in the figure, dx' is related to dx as ct' to ct. In a more general Lorentz transformation, we have over the light cone

$$\frac{dx_1\,dx_2\,dx_3}{x_4} = \frac{dx_1'\,dx_2'\,dx_3'}{x_4'}. \tag{9.43}$$

A similar invariance holds in any other four-dimensional space in which a Lorentz transformation takes place. For instance, let k_μ be a propagation vector of a plane wave

$$k_\mu = \left(\mathbf{k}, \frac{i\omega}{c}\right), \tag{9.44}$$

where $|\mathbf{k}| = 2\pi/\lambda$ and λ and ω are wavelength and frequency. In the case of electromagnetic waves

$$k_\mu^2 = \mathbf{k}^2 - \frac{\omega^2}{c^2} = 0. \tag{9.45}$$

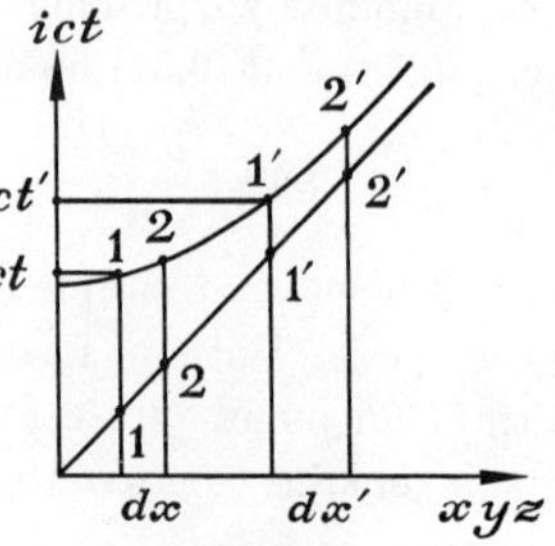

FIG. 24

This equation determines an invariant cone and over this cone we have

$$\frac{dk_1\,dk_2\,dk_3}{\omega c^{-1}} = \frac{dk_1\,dk_2\,dk_3}{(k_1^2+k_2^2+k_3^2)^{\frac{1}{2}}} = \text{an invariant}. \tag{9.46}$$

This result can be extended further to include a wave which is governed by a Klein–Gordon equation

$$\Box\psi - m^2\psi = 0. \tag{9.47}$$

In this case the propagation vector of a plane wave is given by a vector p_μ obeying the condition

$$p_\mu^2 = \mathbf{p}^2 - \frac{p_0^2}{c^2} = -m^2, \tag{9.48}$$

or

$$\frac{p_0^2}{c^2} = \mathbf{p}^2 + m^2. \tag{9.49}$$

Eqn. (9.48) determines an invariant hyperboloid in the space $p_1, p_2, p_3, ip_0 c^{-1}$ which replaces the previous cone. It leads to

$$\frac{dp_1\,dp_2\,dp_3}{p_0} = \frac{dp_1\,dp_2\,dp_3}{c(p_1^2+p_2^2+p_3^2+m^2)^{\frac{1}{2}}} = \text{an invariant}. \tag{9.50}$$

This is also illustrated in Fig. 24.

Another method of proving the four-vector character of A_μ^r or A_μ^a is to show that the respective integrals can be written in covariant form. As a first step towards this, it may be noted that with the help of the δ-function the integrals (9.28) and (9.30) can be expressed as fourfold integrals in the four-dimensional space as follows:†

$$A^r_\mu(x_\kappa) = \frac{1}{4\pi c}\iiiint \frac{J_\mu(x^s_\kappa)\delta\{i^{-1}(x_4-x^s_4)-r\}}{r}\,d^4x^s, \tag{9.51}$$

$$A^a_\mu(x_\kappa) = \frac{1}{4\pi c}\iiiint \frac{J_\mu(x^s_\kappa)\delta\{i^{-1}(x_4-x^s_4)+r\}}{r}\,d^4x^s. \tag{9.52}$$

In this form it is made automatically certain that r is taken for a field point and a source point which lie on the light cone. The element of integration is now an invariant, and it remains to prove that $r^{-1}\delta(r-ct^r)$ and $r^{-1}\delta(r+ct^a)$ are invariants. One way of proving this is to apply an infinitesimal Lorentz transformation, involving a velocity dv between two observers and neglecting terms involving (dv^2) and higher powers. The other and shorter way is to make use of the formula

$$\delta(x^2_\mu) \equiv \delta(x_1^2+x_2^2+x_3^2-c^2t^2) \equiv \delta(r^2-c^2t^2)$$
$$= \frac{1}{2r}[\delta(r-ct)+\delta(r+ct)]. \tag{9.53}$$

Since x^2_μ is an invariant, this also applies to any function of x^2_μ and consequently also to $\delta(x^2_\mu)$. Invariant functions of this type are characterized by having a constant value along each hyperboloid $x^2_\mu = \text{const}$. It follows now that the right-hand side of (9.53) is also an invariant. This consists of two terms, one representing a spherical pulse wave travelling with light velocity away from the source point, and the other a spherical pulse wave converging with the velocity of light towards the source. Such pulses are, as easily verified for any function $r^{-1}f(r\pm ct)$, solutions of the wave equation, with a singularity at $r = 0$, and appear as such waves in each inertial frame of reference. Consequently, each term in itself is an invariant. Formally, the invariance of each of these terms can be expressed by means of the step function $\Theta(x_\mu)$ which is defined as follows:

$$\begin{aligned} \Theta(x_\mu) &= 1 \quad \text{for } i^{-1}x_4 > 0,\\ \Theta(x_\mu) &= 0 \quad \text{for } i^{-1}x_4 < 0,\\ \Theta(-x_\mu) &= 0 \quad \text{for } i^{-1}x_4 > 0,\\ \Theta(-x_\mu) &= 1 \quad \text{for } i^{-1}x_4 < 0. \end{aligned} \tag{9.54}$$

We have

$$r^{-1}\delta(r-ct) = 2\delta(x^2_\mu)\Theta(x_\mu), \tag{9.55}$$

$$r^{-1}\delta(r+ct) = 2\delta(x^2_\mu)\Theta(-x_\mu). \tag{9.56}$$

† $d^4x = i^{-1}\,dx_1\,dx_2\,dx_3\,dx_4 = dx\,dy\,dz\,cdt$.

It is true that $\Theta(x_\mu)$ is not an invariant, because it refers to the hyperplane $t = 0$. However, the product of an invariant which is zero everywhere except on the invariant light cone with $\Theta(x_\mu)$ is also an invariant. A hyperplane $t' = 0$ may also be used to define $\Theta(x_\mu)$ so that in every frame of reference the multiplication with $\Theta(x_\mu)$ abolishes anything which is on the backward light cone, and multiplication with $\Theta(-x_\mu)$ abolishes anything which is on the forward light cone. We now get

$$A_\mu^r(x_\kappa) = \frac{1}{2\pi c}\iiiint J_\mu(x_\kappa^s)\delta[(x_\kappa - x_\kappa^s)^2]\Theta(x_\kappa - x_\kappa^s)\, d^4x_\kappa^s, \tag{9.57}$$

$$A_\mu^a(x_\kappa) = \frac{1}{2\pi c}\iiiint J_\mu(x_\kappa^s)\delta[(x_\kappa - x_\kappa^s)^2]\Theta(x_\kappa^s - x_\kappa)\, d^4x_\kappa^s. \tag{9.58}$$

Particular solutions of (9.16) and (9.17) with appropriate boundary conditions can be expressed by means of Green's functions which obey similar conditions. Let the Green function be denoted by $D(x_\kappa - x_\kappa^s)$. Then it must obey the inhomogeneous equation of a point source

$$\partial_\kappa^2 D(x_\kappa - x_\kappa^s) = -\delta^4(x_\kappa - x_\kappa^s) = \delta(x_1 - x_1^s)\delta(x_2 - x_2^s)\delta(x_3 - x_3^s)\delta(x_4 - x_4^s). \tag{9.59}$$

The corresponding particular solution of (9.16) and (9.17) is given by

$$A_\mu^D(x) = \frac{1}{c}\iiiint J_\mu(x_\kappa^s)D(x_\kappa - x_\kappa^s)\, d^4x_\kappa^s. \tag{9.60}$$

From (9.57) and (9.58) we see that

$$D^r(x_\kappa - x_\kappa^s) = \frac{1}{2\pi}\delta[(x_\kappa - x_\kappa^s)^2]\Theta(x_\kappa - x_\kappa^s) = \frac{1}{4\pi}r^{-1}\delta[r - c(t - t^s)], \tag{9.61}$$

$$D^a(x_\kappa - x_\kappa^s) = \frac{1}{2\pi}\delta[(x_\kappa - x_\kappa^s)^2]\Theta(x_\kappa^s - x_\kappa) = \frac{1}{4\pi}r^{-1}\delta[r + c(t - t^s)]. \tag{9.62}$$

D^r vanishes everywhere except on the backward cone through the field point, while D^a vanishes everywhere except on the forward cone through the field point. We also note that

$$\tfrac{1}{2}D^r(x_\kappa) + \tfrac{1}{2}D^a(x_\kappa) = \frac{1}{4\pi}\delta(x_\kappa^2), \tag{9.63}$$

$$A_\mu^+(x_\kappa) = \frac{1}{c}\iiiint J_\mu(x_\kappa^s)\delta\{(x_\kappa - x_\kappa^s)^2\}\, d^4x^s. \tag{9.64}$$

CHAPTER 4

GENERAL FIELD THEORY

1. The Variation Principle

The electromagnetic field is the best-known example of a classical field and the one of greatest practical importance, even without subjection to quantum theory. In recent years other types of fields have been extensively considered in nuclear physics. Like the Maxwell field they can be studied from a classical point of view, although in their case, the full application can be achieved after quantization. In this chapter we shall describe, from a classical point of view, only some general features of such fields in as far as they depend on relativistic invariance. The results of this chapter apply also to the electromagnetic field.

The first task in any field theory is to derive differential equations (field equations) which determine its development in space-time. The next task is to define all the entities of physical interest, such as energy, momentum, and angular momentum and to prove their conservation in conservative systems.

In a non-relativistic field theory it would be natural to consider a field at a given time and to investigate what happens to the field at future times. While this method could also be applied in the theory of relativity, it is not completely in line with it, and can therefore not be expected to yield results which are relativistic in form. When a field is considered at a given time relative to one observer this field will appear at a range of times to another observer who is in motion relative to the first one. Each field point will appear at a different time. Hence, the concept of a field at a given time is not sufficiently general when Lorentz transformations are envisaged. One feature which is common to both observers is the following: the 'surfaces' (three-dimensional spaces) t = constant and $t' = \sigma(x', y', z')$, into which the former is transformed, are both space-like. It is, therefore, more appropriate to consider a field over a three-dimensional space-like region σ which is not necessarily t = constant. However, it will not involve great loss of generality if we assume the surface σ as flat. This property is Lorentz-invariant.

The history of the field, both in its spatial distribution as well as its temporal development, will involve the four-dimensional space. We consider a region which is bounded by two flat space-like surfaces σ_1 and σ_2 and by a three-dimensional 'cylindrical' surface over which the field is sufficiently small to be negligible (as in Fig. 22). Next we postulate the existence of an invariant Lagrangian L which depends on the field components ψ_μ (a tensor with n components so that $\mu = 1,..., n$),† and the first derivatives $\partial_\kappa \psi_\mu$. The action integral

$$I = \iiiint_{\sigma_1}^{\sigma_2} L(\psi_\mu, \partial_\kappa \psi_\mu)\, d^4x \tag{1.1}$$

then yields
$$\delta I = 0 \tag{1.2}$$

if arbitrary variations of ψ_μ inside the considered region are performed which vanish over the surfaces σ_1 and σ_2. Eqn. (1.2) expresses Hamilton's principle. If L is to be a scalar it can only involve absolute operations on the tensors ψ_μ and $\partial_\kappa \psi_\mu$ which yield a scalar. Hence we must have

$$L(\psi_\mu, \partial_\kappa \psi_\mu) = L(\psi'_\mu, \partial'_\kappa \psi'_\mu). \tag{1.3}$$

For instance, for a vector field ψ_λ, L might be

$$L(\psi_\lambda, \partial_\kappa \psi_\lambda) = \psi_\lambda^2 + \partial_\lambda \psi_\lambda = \psi_\lambda'^2 + \partial'_\lambda \psi'_\lambda = L(\psi'_\lambda, \partial'_\kappa \psi'_\lambda)$$

or
$$L(\psi_\lambda, \partial_\kappa \psi_\lambda) = \psi_\kappa \partial_\kappa \partial_\lambda \psi_\lambda = \psi'_\kappa \partial'_\kappa \partial'_\lambda \psi'_\lambda = L(\psi'_\lambda, \partial'_\kappa \psi'_\lambda),$$

but L could not be

$$L(\psi_\lambda, \partial_\kappa \psi_\lambda) = \sum \psi_\lambda^3 + ... \quad \text{or} \quad L(\psi_\lambda, \partial_\kappa \psi_\lambda) = \psi_\lambda + \partial_\lambda \partial_\kappa \psi_\kappa.$$

In the first of these last two cases L is not an absolute operator; in the second, although the operation is absolute, it yields a vector and not a scalar as required.

In (1.1) the integral is over the four-dimensional region considered and for convenience the integration is indicated as between σ_1 and σ_2. The cylindrical surface over which the field is assumed as vanishing completes the boundary of the region of integration, but there is no need to exhibit this boundary explicitly.

From the variation principle (1.2) we can derive the Euler–Lagrange equations, i.e. the field equations. However, before this is done it will be useful to get an expression for δI for arbitrary variations of ψ_μ which do not necessarily vanish over σ_1 and σ_2 and assume that the boundaries σ_1 and σ_2 may themselves also be varied infinitesimally. The actual field is thus considered not only in the original region but also in a

† In this chapter ψ_μ stands for any tensor, not necessarily a four-vector. When μ is attached to ψ it runs from 1 to n, otherwise from 1 to 4. When attached to ψ and appearing twice, we must sum from 1 to n, otherwise as usual from 1 to 4.

slightly modified region. The displacement from σ_1 to $\sigma_1+\delta\sigma_1$ may be specified by an infinitesimal displacement vector $\delta x_{\kappa 1}$ which may vary from point to point over σ_1 (Fig. 25). The increase of the considered region as a result of the displacement of the boundaries is given by

$$\iiint_{\sigma_2} \delta x_{\kappa 2}\, d\sigma_{\kappa 2} - \iiint_{\sigma_1} \delta x_{\kappa 1}\, d\sigma_{\kappa 1}, \tag{1.4}$$

where $d\sigma_{\kappa 1}$ is a vector which is perpendicular to the surface element $d\sigma_1$ and has the latter's magnitude. The same applies to $d\sigma_{\kappa 2}$. The change

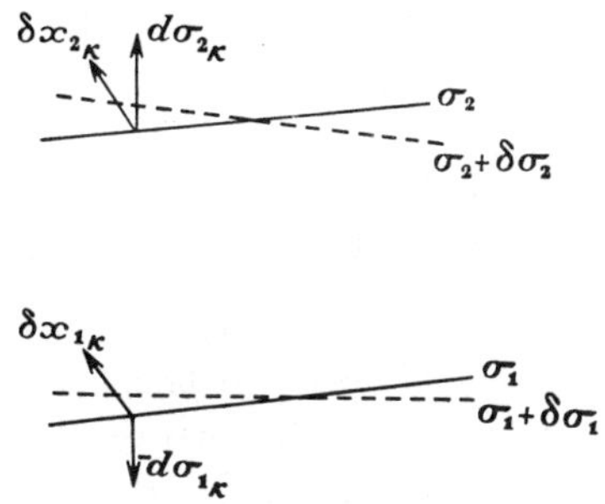

FIG. 25

in the value of $I = \iiiint L\, d^4x$ as a result of the displacement of the boundaries will thus be given simply by

$$\delta I = \iiint_{\sigma_2} L\, \delta x_{\kappa 2}\, d\sigma_{\kappa 2} - \iiint_{\sigma_1} L\, \delta x_{\kappa 1}\, d\sigma_{\kappa 1}. \tag{1.5}$$

As to the contribution to δI which arises from varying the field, we have

$$\delta L = \frac{\partial L}{\partial \psi_\mu}\, \delta\psi_\mu + \frac{\partial L}{\partial(\partial\psi_\mu/\partial x_\kappa)}\, \delta\left(\frac{\partial \psi_\mu}{\partial x_\kappa}\right). \tag{1.6}$$

Hence, altogether we have

$$\delta I = \delta \iiiint_{\sigma_1}^{\sigma_2} L\, d^4x = \iiiint_{\sigma_1}^{\sigma_2} \left[\frac{\partial L}{\partial \psi_\mu}\, \delta\psi_\mu + \frac{\partial L}{\partial(\partial_\kappa \psi_\mu)}\, \delta(\partial_\kappa \psi_\mu)\right] d^4x +$$
$$+ \iiint_{\sigma_2} L\, \delta x_{\kappa 2}\, d\sigma_{\kappa 2} - \iiint_{\sigma_1} L\, \delta x_{\kappa 1}\, d\sigma_{\kappa 1}. \tag{1.7}$$

Since the variation of $\partial_\kappa \psi_\mu$ is obtained from the variation of ψ_μ we have

$$\delta\left(\frac{\partial \psi_\mu}{\partial x_\kappa}\right) = \frac{\partial}{\partial x_\kappa}(\delta\psi_\mu). \tag{1.8}$$

Consequently, we can carry out a partial integration in the second term of the right-hand side of (1.7):

$$\delta I$$

$$= \int\!\!\int\!\!\int\!\!\int_{\sigma_1}^{\sigma_2} \frac{\partial L}{\partial \psi_\mu} \delta\psi_\mu \, d^4x + \int\!\!\int\!\!\int_{\sigma_2} \frac{\partial L}{\partial(\partial_\kappa \psi_\mu)} \delta\psi_\mu \, d\sigma_{\kappa 2} - \int\!\!\int\!\!\int_{\sigma_1} \frac{\partial L}{\partial(\partial_\kappa \psi_\mu)} \delta\psi_\mu \, d\sigma_{\kappa 1} -$$

$$- \int\!\!\int\!\!\int\!\!\int_{\sigma_1}^{\sigma_2} \partial_\kappa\left(\frac{\partial L}{\partial(\partial_\kappa \psi_\mu)}\right) \delta\psi_\mu \, d^4x + \int\!\!\int\!\!\int_{\sigma_2} L \, \delta x_{\kappa 2} \, d\sigma_{\kappa 2} - \int\!\!\int\!\!\int_{\sigma_1} L \, \delta x_{\kappa 1} \, d\sigma_{\kappa 1}$$

$$= \int\!\!\int\!\!\int\!\!\int_{\sigma_1}^{\sigma_2} \left[\frac{\partial L}{\partial \psi_\mu} - \partial_\kappa \frac{\partial L}{\partial(\partial_\kappa \psi_\mu)}\right] \delta\psi_\mu \, d^4x + F(\sigma_2) - F(\sigma_1), \tag{1.9}$$

where
$$F(\sigma) = \int\!\!\int\!\!\int_{\sigma} \left[\frac{\partial L}{\partial(\partial_\kappa \psi_\mu)} \delta\psi_\mu + L \, \delta x_\kappa\right] d\sigma_\kappa. \tag{1.10}$$

Let us now return to the variational principle which states that $\delta I = 0$ for an actual field, provided the boundaries of the considered region are kept constant and the variation over them is zero. Then, in (1.10) we must let $\delta\psi_\mu$ and δx_κ equal zero, so that $F(\sigma) = 0$. Eqn. (1.9) becomes

$$\delta I = \int\!\!\int\!\!\int\!\!\int_{\sigma_1}^{\sigma_2} \left[\frac{\partial L}{\partial \psi_\mu} - \partial_\kappa \left(\frac{\partial L}{\partial(\partial_\kappa \psi_\mu)}\right)\right] \delta\psi_\mu \, d^4x. \tag{1.11}$$

If $\delta I = 0$ for arbitrary variations $\delta\psi_\mu$, it follows further that this can be the case only if the factor in square brackets in (1.11) vanishes:

$$\frac{\partial L}{\partial \psi_\mu} - \partial_\kappa \left(\frac{\partial L}{\partial(\partial_\kappa \psi_\mu)}\right) = 0 \quad (\mu = 1,\ldots, n; \ \kappa = 1,\ldots, 4). \tag{1.12}$$

These are the Euler–Lagrange equations or the field equations. From the invariance of L it follows that these equations retain their form under Lorentz transformations. Provided we know the Lagrangian of a given field, its field equations can be written immediately according to (1.12).

As an example we mention that for the electromagnetic field a suitable Lagrangian is given by

$$L = \tfrac{1}{2}(H^2 - E^2) = \tfrac{1}{4}F_{\kappa\lambda}^2 = \tfrac{1}{4}(\partial_\kappa A_\lambda - \partial_\lambda A_\kappa)^2, \tag{1.13}$$

where
$$F_{\kappa\lambda} = \partial_\kappa A_\lambda - \partial_\lambda A_\kappa, \tag{1.14}$$

is expressed in terms of the vector potential A_κ. In general it is necessary to subject the vector potential to an additional condition, say the Lorentz condition,

$$\partial_\kappa A_\kappa = 0. \tag{1.15}$$

From (1.12), (1.13), and (1.15) we get

$$\partial_{\kappa}^2 A_{\lambda} = 0. \tag{1.16}$$

These are four wave equations for the components A_{λ} which, together with (1.15) and (1.14), are equivalent to Maxwell's equations.

2. The Canonical Energy-Momentum Tensor and the Laws of Conservation of Energy and Momentum

From the invariance of the Lagrangian it follows that

$$I = \iiiint L\, d^4x \tag{2.1}$$

is also an invariant against Lorentz transformation. We shall now make use of this invariance which holds not only with respect to homogeneous Lorentz transformations but also with respect to inhomogeneous transformations involving shifts of origin of the cartesian axes x_1, x_2, x_3 and shifts of time origin in x_4. Such shifts are given by

$$x'_{\kappa} = x_{\kappa} + \epsilon_{\kappa}, \tag{2.2}$$

where ϵ_{κ} is an arbitrary constant four-vector.

With respect to such transformations, vectors and indeed all tensors behave like scalars:

$$\psi'_{\mu}(x') = \psi_{\mu}(x). \tag{2.3}$$

Let us consider any infinitesimal Lorentz transformation. When such a transformation is performed each field quantity undergoes a small change as a result of its re-expression in terms of the new coordinates. Such a transformation has therefore the same effect as a corresponding variation in the field and we can make use of the general formula (1.9) of the previous section. Fom the invariance of I we have

$$I = \iiiint L(\psi_{\mu}, \partial_{\kappa}\psi_{\mu})\, d^4x = \iiiint L(\psi'_{\mu}, \partial'_{\kappa}\psi'_{\mu})\, d^4x'$$

or

$$\delta I = 0. \tag{2.4}$$

Hence, if in (1.10) δx_{κ} corresponds mathematically to an infinitesimal Lorentz transformation and $\delta\psi_{\mu}$ is the corresponding infinitesimal transformation of the tensor ψ_{μ}, by using (2.4) and the field equations (1.12), we get

$$F(\sigma_2) = F(\sigma_1). \tag{2.5}$$

In (2.5) the infinitesimal changes δx_{κ} and the corresponding $\delta\psi_{\mu}$ over σ_1 are correlated to these quantities over σ_2 since they are all generated by the same infinitesimal Lorentz transformation of the whole space. (2.5) expresses a law of conservation for such $F(\sigma)$. In this section we shall consider infinitesimal shifts of axes and we shall find that (2.5) yields the laws of conservation of energy and momentum. In the next section we shall consider infinitesimal Lorentz transformations (rotations of the four-space) and (2.5) will yield the laws of conservation of angular momentum. More generally, should L be invariant against additional groups of transformations additional laws of conservation will result.

Let
$$\delta x_\kappa = \epsilon_\kappa. \tag{2.6}$$

For such a transformation we have from (2.3)

$$\psi'_\mu(x'_\kappa)-\psi_\mu(x_\kappa) = 0. \tag{2.7}$$

For this change we cannot use the symbol $\delta\psi_\mu$ since in (1.10) $\delta\psi_\mu$ corresponds to a variation at a given point:

$$\delta\psi_\mu = \psi'_\mu(x)-\psi_\mu(x). \tag{2.8}$$

We shall use for the difference in (2.7) the symbol $\delta^*\psi_\mu$. The relationship between $\delta^*\psi_\mu$ and $\delta\psi_\mu$ is obtained from

$$\psi'_\mu(x'_\kappa)-\psi_\mu(x_\kappa) = \psi'_\mu(x'_\kappa)-\psi_\mu(x'_\kappa)+\psi_\mu(x'_\kappa)-\psi_\mu(x_\kappa),$$

therefore
$$\delta^*\psi_\mu = \delta\psi_\mu+\frac{\partial\psi_\mu}{\partial x_\kappa}\delta x_\kappa. \tag{2.9}$$

From (2.7), which applies to any shifts, we get in our present case $\delta^*\psi_\mu = 0$, and therefore

$$\delta\psi_\mu = -\frac{\partial\psi_\mu}{\partial x_\lambda}\delta x_\lambda = -\epsilon_\lambda\frac{\partial\psi_\mu}{\partial x_\lambda}. \tag{2.10}$$

$F(\sigma)$ becomes

$$\begin{aligned} F(\sigma) &= -\epsilon_\lambda\int\limits_\sigma \frac{\partial L}{\partial(\partial_\kappa\psi_\mu)}\partial_\lambda\psi_\mu\,d\sigma_\kappa+\epsilon_\kappa\int\limits_\sigma L\,d\sigma_\kappa \\ &= -\epsilon_\kappa\int\limits_\sigma \frac{\partial L}{\partial(\partial_\kappa\psi_\mu)}\partial_\lambda\psi_\mu\,d\sigma_\kappa+\delta_{\kappa\lambda}\epsilon_\lambda\int\limits_\sigma L\,d\sigma_\kappa \\ &= -\epsilon_\lambda\int\limits_\sigma \left[\frac{\partial L}{\partial(\partial_\kappa\psi_\mu)}\partial_\lambda\psi_\mu-\delta_{\kappa\lambda}L\right]d\sigma_\kappa. \end{aligned} \tag{2.11}$$

Since ϵ_λ is constant and arbitrary it follows from the conservation of $F(\sigma)$ that the four-vector P_λ,

$$P_\lambda = \int\limits_\sigma \left[\frac{\partial L}{\partial(\partial_\kappa\psi_\mu)}\partial_\lambda\psi_\mu-\delta_{\kappa\lambda}L\right]d\sigma_\kappa, \tag{2.12}$$

is conserved:
$$P_\lambda(\sigma_1) = P_\lambda(\sigma_2). \tag{2.13}$$

This vector is taken to represent the total momentum and the total energy of the field, the latter being given by $i^{-1}cP_4$.

The integrand, which is a tensor of second rank, is called the canonical energy-momentum tensor and denoted by $T_{\kappa\lambda}$. We have

$$T_{\kappa\lambda} = \frac{\partial L}{\partial(\partial_\kappa \psi_\mu)} \partial_\lambda \psi_\mu - \delta_{\kappa\lambda} L. \tag{2.14}$$

The law of conservation of energy and momentum has been obtained above in integral form. We can also express it in differential form. For this purpose we consider two surfaces σ_1 and σ_2 which differ only slightly in the neighbourhood of a point x_κ. $P_\lambda(\sigma)$ being a functional of σ, the change of P_λ through change of σ will be given by the functional derivative

$$\frac{\delta P_\lambda(\sigma)}{\delta\sigma(x_\kappa)} \equiv \lim_{\sigma_2 \to \sigma_1} \frac{P_\lambda(\sigma_2) - P_\lambda(\sigma_1)}{\Delta^4 x}, \tag{2.15}$$

where $\Delta^4 x$ is the four-dimensional volume enclosed between σ_2 and σ_1 around the point x_κ.

Since the functional P_λ is given by

$$P_\lambda(\sigma) = \int_\sigma T_{\kappa\lambda}\, d\sigma_\kappa, \tag{2.16}$$

the functional derivative is given by†

$$\frac{\delta P_\lambda(\sigma)}{\delta\sigma(x_\kappa)} = \frac{\partial T_{\kappa\lambda}}{\partial x_\kappa}. \tag{2.17}$$

From (2.13) we know that the functional derivative of P_λ must vanish, since P_λ is independent of σ. Hence

$$\frac{\partial T_{\kappa\lambda}}{\partial x_\kappa} = 0 \quad \left(\text{also: } \frac{\partial T_{\lambda\kappa}}{\partial x_\kappa} = 0\right). \tag{2.18}$$

This relation expresses the law of conservation of momentum and energy in differential form.

† This follows from Gauss's theorem,

$$\int \frac{\partial T_{\kappa\lambda}}{\partial x_\kappa} d^4x = \int_{\Sigma,\sigma_1,\sigma_2} T_{\kappa\lambda}\, d\sigma_\kappa = \int_{\sigma_2} T_{\kappa\lambda}\, d\sigma_{\kappa 2} - \int_{\sigma_1} T_{\kappa\lambda}\, d\sigma_{\kappa 1}. \tag{2.17 a}$$

In the limit where $\sigma_2 \to \sigma_1$ and differs from it only in the neighbourhood of x_κ, we have

$$\frac{\partial T_{\kappa\lambda}}{\partial x_\kappa} \Delta^4 x = \lim_{\sigma_2 \to \sigma_1} \left[\int_{\sigma_2} T_{\kappa\lambda}\, d\sigma_{\kappa 2} - \int_{\sigma_1} T_{\kappa\lambda}\, d\sigma_{\kappa 1}\right]. \tag{2.17 b}$$

Therefore

$$\frac{\partial T_{\kappa\lambda}}{\partial x_\kappa} = \frac{\delta P_\lambda(\sigma)}{\delta\sigma(x_\kappa)}. \tag{2.17 c}$$

The components of $T_{\kappa\lambda}$ could be identified with the density of energy H, the flow of energy S_k, and the momentum density G_k as follows:

$$-H = T_{44} = \frac{\partial L}{\partial(\partial_4 \psi_\mu)} \partial_4 \psi_\mu - L, \tag{2.19}$$

$$\frac{i}{c} S_k = T_{k4} = \frac{\partial L}{\partial(\partial_k \psi_\mu)} \partial_4 \psi_\mu \quad (k = 1, 2, 3, \text{or } x, y, z), \tag{2.20}$$

$$icG_k = T_{4k} = \frac{\partial L}{\partial(\partial_4 \psi_\mu)} \partial_k \psi_\mu. \tag{2.21}$$

The remaining components of the tensor $T_{\kappa\lambda}$ would then represent the stress tensor T_{kl} $(k, l = x, y, z)$. (2.18) would thus be equivalent to

$$\frac{\partial H}{\partial t} + \operatorname{div} \mathbf{S} = 0, \tag{2.22}$$

$$\frac{\partial G_k}{\partial t} + \operatorname{div} \mathbf{T}_k = 0, \tag{2.23}$$

where $\mathbf{T}_k$ is a vector with three components T_{lk} $(l = x, y, z)$. We are dealing here with three vectors $\mathbf{T}_k$ according to $k = x, y, z$.

In section 4 we shall advance reasons for not accepting the canonical energy-momentum tensor (2.14) as yielding the correct distribution of energy and momentum in fields with more than one component, i.e. in non-scalar fields. This does not relate to the total energy-momentum vector P_λ. We have obtained this vector from the invariance of the action integral to shifts of coordinates, a procedure which not only defines P_λ but also yields its conservation. It will turn out that the modification of $T_{\kappa\lambda}$ will not have any effect on the integrals P_λ. The reasons why $T_{\kappa\lambda}$ is the correct energy-momentum tensor for scalar fields, which have only one component, is that in this case there exists complete analogy with classical particle dynamics, so that, to mention just the energy, $(\partial L/\partial\dot{\psi})\dot{\psi} - L$ is the analogue of $p\dot{q} - L$. Fields with more than one component have no complete analogy with classical particles and the analogy extends only over orbital properties. What corresponds to the orbital angular momentum in particle dynamics does not yield the total angular momentum in fields with more than one component. The additional angular momentum is called spin. Classical particles have no spin.

3. The Angular Momentum

We shall now return to the invariance of the action integral and consider the implications of the conservation of $F(\sigma)$ when δx_κ and

$\delta\psi_\mu$ correspond to infinitesimal homogeneous Lorentz transformations. Such transformations are given by

$$x'_\kappa = x_\kappa + \delta x_\kappa(x_1,\ldots,x_4) = x_\kappa + \sum \omega_{\kappa\lambda} x_\lambda, \tag{3.1}$$

where $\omega_{\kappa\lambda}$ is an anti-symmetrical tensor with infinitesimal components

$$\omega_{\kappa\lambda} = -\omega_{\lambda\kappa}. \tag{3.2}$$

(3.1) is the four-dimensional analogue of rotations in three- and two-dimensional spaces. In the latter case we have for a rotation through a small angle $d\phi$,

$$x' = \cos d\phi . x + \sin d\phi . y \approx x + d\phi . y,$$
$$y' = -\sin d\phi . x + \cos d\phi . y \approx -d\phi . x + y, \tag{3.3}$$

so that
$$\{\omega_{\kappa\lambda}\} = \begin{bmatrix} 0 & d\phi \\ -d\phi & 0 \end{bmatrix} = \begin{bmatrix} 0 & 1 \\ -1 & 0 \end{bmatrix} d\phi. \tag{3.4}$$

The $\omega_{\kappa\lambda}$ depend on the three components of infinitesimal velocity between the frames K and K' and the slight inclination of axes x, y, z relative to each other. This relationship requires six parameters and this is also the number of components $\omega_{\kappa\lambda}$.

It will be more convenient to express (3.1) in matrix form as follows, where X is a column matrix:

$$X' = X + \delta X = X + \tfrac{1}{2} \sum_{\kappa,\lambda} M_{(\kappa\lambda)} \omega_{\kappa\lambda} X.\dagger \tag{3.5}$$

The $M_{(\kappa\lambda)}$ are six matrices which are given as follows:

$$M_{(11)} = M_{(22)} = M_{(33)} = M_{(44)} = 0, \qquad M_{(\kappa\lambda)} = -M_{(\lambda\kappa)}, \tag{3.6}$$

$$M_{(23)} = \begin{bmatrix} 0 & 0 & 0 & 0 \\ 0 & 0 & 1 & 0 \\ 0 & -1 & 0 & 0 \\ 0 & 0 & 0 & 0 \end{bmatrix}, \qquad M_{(13)} = \begin{bmatrix} 0 & 0 & 1 & 0 \\ 0 & 0 & 0 & 0 \\ -1 & 0 & 0 & 0 \\ 0 & 0 & 0 & 0 \end{bmatrix},$$

$$M_{(12)} = \begin{bmatrix} 0 & 1 & 0 & 0 \\ -1 & 0 & 0 & 0 \\ 0 & 0 & 0 & 0 \\ 0 & 0 & 0 & 0 \end{bmatrix}, \qquad M_{(34)} = \begin{bmatrix} 0 & 0 & 0 & 0 \\ 0 & 0 & 0 & 0 \\ 0 & 0 & 0 & 1 \\ 0 & 0 & -1 & 0 \end{bmatrix},$$

$$M_{(24)} = \begin{bmatrix} 0 & 0 & 0 & 0 \\ 0 & 0 & 0 & 1 \\ 0 & 0 & 0 & 0 \\ 0 & -1 & 0 & 0 \end{bmatrix}, \qquad M_{(14)} = \begin{bmatrix} 0 & 0 & 0 & 1 \\ 0 & 0 & 0 & 0 \\ 0 & 0 & 0 & 0 \\ -1 & 0 & 0 & 0 \end{bmatrix}. \tag{3.7}$$

† The brackets on $(\kappa\lambda)$ are to indicate that $M_{(\kappa\lambda)}$ is a matrix and not the κ, λ component. $M_{(\kappa\lambda)}$ has components $M_{(\kappa\lambda)1,1}$, $M_{(\kappa\lambda)1,2}$, etc.

(3.5) is identical with

$$x'_\mu = \Big(\delta_{\mu\nu} + \tfrac{1}{2} \sum_{\kappa,\lambda} M^{\mu\nu}_{(\kappa\lambda)} \omega_{\kappa\lambda}\Big) x_\nu.$$

As is easily verified, the matrix $\frac{1}{2}\sum_{\kappa,\lambda} M_{(\kappa\lambda)}\omega_{\kappa\lambda}$ is identical with the matrix $\{\omega_{\kappa\lambda}\}$, so that (3.5) is a re-expression of (3.1). By letting the $\omega_{\kappa\lambda}$ in (3.5) take all possible infinitesimal values, we get all the possible infinitesimal Lorentz transformations. The $M_{(\kappa\lambda)}$ are called the operators of the infinitesimal transformations. Each group of continuous transformations is characterized by a set of such operators which yield the infinitesimal transformations. In the case of the Lorentz group the operators are the above $M_{(\kappa\lambda)}$. In the ordinary three-dimensional space the infinitesimal rotations depend on three infinitesimal angles $d\phi_x = \omega_{23}$, $d\phi_y = \omega_{13}$, and $d\phi_z = \omega_{12}$, around the x-, y-, z-axes. In each rotation only two axes get transformed. This holds also in the four-dimensional space. Thus the rotation $M_{(34)}\omega_{34}$ involves only the z- and ict-axes. The x- and y-axes are unchanged and they are the axes around which the rotation takes place. In a three-dimensional space only one real axis of rotation exists. In this case we have

$$\delta X = \tfrac{1}{2}\sum_{a,b} R_{(ab)}\omega_{ab} X, \tag{3.8}$$

where

$$R_{(ab)} = -R_{(ba)} \tag{3.9}$$

and

$$R_{(23)} = \begin{bmatrix} 0 & 0 & 0 \\ 0 & 0 & 1 \\ 0 & -1 & 0 \end{bmatrix}, \quad R_{(31)} = \begin{bmatrix} 0 & 0 & -1 \\ 0 & 0 & 0 \\ 1 & 0 & 0 \end{bmatrix}, \quad R_{(12)} = \begin{bmatrix} 0 & 1 & 0 \\ -1 & 0 & 0 \\ 0 & 0 & 0 \end{bmatrix}. \tag{3.10}$$

The $M_{(\kappa\lambda)}$ and similarly the $R_{(ab)}$ are not independent of each other. In fact the following relations between the different matrices hold:

$$\begin{aligned}[M_{(\kappa\lambda)}, M_{(\mu\nu)}] &\equiv M_{(\kappa\lambda)}M_{(\mu\nu)} - M_{(\mu\nu)}M_{(\kappa\lambda)} \\ &= \delta_{\lambda\mu}M_{(\kappa\nu)} - \delta_{\kappa\mu}M_{(\lambda\nu)} + \delta_{\kappa\nu}M_{(\lambda\mu)} - \delta_{\lambda\nu}M_{(\kappa\mu)}.\end{aligned} \tag{3.11}$$

(3.11) may be verified directly by using the expressions (3.7). In the three-dimensional space the corresponding relations simplify and are more familiar from the connexion with the angular momenta operators in quantum theory:

$$[R_x, R_y] \equiv R_x R_y - R_y R_x = -R_z \quad \text{(and cyclically)},$$

where

$$R_x = R_{(23)}, \quad R_y = R_{(31)}, \quad R_z = R_{(12)}. \tag{3.12}$$

The relations (3.11) (and (3.12) which result from them), are known as the integrability conditions. Because of these relations any finite transformation which is obtained by a succession of infinitesimal transformations is independent of the path chosen in the space of parameters $\omega_{\kappa\lambda}$.

The infinitesimal transformation (3.5) refers to the four-vector $X = (x_1 x_2 x_3 x_4)$ and will apply to any four-vector. However, to evaluate (1.10) we must calculate the infinitesimal change of any tensor $\delta\psi_\mu$ when an infinitesimal Lorentz transformation is performed. Again, it will be convenient to order all the components of ψ_μ as a column matrix Ψ and express the change $\delta^*\psi_\mu = \psi'_\mu(x') - \psi_\mu(x)$ as follows:

$$\delta^*\Psi = \tfrac{1}{2}\sum_{\kappa\lambda} P_{(\kappa\lambda)}\,\omega_{\kappa\lambda}\,\Psi, \tag{3.13}$$

where the $P_{(\kappa\lambda)}$ are matrices of n dimensions, n being the number of components of ψ_μ. Like δX, $\delta^*\Psi$ must be proportional to the infinitesimal parameters. The six matrices $P_{(\kappa\lambda)}$ produce infinitesimal tensor transformations which represent the infinitesimal Lorentz transformations. They may be obtained from the finite transformations of the tensors in the same way as the $M_{(\kappa\lambda)}$ are obtainable from the finite Lorentz transformations. Since the ψ_μ transform according to a representation of the Lorentz group, the $P_{(\kappa\lambda)}$ must satisfy the same relations as the $M_{(\kappa\lambda)}$:

$$[P_{(\kappa\lambda)}, P_{(\mu\nu)}] = \delta_{\lambda\mu} P_{(\kappa\nu)} - \delta_{\kappa\mu} P_{(\lambda\nu)} + \delta_{\kappa\nu} P_{(\lambda\mu)} - \delta_{\lambda\nu} P_{(\kappa\mu)}. \tag{3.14}$$

Explicit expressions for the $P_{(\kappa\lambda)}$ for any tensor ψ_μ will be given at the end of this section.

For $\delta\Psi$ we get

$$\delta\Psi = \Psi'(x) - \Psi(x) = \Psi'(x') - \Psi(x) - (\Psi'(x') - \Psi'(x)). \tag{3.15}$$

Consequently,
$$\delta\Psi = \delta^*\Psi - \sum_\kappa \frac{\partial\Psi}{\partial x_\kappa}\,\delta x_\kappa. \tag{3.16}$$

Using (3.13) and (3.1), we get

$$\delta\Psi = \frac{1}{2}\sum_{\kappa\lambda}\left(P_{(\kappa\lambda)}\,\omega_{\kappa\lambda}\,\Psi - 2\frac{\partial\Psi}{\partial x_\kappa}\,\omega_{\kappa\lambda}\,x_\lambda\right) = \tfrac{1}{2}\sum_{\kappa\lambda}\omega_{\kappa\lambda}(P_{(\kappa\lambda)} + x_\kappa\,\partial_\lambda - x_\lambda\,\partial_\kappa)\Psi. \tag{3.17}$$

In (1.10) we replace the index κ by μ so as not to confuse this index with κ of (3.17) (from now on μ is an ordinary index $= 1, 2, 3, 4$). We have

$$\begin{aligned} F(\sigma) = \int_\sigma \left(\frac{\partial L}{\partial(\partial_\mu\Psi)}\,\delta\Psi + L\,\delta x_\mu\right) d\sigma_\mu &= \sum_{\kappa\lambda}\int_\sigma \frac{\partial L}{\partial(\partial_\mu\Psi)}\,(\tfrac{1}{2}P_{(\kappa\lambda)}\Psi)\omega_{\kappa\lambda}\,d\sigma_\mu + \\ &+ \sum_{\kappa\lambda}\int_\sigma \left\{\frac{\partial L}{\partial(\partial_\mu\Psi)}\,(\tfrac{1}{2}(x_\kappa\,\partial_\lambda - x_\lambda\,\partial_\kappa)\Psi)\omega_{\kappa\lambda} + \right. \\ &\left. + \tfrac{1}{2}\left(\sum_{\kappa\lambda}\delta_{\mu\kappa}\,L\omega_{\kappa\lambda}\,x_\lambda - \sum_{\kappa\lambda}\delta_{\mu\lambda}\,L\omega_{\kappa\lambda}\,x_\kappa\right)\right\} d\sigma_\mu. \end{aligned} \tag{3.18}$$

In terms of the canonical energy-momentum tensor $T_{\kappa\lambda}$ we can express $F(\sigma)$ as follows:

$$F(\sigma) = \sum_{\kappa\lambda}\left\{\int\limits_{\sigma} \frac{\partial L}{\partial(\partial\Psi/\partial x_{\mu})}(\tfrac{1}{2}P_{(\kappa\lambda)}\Psi)\omega_{\kappa\lambda}\,d\sigma_{\mu}+\tfrac{1}{2}\int\limits_{\sigma}(T_{\mu\lambda}x_{\kappa}-T_{\mu\kappa}x_{\lambda})\omega_{\kappa\lambda}\,d\sigma_{\mu}\right\}. \tag{3.19}$$

Since the $\frac{1}{2}\omega_{\kappa\lambda}$ are constants it follows that

$$-\int\limits_{\sigma}\left(T_{\mu\kappa}x_{\lambda}-T_{\mu\lambda}x_{\kappa}-\frac{\partial L}{\partial(\partial_{\mu}\Psi)}P_{(\kappa\lambda)}\Psi\right)d\sigma_{\mu} = J_{\kappa\lambda} \tag{3.20}$$

is a quantity which is conserved. Expressed more explicitly, $J_{\kappa\lambda}$ is given by the following formula:

$$J_{\kappa\lambda} = -\int\limits_{\sigma}\left(T_{\mu\kappa}x_{\lambda}-T_{\mu\lambda}x_{\kappa}-\frac{\partial L}{\partial(\partial_{\mu}\psi_{a})}P_{(\kappa\lambda)ab}\psi_{b}\right)d\sigma_{\mu} \quad (a,b = 1,\ldots,n). \tag{3.21}$$

The physical significance of $J_{\kappa\lambda}$ may be obtained as follows. The canonical tensor $T_{\mu\kappa}$ is interpreted as an energy-momentum tensor from an analogy between a scalar field and spin-less particles. Hence, in more general fields with more than one component we may look upon it as the contribution to the energy-momentum tensor which arises from what may be termed the orbital properties of the field, such as flow of energy and momentum. The term

$$L_{\mu\kappa\lambda} = T_{\mu\lambda}x_{\kappa}-T_{\mu\kappa}x_{\lambda} \tag{3.22}$$

will then represent the density of the orbital angular momentum $L_{\kappa\lambda} = \int L_{\mu\kappa\lambda}\,d\sigma_{\mu}$. This term is similar to the one obtained in ordinary particle dynamics for orbital angular momentum,

$$L_{kl} = P_{l}x_{k}-P_{k}x_{l} \quad (k,l = x,y,z), \tag{3.23}$$

where P_k is the momentum of the particle and x_k the radius vector with respect to some given point. L_{kl} is an anti-symmetrical tensor in a space of three dimensions. In the theory of relativity we must raise the number of dimensions from three to four and, when dealing with fields, we have to consider densities, since in each volume element we have energy, momentum, and moment of momentum with respect to some given point. To be more specific, if $T_{4k} = icG_k$ are the spatial components of the momentum density, the relation corresponding to (3.23) is

$$L_{4kl} = T_{4l}x_{k}-T_{4k}x_{l} = ic(G_{l}x_{k}-G_{k}x_{l}). \tag{3.24}$$

The other components contained in $L_{\mu\kappa\lambda}$ are those obtained through changing from three to four dimensions.

Apart from the orbital part, the density $J_{\mu\kappa\lambda}$ of angular momentum (i.e. the integrand in (3.21) $J_{\kappa\lambda} = \int J_{\mu\kappa\lambda}\, d\sigma_\mu$) contains another part which is called the spin density $S_{\mu\kappa\lambda}$,

$$S_{\mu\kappa\lambda} = \sum_{a,b=1}^{n} \frac{\partial L}{\partial(\partial_\mu \psi_a)} P_{(\kappa\lambda)ab}\psi_b. \tag{3.25}$$

According to (3.21) neither the orbital momentum alone nor the spin angular momentum alone need be conserved, since the conservation law applies to their sum.

From (3.25) it follows that a scalar field possesses no spin. This is due to the vanishing of all the matrices. In the case of a scalar field they must be six identical one-dimensional matrices satisfying the relations (3.14). The only solution is $P_{(\kappa\lambda)} \equiv 0$. It follows further that since $L_{\mu\kappa\lambda}$ then represents the density of total angular momentum, the canonical tensor of a scalar field is necessarily symmetrical. This is proved as follows. From

$$\frac{\partial L_{\mu\lambda\kappa}}{\partial x_\mu} = 0 \tag{3.26}$$

we get, using (3.22),

$$\partial_\mu L_{\mu\lambda\kappa} = \{(\partial_\mu T_{\mu\kappa})x_\lambda - T_{\mu\kappa}(\partial_\mu x_\lambda) - (\partial_\mu T_{\mu\lambda})x_\kappa - T_{\mu\lambda}(\partial_\mu x_\kappa)\} = 0. \tag{3.27}$$

On the other hand, from (2.18), we have

$$\partial_\mu T_{\mu\lambda} \equiv \partial_\mu T_{\mu\kappa} = 0. \tag{3.28}$$

We also have $$\partial_\mu x_\lambda = \delta_{\mu\lambda}, \qquad \partial_\mu x_\kappa = \delta_{\mu\kappa}; \tag{3.29}$$

(3.27) therefore yields

$$\partial_\mu L_{\mu\lambda\kappa} = T_{\lambda\kappa} - T_{\kappa\lambda} = 0, \tag{3.30}$$

so that $$T_{\kappa\lambda} = T_{\lambda\kappa}. \tag{3.31}$$

Thus, in a scalar field, if the orbital angular momentum is to be conserved, the canonical energy-momentum tensor is necessarily symmetrical and vice versa. More generally, the above calculation shows that for any symmetrical tensor $\Theta_{\kappa\lambda}$ with vanishing divergence

$$\partial_\kappa \Theta_{\kappa\lambda} = 0, \tag{3.32}$$

the quantity $\Theta_{\kappa\lambda} x_\mu - \Theta_{\kappa\mu} x_\lambda$ obeys the relation

$$\partial_\kappa(\Theta_{\kappa\lambda} x_\mu - \Theta_{\kappa\mu} x_\lambda) = 0. \tag{3.33}$$

Hence, by Gauss's theorem,

$$\int (\Theta_{\kappa\lambda} x_\mu - \Theta_{\kappa\mu} x_\lambda)\, d\sigma_\kappa \quad \text{is conserved.} \tag{3.34}$$

It may be added that if in a scalar field we introduce an energy-momentum tensor which is non-symmetrical, the angular momentum based on it will not be conserved and we shall have to add to it a non-orbital angular momentum such that the total is conserved. The additional angular momentum may again be attributed to some spin although it is of different origin from the one introduced in this chapter. The spin, in the proper sense of the word, is connected with the existence of several components and vanishes for a scalar field.

The explicit expressions for the $P_{(\kappa\lambda)}$ in the case of any tensor may be obtained as follows. The general transformation law of a tensor $F_{\kappa_1 \dots \kappa_n}$ of rank n is given by

$$F'_{\kappa_1 \dots \kappa_n} = \sum_{\lambda_1 \dots \lambda_n = 1}^{4} L_{\kappa_1 \lambda_1} L_{\kappa_2 \lambda_2} \dots L_{\kappa_n \lambda_n} F_{\lambda_1 \dots \lambda_n}, \tag{3.35}$$

where $L_{\kappa\lambda}$ is the matrix of a Lorentz transformation. $L_{\kappa\lambda}$ is identical with $\partial x_\lambda / \partial x'_\kappa$. Hence we may express (3.35) as follows:

$$F'_{\kappa_1 \dots \kappa_n} = \sum_{\lambda_1 \dots \lambda_n = 1}^{4} \frac{\partial x_{\lambda_1}}{\partial x'_{\kappa_1}} \frac{\partial x_{\lambda_2}}{\partial x'_{\kappa_2}} \dots \frac{\partial x_{\lambda_n}}{\partial x'_{\kappa_n}} F_{\lambda_1 \dots \lambda_n}. \tag{3.36}$$

For an infinitesimal transformation

$$x_\lambda = x'_\lambda - \epsilon_\lambda(x'_1, \dots, x'_4), \tag{3.37}$$

we get to first order

$$\frac{\partial x_\lambda}{\partial x'_\kappa} = \delta_{\kappa\lambda} - \frac{\partial \epsilon_\lambda}{\partial x'_\kappa} \approx \delta_{\kappa\lambda} - \frac{\partial \epsilon_\lambda}{\partial x_\kappa}. \tag{3.38}$$

Consequently,

$$F'_{\kappa_1 \dots \kappa_n} = \sum_{\lambda_1 \dots \lambda_n = 1}^{4} F_{\lambda_1 \dots \lambda_n} \prod_{r=1}^{n} \left(\delta_{\kappa_r \lambda_r} - \frac{\partial \epsilon_{\lambda_r}}{\partial x_{\kappa_r}} \right), \tag{3.39}$$

where $\prod$ indicates the product over the n terms. To first order we get further

$$\begin{aligned} F'_{\kappa_1 \dots \kappa_n} &= \sum_{\lambda_1 \dots \lambda_n = 1}^{4} \left\{ \prod_{r=1}^{n} \delta_{\kappa_r \lambda_r} - \sum_{r=1}^{n} \delta_{\lambda_1 \kappa_1} \delta_{\lambda_2 \kappa_2} \dots \delta_{\lambda_{r-1} \kappa_{r-1}} \frac{\partial \epsilon_{\lambda_r}}{\partial x_{\kappa_r}} \delta_{\lambda_{r+1} \kappa_{r+1}} \dots \delta_{\lambda_n \kappa_n} \right\} F_{\lambda_1 \dots \lambda_n} \\ &= F_{\kappa_1 \dots \kappa_n} - \sum F_{\kappa_1 \kappa_2 \dots \kappa_{r-1} \lambda_r \kappa_{r+1} \dots \kappa_n} \frac{\partial \epsilon_{\lambda_r}}{\partial x_{\kappa_r}} \\ &= F_{\kappa_1 \dots \kappa_n} - \sum_{\sigma, \tau = 1}^{4} \sum_{r=1}^{n} \delta_{\tau \kappa_r} F_{\kappa_1 \kappa_2 \dots \kappa_{r-1} \sigma \kappa_{r+1} \dots \kappa_n} \frac{\partial \epsilon_\sigma}{\partial x_\tau}. \end{aligned} \tag{3.40}$$

Thus we get

$$\delta^* F_{\kappa_1 \dots \kappa_n} \equiv F'_{\kappa_1 \dots \kappa_n}(x') - F_{\kappa_1 \dots \kappa_n}(x) = - \sum_{\sigma, \tau = 1}^{4} \sum_{r=1}^{n} \delta_{\tau \kappa_r} F_{\kappa_1 \dots \kappa_{r-1} \sigma \kappa_{r+1} \dots \kappa_n} \frac{\partial \epsilon_\sigma}{\partial x_\tau}. \tag{3.41}$$

For an infinitesimal Lorentz transformation

$$\epsilon_\sigma = \sum_{\tau=1}^{4} \omega_{\sigma\tau} x_\tau = - \sum_{\tau=1}^{4} \omega_{\tau\sigma} x_\tau \tag{3.42}$$

and (3.41) yields

$$\begin{aligned}\delta^* F_{\kappa_1\ldots\kappa_n} &= - \sum_{\sigma,\tau=1}^{4} \sum_{r=1}^{n} \omega_{\sigma\tau} \delta_{\tau\kappa_r} F_{\kappa_1\ldots\kappa_{r-1}\sigma\kappa_{r+1}\ldots\kappa_n} \\ &= -\tfrac{1}{2} \sum_{\lambda_1\ldots\lambda_n=1}^{4} \sum_{\sigma,\tau=1}^{4} \omega_{\sigma\tau} \Big(\sum_{r=1}^{n} (\delta_{\tau\kappa_r}\delta_{\sigma\lambda_r} - \delta_{\sigma\kappa_r}\delta_{\tau\lambda_r}) \times \\ &\qquad \times \delta_{\kappa_1\lambda_1}\delta_{\kappa_2\lambda_2}\ldots\delta_{\kappa_{r-1}\lambda_{r-1}}\delta_{\kappa_{r+1}\lambda_{r+1}}\ldots\delta_{\kappa_n\lambda_n} F_{\lambda_1\ldots\lambda_n}\Big).\end{aligned} \tag{3.43}$$

Thus we get

$$\begin{aligned}&P_{(\sigma\tau)\kappa_1\ldots\kappa_n\lambda_1\ldots\lambda_n} \\ &\quad = - \sum_{r=1}^{n} (\delta_{\tau\kappa_r}\delta_{\sigma\lambda_r} - \delta_{\sigma\kappa_r}\delta_{\tau\lambda_r}) \delta_{\kappa_1\lambda_1}\delta_{\kappa_2\lambda_2}\ldots\delta_{\kappa_{r-1}\lambda_{r-1}}\delta_{\kappa_{r+1}\lambda_{r+1}}\ldots\delta_{\kappa_n\lambda_n}.\end{aligned} \tag{3.44}$$

In particular $P_{(\sigma\tau)\kappa_1\lambda_1} \equiv P_{(\sigma\tau)\kappa\lambda} = \delta_{\sigma\kappa}\delta_{\tau\lambda} - \delta_{\tau\kappa}\delta_{\sigma\lambda}$. This yields (3.6) and (3.7).

4. The Symmetrical Energy-Momentum Tensor

In general, the canonical energy-momentum tensor is not symmetrical and whenever this is the case $\iiint (T_{\kappa\lambda} x_\mu - T_{\kappa\mu} x_\lambda)\, d\sigma_\kappa$ is not conserved. Now, although $T_{\kappa\lambda}$ yields the total energy and momentum (by virtue of its connexion with the invariance of the action integral with respect to shifts of axes), it cannot represent the actual distribution of energy and momentum in the field when a spin is also present. Let the actual energy-momentum tensor of the field be denoted by $\Theta_{\kappa\lambda}$. Then from the law of conservation of energy and momentum we have

$$\partial_\kappa \Theta_{\kappa\lambda} = 0. \tag{4.1}$$

Moreover, since the energy-momentum tensor is now given by $\Theta_{\kappa\lambda}$, the total angular momentum, as already derived in (3.21), should be obtainable also from the formula

$$J_{\lambda\mu} = \iiint (\Theta_{\kappa\mu} x_\lambda - \Theta_{\kappa\lambda} x_\mu)\, d\sigma_\kappa. \tag{4.2}$$

From the law of conservation of angular momentum and (4.1) it follows that $\Theta_{\kappa\lambda}$ is necessarily a symmetrical tensor:

$$\Theta_{\kappa\lambda} = \Theta_{\lambda\kappa}. \tag{4.3}$$

That the energy-momentum tensor must be symmetrical follows also from (2.22) and (2.23). Depending on whether we use columns or rows of $T_{\kappa\lambda}$ we shall get different expressions for flow of energy and momentum when $T_{\kappa\lambda}$ is not symmetrical.

Our task now is to derive an expression for $\Theta_{\kappa\lambda}$ which is symmetrical and yields the same total energy and momentum and the same total angular momentum as already derived in the previous section from symmetry considerations.

Let the difference between $\Theta_{\kappa\lambda}$ and $T_{\kappa\lambda}$ be denoted by $T^S_{\kappa\lambda}$, so that

$$\Theta_{\kappa\lambda} = T_{\kappa\lambda}+T^S_{\kappa\lambda}. \tag{4.4}$$

While $T_{\kappa\lambda}$ may be looked upon as the orbital contribution to the energy-momentum tensor, $T^S_{\kappa\lambda}$ is the spin contribution.

From (4.1) and the vanishing of the divergence of $T_{\kappa\lambda}$ it follows that the divergence of $T^S_{\kappa\lambda}$ must also vanish:

$$\partial_\kappa T^S_{\kappa\lambda} = 0. \tag{4.5}$$

Now, from (3.21) and the law of conservation of $J_{\kappa\lambda}$ in differential form we have (using also (3.25))

$$-\partial_\mu J_{\mu\kappa\lambda} = \partial_\mu(T_{\mu\kappa}x_\lambda - T_{\mu\lambda}x_\kappa - S_{\mu\kappa\lambda}) = 0. \tag{4.6}$$

Consequently,

$$-\partial_\mu S_{\mu\kappa\lambda}+x_\lambda\partial_\mu T_{\mu\kappa}+T_{\lambda\kappa}-x_\kappa\partial_\mu T_{\mu\lambda}-T_{\kappa\lambda} = 0. \tag{4.7}$$

Now, since $\partial_\mu T_{\mu\kappa}$ and $\partial_\mu T_{\mu\lambda}$ vanish, we get

$$T_{\kappa\lambda}-T_{\lambda\kappa} = -\partial_\mu S_{\mu\kappa\lambda}. \tag{4.8}$$

This relation shows that for a symmetrical $T_{\kappa\lambda}$, the divergence of $S_{\mu\kappa\lambda}$ vanishes. This is, of course, as it should be, because if orbital angular momentum is conserved, the total angular momentum will be conserved only if the spin angular momentum is also conserved. We note that $S_{\mu\kappa\lambda}$ is antisymmetric in the indices κ, λ:

$$S_{\mu\kappa\lambda} = -S_{\mu\lambda\kappa}. \tag{4.9}$$

This also follows from (3.25). Hence, we introduce a new tensor $A_{\mu\lambda\kappa}$ which makes this antisymmetry more explicit:

$$S_{\mu\kappa\lambda} = A_{\mu\kappa\lambda}-A_{\mu\lambda\kappa}. \tag{4.10}$$

(4.8) and (4.10) yield

$$T_{\lambda\kappa}+\partial_\mu A_{\mu\lambda\kappa} = T_{\kappa\lambda}+\partial_\mu A_{\mu\kappa\lambda}, \tag{4.11}$$

so that the tensor

$$\Theta_{\kappa\lambda} = T_{\kappa\lambda}+\partial_\mu A_{\mu\kappa\lambda} \tag{4.12}$$

is a symmetrical tensor. Although $S_{\mu\kappa\lambda}$ is given in terms of the Lagrangian L this cannot yet be claimed of $A_{\mu\kappa\lambda}$ because (4.10) does not define $A_{\mu\kappa\lambda}$ completely; the addition of any symmetrical tensor to $A_{\mu\kappa\lambda}$ will not upset the relation (4.10). However, if $\partial_\mu A_{\mu\kappa\lambda}$ is to be identified with $T^S_{\kappa\lambda}$ its divergence must vanish:

$$\partial_\kappa T^S_{\kappa\lambda} = \partial_\kappa\partial_\mu A_{\mu\kappa\lambda} = 0. \tag{4.13}$$

Since $\partial_\kappa \partial_\mu$ is symmetrical in κ, μ it follows that $A_{\mu\kappa\lambda}$ is anti-symmetrical in κ, μ:

$$A_{\mu\kappa\lambda} = -A_{\kappa\mu\lambda}. \tag{4.14}$$

From (4.10) and (4.14) we can now obtain $A_{\mu\kappa\lambda}$ in a unique way on using the following relations which are obtained by permutation of indices:

$$\begin{aligned} S_{\mu\kappa\lambda} &= A_{\mu\kappa\lambda} - A_{\mu\lambda\kappa}, \\ S_{\kappa\lambda\mu} &= A_{\kappa\lambda\mu} - A_{\kappa\mu\lambda}, \\ S_{\lambda\kappa\mu} &= A_{\lambda\kappa\mu} - A_{\lambda\mu\kappa}. \end{aligned} \tag{4.15}$$

On adding these equations and taking (4.14) into account some of the terms cancel and we get

$$A_{\mu\kappa\lambda} = \tfrac{1}{2}(S_{\mu\kappa\lambda} + S_{\kappa\lambda\mu} + S_{\lambda\kappa\mu}). \tag{4.16}$$

Thus we get finally

$$\Theta_{\kappa\lambda} = T_{\kappa\lambda} + \tfrac{1}{2}\partial_\mu(S_{\mu\kappa\lambda} + S_{\kappa\lambda\mu} + S_{\lambda\kappa\mu}). \tag{4.17}$$

The interpretation of this tensor as an energy-momentum tensor is still dependent on whether or not it yields the same value for the total energy and momentum vector P_λ as our former $T_{\kappa\lambda}$, since otherwise the relationship between the energy-momentum vector P_λ and the shift of axes will be destroyed. However, this is the case, for we can prove that

$$P_\lambda = \iiint\limits_\sigma T_{\kappa\lambda}\, d\sigma_\kappa = \iiint\limits_\sigma \Theta_{\kappa\lambda}\, d\sigma_\kappa. \tag{4.18}$$

The proof of (4.18) follows from

$$\iiint \frac{\partial A_{\mu\kappa\lambda}}{\partial x_\mu}\, d\sigma_\kappa = \iint A_{\mu\kappa\lambda}\, d\sigma_{\mu\kappa}, \tag{4.19}$$

where $d\sigma_{\kappa\lambda}$ is a surface element of the boundary of the space-like region σ in which the field is considered. Now, if the field is contained in a finite region only, we may take the boundary sufficiently far away for the field on it to be negligible. The r.h.s. of (4.19) thus vanishes and (4.18) follows. We also see that the total energy-momentum due to spin is zero. The spin only modifies the distribution of energy-momentum.

Furthermore, if we accept $\Theta_{\kappa\lambda}$ as the energy-momentum tensor, we must redefine the density of total angular momentum as indicated in the integrand of (4.2). Let us denote this modified density by $J^{(\mathrm{mod})}_{\mu\kappa\lambda}$, then

$$J^{(\mathrm{mod})}_{\mu\kappa\lambda} = \Theta_{\mu\lambda} x_\kappa - \Theta_{\mu\kappa} x_\lambda. \tag{4.20}$$

From the symmetry of $\Theta_{\kappa\lambda}$ and the vanishing of its divergence the conservation of the total angular momentum $J^{(\mathrm{mod})}_{\kappa\lambda}$ is guaranteed. However, if the relationship between the total angular momentum and the homogeneous Lorentz transformations is to be satisfied, we can accept $\Theta_{\kappa\lambda}$ as an energy-momentum tensor only if $J^{(\text{mod})}_{\kappa\lambda} = J_{\kappa\lambda}$. Here again we can prove that

$$J_{\kappa\lambda} = \int\limits_{\sigma} J_{\mu\kappa\lambda}\,d\sigma_{\mu} = \int\limits_{\sigma} J^{(\text{mod})}_{\mu\kappa\lambda}\,d\sigma_{\mu} = J^{(\text{mod})}_{\kappa\lambda}, \tag{4.21}$$

so that although the densities of angular momentum are affected by the new definition, the total angular momentum is not affected. To prove (4.21), we note firstly that the density of orbital angular momentum remains unchanged by the new energy-momentum tensor and is given by

$$L_{\mu\kappa\lambda} = T_{\mu\lambda}x_{\kappa} - T_{\mu\kappa}x_{\lambda}. \tag{4.22}$$

As to the spin angular momentum, it is now given by

$$S^{(\text{mod})}_{\mu\kappa\lambda} = T^{S}_{\mu\lambda}x_{\kappa} - T^{S}_{\mu\kappa}x_{\lambda} \tag{4.23}$$

where

$$T^{S}_{\kappa\lambda} = \partial_{\nu}A_{\nu\kappa\lambda}. \tag{4.24}$$

From (4.23) and (4.24) we get

$$\begin{aligned} S^{(\text{mod})}_{\mu\kappa\lambda} &= A_{\mu\kappa\lambda} - A_{\mu\lambda\kappa} + \partial_{\nu}(A_{\nu\mu\lambda}x_{\kappa} - A_{\nu\mu\kappa}x_{\lambda}) \\ &= S_{\mu\kappa\lambda} + \partial_{\nu}(A_{\nu\mu\lambda}x_{\kappa} - A_{\nu\mu\kappa}x_{\lambda}). \end{aligned} \tag{4.25}$$

We can now conclude, as in the case of (4.19) that the second part on the r.h.s. of (4.25) will give no contribution to the integral over the space-like surface, so that

$$\iiint\limits_{\sigma} S^{(\text{mod})}_{\mu\kappa\lambda}\,d\sigma_{\mu} = \iiint\limits_{\sigma} S_{\mu\kappa\lambda}\,d\sigma_{\mu} = S_{\kappa\lambda}. \tag{4.26}$$

Since neither orbital angular momentum nor the spin is changed by the new definition of the energy-momentum tensor, it follows that the total angular momentum also remains unchanged and so (4.21) is proved.

The symmetrical tensor $\Theta_{\kappa\lambda}$ has been introduced by J. F. Belinfante and by L. Rosenfeld. Recently a certain criticism has been raised against the manner of decomposition of the angular momentum into the spin part and the orbital part as it follows from it. Before dealing with this, in section 6, it will be useful to consider a few examples.

5. Examples

(*a*) *The scalar field.* As an example, let us consider a field which has only one component $\psi(x, y, z, t)$ and is thus a scalar field. The simplest Lagrangian which we can form, involving ψ and $\partial_k \psi$, may be given by

$$L = \tfrac{1}{2}\{(\partial_{\kappa}\psi)^2 + k^2\psi^2\}, \tag{5.1}$$

where k is a constant.

Eqn. (5.1) yields

$$T_{\kappa\lambda} = \partial_\kappa \psi \frac{\partial L}{\partial(\partial_\lambda \psi)} - \delta_{\kappa\lambda} L = \partial_k \psi \partial_\lambda \psi - \delta_{\kappa\lambda} L, \tag{5.2}$$

which is thus a symmetrical tensor.

The field equation is obtained by using (1.12):

$$\frac{\partial L}{\partial \psi} - \partial_\kappa \left\{ \frac{\partial L}{\partial(\partial_\kappa \psi)} \right\} = k^2 \psi - \partial_\kappa^2 \psi = 0,$$

therefore

$$\partial_\kappa^2 \psi - k^2 \psi = 0. \tag{5.3}$$

This is a scalar wave equation, known as the Klein–Gordon equation.

(*b*) *The Proca field.* As the next simplest field we may consider a vector field ψ_λ. In quantum theory such a field can contain particles (mesons) of spin 0 or 1 (in units of $\hbar$). This follows from a consideration of the matrix

$$S^2 = -[P_{(12)}^2 + P_{(23)}^2 + P_{(31)}^2]. \tag{5.4}$$

The eigenvalues of this matrix yield the square $s(s+1)$ of the magnitude of the spin angular momentum of one meson. Now for a vector field the $P_{(\kappa\lambda)}$ are identical with the $M_{(\kappa\lambda)}$ (3.7) of the Lorentz transformations and we get

$$S^2 = \begin{bmatrix} 2 & 0 & 0 & 0 \\ 0 & 2 & 0 & 0 \\ 0 & 0 & 2 & 0 \\ 0 & 0 & 0 & 0 \end{bmatrix} = \begin{bmatrix} 1(1+1) & 0 & 0 & 0 \\ 0 & 1(1+1) & 0 & 0 \\ 0 & 0 & 1(1+1) & 0 \\ 0 & 0 & 0 & 0(0+1) \end{bmatrix}. \tag{5.5}$$

Hence s is either 1 or 0.

The simplest vector field ψ_λ may be expected to satisfy the relativistically invariant wave equation

$$\partial_\kappa^2 \psi_\lambda - k^2 \psi_\lambda = 0. \tag{5.6}$$

The Lagrangian which yields this equation is

$$L = \tfrac{1}{2}[(\partial_\kappa \psi_\lambda)^2 + k^2 \psi_\lambda^2]. \tag{5.7}$$

It is easily verified that (5.6) follows from the Euler–Lagrange equation:

$$\frac{\partial L}{\partial \psi_\lambda} - \partial_\kappa \left\{ \frac{\partial L}{\partial(\partial_\kappa \psi_\lambda)} \right\} = 0. \tag{5.8}$$

The canonical energy-momentum tensor turns out to be again symmetrical:

$$T_{\kappa\lambda} = \frac{\partial L}{\partial(\partial_\kappa \psi_\sigma)} \frac{\partial \psi_\sigma}{\partial x_\lambda} - \delta_{\kappa\lambda} L = (\partial_\kappa \psi_\sigma)(\partial_\lambda \psi_\sigma) - \delta_{\kappa\lambda} L. \tag{5.9}$$

Hence, orbital momentum is conserved.

From this it does not follow that the spin angular momentum vanishes.

In fact in view of (5.5), fields with both spin $s = 0$ and $s = 1$ can occur in the quantum theoretical case. From the conservation of the orbital momentum it necessarily follows that the spin angular momentum is also conserved. For the spin density we get

$$S_{\mu\kappa\lambda} = \frac{\partial L}{\partial(\partial_\mu \psi_\mu)} P_{(\kappa\lambda)\nu\sigma}\psi_\sigma = \frac{\partial L}{\partial(\partial_\mu \psi_\nu)} M_{(\kappa\lambda)\nu\sigma}\psi_\sigma, \tag{5.10}$$

where the $M_{(\kappa\lambda)\nu\sigma}$ are given by (3.6) and (3.7). On using these matrices, we get

$$S_{\mu\kappa\lambda} = 2(\partial_\mu \psi_\kappa . \psi_\lambda - \partial_\mu \psi_\lambda . \psi_\kappa). \tag{5.11}$$

The divergence of this tensor vanishes as can be proved by using (5.6):

$$\begin{aligned} \frac{1}{2}\frac{\partial S_{\mu\kappa\lambda}}{\partial x_\mu} &= \partial_\mu^2 \psi_\kappa . \psi_\lambda - \partial_\mu^2 \psi_\lambda . \psi_\kappa \\ &= k^2 \psi_\kappa \psi_\lambda - k^2 \psi_\lambda \psi_\kappa = 0. \end{aligned} \tag{5.12}$$

From the foregoing we see that by combining several Klein–Gordon equations we cannot hope to get a field which would lead (in quantum theory) to a definite spin. To get a case which corresponds to particles with spin 1, we must reduce the number of components of ψ_λ from four to three by superimposing over the wave equation a relativistically invariant equation. A set of such equations could be

$$\partial_\kappa^2 \psi_\lambda - k^2 \psi_\lambda = 0, \tag{5.13}$$

$$\partial_\kappa \psi_\kappa = 0. \tag{5.14}$$

However, there appears to be no simple Lagrangian which leads directly to these equations. On the other hand, it was shown by Proca that by replacing equations (5.13) and (5.14) by an equivalent set of first-order equations it is possible to form a Lagrangian which yields the first-order equations which in their turn are equivalent to (5.13) and (5.14). The set of first-order equations is obtained by introducing the quantity $\phi_{\kappa\lambda} = -\phi_{\lambda\kappa}$, defined by

$$\partial_\kappa \psi_\lambda - \partial_\lambda \psi_\kappa = \phi_{\kappa\lambda}, \tag{5.15}$$

and using

$$\partial_\kappa \phi_{\kappa\lambda} = k^2 \psi_\lambda. \tag{5.16}$$

From (5.16), we get on multiplying by ∂_λ

$$\partial_\lambda \partial_\kappa \phi_{\kappa\lambda} = k^2 \partial_\lambda \psi_\lambda. \tag{5.17}$$

In the left-hand side $\partial_\lambda \partial_\kappa$ is symmetrical in κ, λ while by (5.15) $\phi_{\kappa\lambda}$ is anti-symmetrical. Hence the left-hand side is zero and we get (5.14). Similarly, if we multiply (5.15) by ∂_κ, we get

$$\partial_\kappa^2 \psi_\lambda - \partial_\kappa \partial_\lambda \psi_\kappa = \partial_\kappa \phi_{\kappa\lambda}, \tag{5.18}$$

therefore

$$\partial_k^2 \psi_\lambda - \partial_\lambda \partial_\kappa \psi_\kappa = k^2 \psi_\lambda. \tag{5.19}$$

Since we have already derived $\partial_\kappa \psi_\kappa = 0$, (5.19) yields the wave equation (5.13).

Equation (5.16) with $\phi_{\kappa\lambda}$ as given in (5.15) may be obtained from the following Lagrangian:

$$L = \tfrac{1}{2}(\tfrac{1}{2}\phi_{\sigma\kappa}^2 + k^2\psi_\kappa^2) = \tfrac{1}{2}\{\tfrac{1}{2}(\partial_\sigma \psi_\kappa - \partial_\kappa \psi_\sigma)^2 + k^2 \psi_\kappa^2\}. \tag{5.20}$$

We get from

$$\frac{\partial L}{\partial \psi_\kappa} - \partial_\sigma \left\{\frac{\partial L}{\partial(\partial_\sigma \psi_\kappa)}\right\} = 0, \quad \partial_\sigma(\partial_\sigma \psi_\kappa - \partial_\kappa \psi_\sigma) - k^2\psi_\kappa = 0, \tag{5.21}$$

which is identical with (5.16) in view of (5.15).

The canonical energy-momentum tensor is given by

$$T_{\kappa\lambda} = \partial_\lambda \psi_\sigma \frac{\partial L}{\partial(\partial_\kappa \psi_\sigma)} - \delta_{\kappa\lambda} L = \partial_\lambda \psi_\sigma(\partial_\kappa \psi_\sigma - \partial_\sigma \psi_\kappa) - \delta_{\kappa\lambda} L = \partial_\lambda \psi_\sigma \phi_{\kappa\sigma} - \delta_{\kappa\lambda} L. \tag{5.22}$$

This tensor is not symmetrical and consequently a non-vanishing spin angular momentum must exist. Using (3.25) and (3.7), we get for the $S_{\mu\kappa\lambda}$

$$\begin{aligned} S_{\mu\kappa\lambda} &= \partial_\mu \psi_\kappa . \psi_\lambda - \partial_\mu \psi_\lambda . \psi_\kappa + \partial_\lambda \psi_\mu . \psi_\kappa - \partial_\kappa \psi_\mu . \psi_\lambda \\ &= (\partial_\mu \psi_\kappa - \partial_\kappa \psi_\mu)\psi_\lambda - (\partial_\mu \psi_\lambda - \partial_\lambda \psi_\mu)\psi_\kappa = \phi_{\mu\kappa}\psi_\lambda - \phi_{\mu\lambda}\psi_\kappa. \end{aligned} \tag{5.23}$$

On evaluating the corresponding $S_{\kappa\lambda\mu}$ and $S_{\kappa\mu\lambda}$ and using (4.17) for the symmetrized energy-momentum tensor $\Theta_{\kappa\lambda}$, we get

$$\Theta_{\kappa\lambda} = T_{\kappa\lambda} + \partial_\mu[(\partial_\mu \psi_\kappa - \partial_\kappa \psi_\mu)\psi_\lambda] = T_{\kappa\lambda} + \partial_\mu(\phi_{\mu\kappa}\psi_\lambda). \tag{5.24}$$

On using (5.16) and (5.22), we get further

$$\Theta_{\kappa\lambda} = \partial_\lambda \psi_\mu . \phi_{\kappa\mu} + \partial_\mu \phi_{\mu\kappa} . \psi_\lambda + \phi_{\mu\kappa} \partial_\mu \psi_\lambda - \delta_{\kappa\lambda} L,$$

so that $$\Theta_{\kappa\lambda} = \partial_\lambda \psi_\mu . \phi_{\kappa\mu} - \partial_\mu \psi_\lambda . \phi_{\kappa\mu} + k^2\psi_\kappa \psi_\lambda - \delta_{\kappa\lambda} L; \tag{5.25}$$

therefore $$\Theta_{\kappa\lambda} = \phi_{\kappa\mu}\phi_{\lambda\mu} + k^2\psi_\kappa \psi_\lambda - \delta_{\kappa\lambda} L. \tag{5.26}$$

In this expression $\Theta_{\kappa\lambda}$ is manifestly symmetrical. To obtain the density of spin we have to use (see 5.24)

$$T^S_{\kappa\lambda} = \partial_\mu(\phi_{\mu\kappa}\psi_\lambda) \tag{5.27}$$

and $$S^{(\text{mod})}_{\mu\kappa\lambda} = T^S_{\mu\lambda} x_\kappa - T^S_{\mu\kappa} x_\lambda = \partial_\sigma(\phi_{\sigma\mu}\psi_\lambda)x_\kappa - \partial_\sigma(\phi_{\sigma\mu}\psi_\kappa)x_\lambda. \tag{5.28}$$

To obtain the total spin angular momentum we need not use the density $S^{(\text{mod})}_{\mu\kappa\lambda}$, we can use $S_{\mu\kappa\lambda}$ since both expressions yield the same result. (5.23) yields

$$S_{\kappa\lambda} = \iiint_\sigma S_{\mu\kappa\lambda}\, d\sigma_\mu = \iiint_\sigma (\phi_{\mu\kappa}\psi_\lambda - \phi_{\mu\lambda}\psi_\kappa)\, d\sigma_\mu. \tag{5.29}$$

The vector field of Proca is closely related to the electromagnetic field. If we make $k = 0$ and use the notation $\psi_\lambda \to A_\lambda$, $\phi_{\kappa\lambda} \to F_{\kappa\lambda}$, we get the equations

$$\partial_\lambda A_\lambda = 0, \tag{5.30}$$

$$\partial_\kappa^2 A_\lambda = 0, \tag{5.31}$$

$$F_{\kappa\lambda} = \partial_\kappa A_\lambda - \partial_\lambda A_\kappa, \tag{5.32}$$

which characterize the electromagnetic field $F_{\kappa\lambda}$ in terms of the vector potential and the Lorentz condition. (5.29) yields for the spin of the electromagnetic field

$$S_{\kappa\lambda} = \iiint (F_{\mu\kappa} A_\lambda - F_{\mu\lambda} A_\kappa)\, d\sigma_\mu. \tag{5.33}$$

Relative to a surface σ which is given by $t =$ constant, we get

$$S_{\kappa\lambda} = \frac{1}{ic} \iiint (F_{4\kappa} A_\lambda - F_{4\lambda} A_\kappa)\, dxdydz. \tag{5.34}$$

In particular, with

$$\mathbf{S} = (S_{23}, S_{31}, S_{12}), \quad \mathbf{A} = (A_1, A_2, A_3), \quad \text{and} \quad i\mathbf{E} = (F_{41}, F_{42}, F_{43}) \tag{5.35}$$

where $\mathbf{E}$ is the electric field, we get

$$\mathbf{S} = \frac{1}{c} \iiint [\mathbf{E}, \mathbf{A}]\, dxdydz. \tag{5.36}$$

As a result of this relationship between the vector field of Proca and the Maxwell field, vector-mesons and photons have the same spin, which according to quantum mechanics is equal to 1. The essential difference between photons and vector-mesons is in the mass of the particles. According to quantum theory $k = mc/\hbar$, where m is the mass of the field particles. Hence for photons, where $k = 0$, the mass is zero. This is in accord with their moving always with the velocity of light. If instead of using the anti-symmetrical tensor $\phi_{\kappa\lambda}$ we use the vectors $\mathbf{E}$ and $\mathbf{H}$, we get the set of first-order equations

$$\operatorname{curl} \mathbf{E} = -\frac{1}{c} \dot{\mathbf{H}}, \qquad \operatorname{div} \mathbf{E} + k^2 \phi = 0,$$

$$\operatorname{div} \mathbf{H} = 0, \qquad \operatorname{curl} \mathbf{H} - \frac{1}{c} \dot{\mathbf{E}} + k^2 \mathbf{A} = 0, \tag{5.37}$$

where ψ_λ is replaced by the vector and scalar potentials $\mathbf{A}$, ϕ. In this way the analogy between the Proca field and the Maxwell field is also expressed when the usual notation is used. Again it can be seen that for $k = 0$ we get Maxwell's equations.

6. The Conservation of the Spin and Orbital Angular Momenta

The division of the total angular momentum into a spin part and an orbital part as described in sections 3 and 4 presents itself in a natural way from a mathematical point of view, but it suffers on physical grounds. On the one hand, the orbital and spin energy-momentum tensors obey separately laws of conservation ($\partial T_{\kappa\lambda}/\partial x_\kappa = 0, \partial T^S_{\kappa\lambda}/\partial x_\kappa = 0$), while on the other hand, only the total angular momentum is conserved. Orbital and spin angular momenta will be conserved separately only when the canonical tensor is symmetrical and in general this is not the case. Now, for free fields, such as we have been considering, with no interaction with other fields, we should expect a law of conservation to hold not only for the separate energy-momentum vectors but also for the separate orbital and spin angular momenta. However, this is not the case with the Belinfante–Rosenfeld spin. This problem has been treated recently on the basis of a Lagrangian formalism (as distinct from a more group-theoretical approach) by J. Hilgevoord and S. A. Wouthuysen† and by J. Hilgevoord and E. A. de Kerf.‡ If the total spin and the total orbital angular momentum are to be defined in each frame of reference in the same way, for instance as integrals over a space-like surface t = const (and t' = const for a second frame of reference, etc.), then these integrals must transform into each other in a covariant way. However, this will be the case only if a law of conservation applies to the spin and orbital angular momenta separately. It may be added, that unlike the case of a classical particle (which is treated in the next chapter), when we deal with a free field we have no means of singling out one particular space-like surface, to be used by all observers for evaluating total quantities. The solution to this problem for a large class of fields was given by Hilgevoord and de Kerf. They showed that for all fields which obey the Klein–Gordon equation and subsidiary conditions, it is possible to define the spin and orbital angular momenta in such a way that they are conserved. Their procedure is as follows: In addition to the usual Lagrangian which leads to the whole system of equations (i.e. Klein–Gordon plus subsidiary conditions), they introduce a Lagrangian which leads to the Klein–Gordon equation only. In this case, Hilgevoord and de Kerf have shown that it is possible to introduce a Lagrangian which leads always to a

† For reference, see list at the end of the book.

‡ I am much indebted to these authors for advance information on their paper: 'The covariant definition of spin in relativistic quantum field theory', to be published in *Physica*.

symmetrical energy-momentum tensor. Essentially, the Lagrangian in question is a generalization of (5.7) which applies for a field with four components and which yielded the symmetrical tensor $T_{\kappa\lambda}$ according to (5.9). If we use this Lagrangian to obtain spin and angular momenta according to (2.14), (3.22), and (3.25), we shall get conserved quantities. All that is required for this to hold, is that the field should obey the Klein–Gordon equation. The fact that the field obeys in addition certain subsidiary conditions, so as to make the spin in quantum theory definite ($0, \frac{1}{2}, 1$, etc.) can make no difference to the property of conservation. We are, of course, interested only in solutions of the whole system of equations. For such solutions and in the case of spin $\frac{1}{2}$ and 1, Hilgevoord and de Kerf were able to show that not only are the angular momenta conserved but that they yield also the same total angular momentum as the Lagrangian which determines the full system of equations. (In each case the proof requires the explicit form of the usual Lagrangian.) The same also applies to the total energy-momentum vector. Hence, only a new splitting of the total angular momentum is achieved. Finally it may be added that in the rest system (where the energy-momentum vector has only a time component, $P_1 = P_2 = P_3 = 0$, the orbital angular momentum has vanishing space components) the spin has only three space components. This is as it should be from a classical analogy. The Belinfante–Rosenfeld spin does not possess this property. For the form of a spin tensor in a rest system, see next chapter, section 9.

CHAPTER 5

RELATIVISTIC PARTICLE DYNAMICS

1. Kinematics

When dealing with Maxwell's equations our task consisted merely in rewriting them in tensor form. This is possible because Maxwell's equations retain their form with respect to Lorentz transformations. In the case of pre-relativistic dynamics we know that the basic equations are covariant with respect to Galilei transformations, but not with respect to Lorentz transformations. Consequently, it is not possible to express the classical equations of motion in tensor form, without suitably modifying the content of the theory. We begin with some purely kinematic aspects.

Let us consider a particle which moves with a velocity

$$\mathbf{u} = (dx/dt, dy/dt, dz/dt)$$

in space. We have already seen that the transformation law of a velocity is rather complicated. Now in the theory of relativity it would be more natural to look for a four-vector which has as space components the vector $\mathbf{u}$ and as a fourth component some other suitable quantity.† It turns out that such a four-vector does not exist, but we can define a four-vector which very nearly yields the vector $\mathbf{u}$ for the space components, and this is more nearly so as $u = |\mathbf{u}|$ becomes smaller compared with c. To derive it let us start with the four-vector

$$dx_\kappa = (dx, dy, dz, ic\,dt). \tag{1.1}$$

If we divide this by dt we shall get

$$\frac{dx_\kappa}{dt} = (u_x, u_y, u_z, ic) \equiv (u_a, ic) \quad (a = x, y, z). \tag{1.2}$$

However, as dt or dt^{-1} are not invariants, dx_κ/dt is not a four-vector. From a given vector we can get another vector only by multiplying it by a scalar or an invariant. Let us therefore divide dx_κ by the proper

† The three space components of a four-vector always constitute a three-dimensional space vector because $t' = t$ are space rotations. The converse does not hold. We may multiply the components of an ordinary space vector by a function of the velocity without affecting its three-vector properties, but this will have an effect on the way this vector will transform under Lorentz transformations, because the velocity is not an invariant.

time $d\tau = \gamma^{-1}dt$ where $\gamma = (1-u^2/c^2)^{-\frac{1}{2}}$ and u is the velocity of the particle as measured in the frame considered

$$u = \left[\left(\frac{dx}{dt}\right)^2+\left(\frac{dy}{dt}\right)^2+\left(\frac{dz}{dt}\right)^2\right]^{\frac{1}{2}}. \tag{1.3}$$

Then
$$U_\kappa = \left(\frac{dx}{d\tau}, \frac{dy}{d\tau}, \frac{dz}{d\tau}, ic\frac{dt}{d\tau}\right) \tag{1.4}$$

is a four-vector.

It turns out that the four-vector U_κ is very useful in the formulation of relativistic dynamics, although the vector $\mathbf{u} = (u_x, u_y, u_z)$ is the one which is more directly obtainable from observation.

The four-vector U_κ was introduced by Minkowski and is called after him. The relation between U_κ and $\mathbf{u} \equiv u_a$ is as follows:

$$U_\kappa = (\gamma u_a, ic\gamma). \tag{1.5}$$

The factor γ is essential. As γ is not an invariant (u_a, ic) is not a four-vector.

In a similar way, we introduce the Minkowski acceleration

$$\frac{dU_\kappa}{d\tau} = A_\kappa = \left(\frac{d^2x}{d\tau^2}, \frac{d^2y}{d\tau^2}, \frac{d^2z}{d^2\tau}, ic\frac{d^2t}{d\tau^2}\right) = \left\{\gamma\frac{d}{dt}(\gamma u_a), ic\gamma\frac{d\gamma}{dt}\right\}. \tag{1.6}$$

The velocity four-vector always has the same length irrespective of the value of u

$$U_\kappa^2 = \gamma^2u^2-\gamma^2c^2 = \gamma^2(u^2-c^2) = -c^2. \tag{1.7}$$

U_κ is thus a time-like vector with constant length.

From (1.7) we find

$$\frac{d}{d\tau}U_\kappa^2 = 2U_\kappa\frac{dU_\kappa}{d\tau} = 0. \tag{1.8}$$

Consequently, U_κ and $dU_\kappa/d\tau$ are two four-vectors which are orthogonal to each other in the four-dimensional Minkowski universe. This can also be confirmed by a direct use of (1.5) and (1.6) and using the formula

$$\mathbf{u}\,d\mathbf{u} = u\,du, \tag{1.9}$$

which follows directly from

$$u^2 = u_x^2+u_y^2+u_z^2, \tag{1.10}$$

so that
$$u\,du = u_x\,du_x+u_y\,du_y+u_z\,du_z = \mathbf{u}\,d\mathbf{u}. \tag{1.11}$$

du is the change of length of u and is different from $|d\mathbf{u}|$.

2. The Energy-Momentum Vector

Let us now consider a particle in translational motion. A method which suggests itself in characterizing its dynamical properties is to ascribe to it a four-vector P_κ which we shall call the energy-momentum vector and which satisfies the following postulates:

(1) P_κ is always parallel to the four-velocity,

(2) P_κ is constant in time as long as no force is acting on the particle. Hence

$$P_\kappa = m_0 U_\kappa, \tag{2.1}$$

where m_0 is an invariant (a scalar) and

$$\frac{dP_\kappa}{d\tau} = 0. \tag{2.2}$$

Since at any moment dt and $d\tau$ differ by a factor $(1-u^2/c^2)^{\frac{1}{2}}$, we can also express (2.2) by

$$\frac{dP_\kappa}{dt} = 0. \tag{2.3}$$

From (2.1) and (2.2) it follows that m_0 is also constant in time. Firstly, we have

$$\frac{dP_\kappa}{dt} = \frac{dm_0}{d\tau} U_\kappa + m_0 \frac{dU_\kappa}{d\tau} = 0.$$

Now, as U_κ is always different from zero we get further

$$\frac{dm_0}{d\tau} U_\kappa^2 + m_0 \frac{dU_\kappa}{d\tau} U_\kappa = 0. \tag{2.4}$$

Using (1.8) and $U_\kappa^2 = -c^2$ (= constant), we get

$$\frac{dm_0}{d\tau} = 0. \tag{2.5}$$

Consequently, we also have

$$\frac{dm_0}{dt} = 0. \tag{2.5 a}$$

Recalling that in classical dynamics

$$\mathbf{p} = m\mathbf{u} \text{ (or } p_a = mu_a) \quad (a = x, y, z), \tag{2.6}$$

where $\mathbf{p}$ is the momentum, $\mathbf{u}$ the velocity, and m the mass of the particle, we find on equating (2.6) to the three space components of (2.1):

$$P_a = \frac{m_0 u_a}{(1-u^2/c^2)^{\frac{1}{2}}} = mu_a \quad (a = x, y, z), \tag{2.7}$$

so that the relation between m and m_0 is given by

$$m = \frac{m_0}{(1-u^2/c^2)^{\frac{1}{2}}} = m_0 \gamma. \tag{2.8}$$

Thus the theory of relativity forces us to the conclusion that what is regarded in the classical theory as mass cannot be assumed (as is done in the classical theory) to be independent of the velocity. In view of this it is better to denote m by m_u. In the theory of relativity it is customary to call m_u the relative mass and m_0 the rest mass. The rest mass is a quantity which is independent of the velocity of the considered particle, as follows from its scalar nature. (For various observers the particle has in general different velocities while m_0 has the same value.)

Let us now consider the fourth component of P_κ:

$$P_4 = \frac{im_0 c}{(1-u^2/c^2)^{\frac{1}{2}}} = \frac{iW}{c}, \tag{2.9}$$

where

$$W = \frac{m_0 c^2}{(1-u^2/c^2)^{\frac{1}{2}}}. \tag{2.10}$$

Because of the invariance of c the relation (2.9) between W and P_4 holds in all frames of reference. W has the dimensions of energy and it thus follows from the theory of relativity that if a particle is associated with a momentum P_1, P_2, P_3, it is necessarily also associated with an energy $W = -icP_4$. We call this energy the total energy of the particle. An essential feature here is that the energy is defined without an arbitrary additive constant. We cannot add a constant without destroying the four-vector character of P_κ. For a given momentum the total energy is given according to (2.10). If we transform to rest, i.e. if we introduce a frame of reference which is at rest relative to the particle, $u = 0$, and we find for the total energy in the rest system

$$W_0 = m_0 c^2. \tag{2.11}$$

This value for the total energy of a particle with rest mass m_0 at rest is a necessary consequence from the definition of the four-momentum.

The length of any vector has the same value in all frames of reference. In the case of P_κ the square of the length is given by

$$P_\kappa^2 = P_1^2 + P_2^2 + P_3^2 - W^2/c^2. \tag{2.12}$$

The value of this invariant may be calculated in the rest system, where, because of the assumed parallelism between P_κ and U_κ,

$$P_1 = P_2 = P_3 = 0, \qquad W = W_0 = m_0 c^2.$$

Hence,

$$P_1^2 + P_2^2 + P_3^2 - W^2/c^2 = -m_0^2 c^2, \tag{2.13}$$

$$W = c(P_1^2 + P_2^2 + P_3^2 + m_0^2 c^2)^{\frac{1}{2}}. \tag{2.14}$$

This equation provides a direct relation between momentum and energy.

Suppose the energy at rest is $W_0 = m_0 c^2$. If the particle or body is given some velocity, its energy is increased, and we call the increase the kinetic energy E. The total energy may therefore also be expressed as

$$W = W_0 + E = m_0 c^2 + E. \tag{2.15}$$

From (2.10) and (2.15) we get for the kinetic energy

$$E = m_0 c^2(\gamma - 1). \tag{2.16}$$

For small velocities $u \ll c$ we can expand the square root of $1-u^2/c^2$ and get approximately

$$E = m_0 c^2(1 + u^2/2c^2 - ... - 1) = \tfrac{1}{2}(m_0 u^2) + ... \tag{2.17}$$

This agrees with the classical value for the kinetic energy.

3. The Equations of Motion in a Given Field

Let us consider the case when the particle is moving in a field of force which is given as a function of x, y, z, t, so that the particle itself has no effect on the field. We then introduce the force according to the following defining equations:

$$\frac{dP_a}{dt} = \frac{d}{dt}(m_0 U_a) = f_a \quad (a = x, y, z). \tag{3.1}$$

These three equations will determine $P_a(t)$ and hence U_a and the orbit of the particle for given initial conditions provided we know how the mass m_0 depends on the field. Moreover, since (2.13) always applies, we can also obtain P_4 and dP_4/dt. Let us denote the last quantity by f_4. Then we can replace (3.1) by

$$\frac{dP_\kappa}{dt} = f_\kappa \quad (\kappa = 1, 2, 3, 4). \tag{3.2}$$

f_κ is not a four-vector and for this reason we use the notation f_κ and not F_κ. f_4 is not to be calculated separately, but follows from (3.1) and (2.13). From a mathematical point of view it is more convenient to introduce a four-vector. This was done by Minkowski who introduced

$$\frac{dP_\kappa}{d\tau} = K_\kappa = \left\{\frac{d}{d\tau}\left(m_0 \frac{dx}{d\tau}\right), ..., ic\frac{d}{d\tau}\left(m_0 \frac{dt}{d\tau}\right)\right\}. \tag{3.3}$$

K_κ is called the Minkowski force while f_κ is often referred to as the Newton force. The first measures the change of relativistic momentum in a unit of proper time, while the second measures it in a unit of ordinary time.

The relationship between K_κ and f_κ is given by

$$K_\kappa = (\gamma f_\kappa) = \left(\gamma f_a, ic\gamma\frac{d}{dt}(m_0 \gamma)\right) = \left(\gamma f_a, ic^{-1}\gamma\frac{d}{dt}(m_0 \gamma c^2)\right). \tag{3.4}$$

Since U_κ and $dU_\kappa/d\tau$ are orthogonal to each other we get

$$P_\kappa \frac{dP_\kappa}{d\tau} = m_0 U_\kappa \frac{d}{d\tau}(m_0 U_\kappa) = m_0^2 U_\kappa \frac{dU_\kappa}{d\tau} + m_0 U_\kappa^2 \frac{dm_0}{d\tau}, \tag{3.5}$$

therefore

$$P_\kappa \frac{dP_\kappa}{d\tau} = P_\kappa K_\kappa = -m_0 c^2 \frac{dm_0}{d\tau}, \tag{3.6}$$

Using (3.4) and

$$P_\kappa = (m_0 \gamma u_a, i m_0 \gamma c), \tag{3.7}$$

we get from (3.6)

$$m_0 \gamma^2 (f_a u_a) - m_0 \gamma^2 \frac{d}{dt}(m_0 c^2 \gamma) = -m_0 c^2 \frac{dm_0}{d\tau}. \tag{3.8}$$

Consequently,

$$f_a u_a = \frac{dW}{dt} - \frac{1}{\gamma} \frac{d(m_0 c^2)}{dt}, \tag{3.9}$$

where W is the energy of the particle as defined in (2.10). $f_a u_a$ is the work done by the force f_a per unit of time. It follows therefore that unless $dm_0/dt = 0$ the law of conservation of energy will not hold. Nor can a modification of definition of the energy be found which will change this result with regard to m_0 because the right-hand side of (3.9) is not a perfect differential:

$$\begin{aligned} dW - \frac{1}{\gamma} d(m_0 c^2) &= d(m_0 \gamma c^2) - \frac{1}{\gamma} d(m_0 c^2) \\ &= m_0 c^2 d\gamma + \left(\gamma - \frac{1}{\gamma}\right) d(m_0 c^2) \neq df(m_0, \gamma) \end{aligned} \tag{3.10}$$

where f is some function of m_0 and γ. Hence, also, when a force is acting on a particle

$$\frac{dm_0}{dt} = \frac{d(m_0 c^2)}{dt} = 0 \tag{3.11}$$

and the mass cannot depend on the field of force.

In the next section we shall see that when two or more particles interact with each other and as a result of the interaction new particles are formed, the total of the rest mass is in general not conserved. The sum of the individual masses must therefore be distinguished from the total mass of a system of which the former are constituents. From (3.11) it follows, however, that in an experiment designed to measure the mass of a particle, the result will be independent of any interaction with other particles.

Using (3.9) and (3.11), we can also express the relationship between K_κ and f_κ as follows, instead of through (3.4):

$$K_\kappa = \{\gamma f_a, i c^{-1} \gamma (f_a u_a)\} = \{\gamma f_a, i c^{-1} (f_a U_a)\}. \tag{3.12}$$

Thus the fourth component of K_κ is the work done by the Newton force per unit of proper time (apart from a factor ic^{-1}). For a Newton force the fourth component is the work done by this force per unit of time. We must remember, however, that the Newton force is not a four-vector.

Relative to a frame in which the particle is momentarily at rest, $u_a = 0$ and $K^0_\kappa = (f^0_a, 0)$, since $\gamma = 1$. If we transform from this system into any other frame, with motion along x-axes with velocity v, the four-vector transformation yields, in terms of the Newton force,

$$f_x = f^0_x \qquad = K^0_x, \tag{3.13}$$

$$f_y = f^0_y\left(1-\frac{v^2}{c^2}\right)^{\frac{1}{2}} = K^0_y\left(1-\frac{v^2}{c^2}\right)^{\frac{1}{2}}, \tag{3.14}$$

$$f_z = f^0_z\left(1-\frac{v^2}{c^2}\right)^{\frac{1}{2}} = K^0_z\left(1-\frac{v^2}{c^2}\right)^{\frac{1}{2}}. \tag{3.15}$$

It is often convenient to consider the forces in a frame in which a mechanical system is at rest and then transform to some arbitrary inertial frame.

4. The Non-conservation of Rest Mass and the Binding Energy of a Compound System

In this section we are going to consider some consequences which follow from the laws of conservation of momentum and energy when a system of particles interact and as a result of the interaction a new system of particles is formed. The laws of conservation of total momentum and total energy will be postulated, but we are not going to use any further features regarding the nature of the interaction. For simplicity we assume that two particles 1, 2 are going to collide and that as a result of the collision two new particles a, b, not necessarily identical with 1, 2, are formed. The laws of conservation of momentum and energy are then summed up in the following relation:

$$P_{1\kappa}+P_{2\kappa} = P_{a\kappa}+P_{b\kappa}. \tag{4.1}$$

To begin with we note that this law is Lorentz invariant because it is a relation between four-vectors. Graphically it can be illustrated by a vector polygon in a four-dimensional momentum space (which transforms like x, y, z, ict). Since a Lorentz transformation only rotates the system of coordinates the properties of the polygon are not affected. Hence, if the laws of conservation of energy and momentum hold in one frame they will hold in any other Lorentz frame as well (in Fig. 26

the vectors are of course complex, and we use it only in an abstract sense).

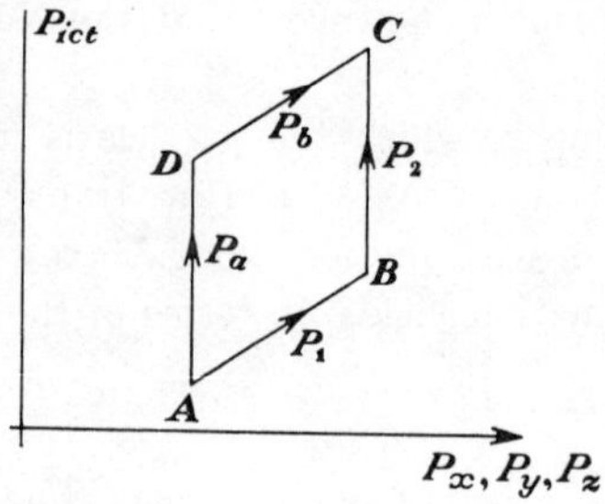

FIG. 26

Eqn. (4.1) stands for the following relations:

$$\frac{M_{01}\,u_{1x}}{(1-\beta_1^2)^{\frac{1}{2}}}+\frac{M_{02}\,u_{2x}}{(1-\beta_2^2)^{\frac{1}{2}}}=\frac{M_{0a}\,u_{ax}}{(1-\beta_a^2)^{\frac{1}{2}}}+\frac{M_{0b}\,u_{bx}}{(1-\beta_b^2)^{\frac{1}{2}}}, \tag{4.2}$$

with similar relations in y, z, and

$$W_1+W_2=W_a+W_b, \tag{4.3}$$

or

$$\frac{M_{01}\,c^2}{(1-\beta_1^2)^{\frac{1}{2}}}+\frac{M_{02}\,c^2}{(1-\beta_2^2)^{\frac{1}{2}}}=\frac{M_{0a}\,c^2}{(1-\beta_a^2)^{\frac{1}{2}}}+\frac{M_{0b}\,c^2}{(1-\beta_b^2)^{\frac{1}{2}}}. \tag{4.4}$$

In all these relations

$$\beta_1=\frac{u_1}{c},\quad \beta_2=\frac{u_2}{c},\quad \beta_a=\frac{u_a}{c},\quad \beta_b=\frac{u_b}{c}. \tag{4.5}$$

To simplify the problem further let us assume that both before and after the collision, u_y, u_z are zero for all particles.

In the non-relativistic theory the corresponding equations are ($u_y=u_z=0$):

$$M_{01}\,u_{1x}+M_{02}\,u_{2x}=M_{0a}\,u_{ax}+M_{0b}\,u_{bx}, \tag{4.6}$$

$$\frac{M_{01}\,u_{1x}^2}{2}+\frac{M_{02}\,u_{2x}^2}{2}=\frac{M_{0a}\,u_{ax}^2}{2}+\frac{M_{0b}\,u_{bx}^2}{2}+Q, \tag{4.7}$$

where Q is some positive or negative energy (for instance internal potential energy) depending on whether kinetic energy was lost or gained in the reaction. In the non-relativistic case there is no difficulty in adding to (4.6) and (4.7) the relation

$$M_{01}+M_{02}=M_{0a}+M_{0b} \tag{4.8}$$

and thus to assume that Lavoisier's law holds and make use of it. Suppose M_{01}, M_{02}, M_{0a}, u_1, u_2, and u_a are given. Then M_{0b} will follow from (4.8) and we shall be left with two equations to determine u_b. This is always possible through proper adjustment of Q. No such possibility exists in the relativistic case, if we add Lavoisier's law (4.8) to (4.2) and (4.3). In this case, if M_{01}, M_{02}, M_{0a}, u_1, u_2, u_a are given, we shall be left with two equations for u_b which are in general not compatible with each other. The situation may be illustrated further, and without restriction to collisions along a line, by again considering the vector diagram in the momentum space. We know already that the laws of conservation of energy and momentum demand that the polygon be closed. On the other hand, since the absolute value of the lengths of the vectors is $M_{01}c$, $M_{02}c$, $M_{0a}c$, $M_{0b}c$, Lavoisier's law demands that the lengths $\overline{AB}+\overline{BC}$ shall be equal to the lengths $\overline{AD}+\overline{DC}$, which is in general not the case. We thus come to the conclusion that Lavoisier's law would forbid certain reactions which are permissible on energy-momentum considerations. Conversely we can say that the theory of relativity shows that cases can exist which though compatible with the laws of conservation of momentum and energy are not compatible with conservation of total rest mass. In such cases rest mass M_0 with its equivalent energy $W = M_0 c^2$ is transformed into other forms of energy.

Lavoisier's law rested solely on chemical observation, where the energies involved are too small to detect any changes in rest masses. On the other hand, nuclear reactions speak against Lavoisier's law and thus testify in favour of the theory of relativity.

Thus, to return to our example, in general we get

$$M_{01}+M_{02} \neq M_{0a}+M_{0b}. \tag{4.9}$$

If $M_{01}+M_{02} > M_{0a}+M_{0b}$ the reaction is said to be exothermic and if $M_{01}+M_{02} < M_{0a}+M_{0b}$ the reaction is endothermic. The kinetic energies before and after the collision are given by

$$\text{K.E.}_{\text{(before collision)}} = \frac{M_{01}c^2}{(1-\beta_1^2)^{\frac{1}{2}}}+\frac{M_{02}c^2}{(1-\beta_2^2)^{\frac{1}{2}}}-(M_{01}+M_{02})c^2, \tag{4.10}$$

$$\text{K.E.}_{\text{(after collision)}} = \frac{M_{0a}c^2}{(1-\beta_a^2)^{\frac{1}{2}}}+\frac{M_{0b}c^2}{(1-\beta_b^2)^{\frac{1}{2}}}-(M_{0a}+M_{0b})c^2. \tag{4.11}$$

Hence the difference in these kinetic energies is given by

$$\Delta(\text{K.E.}) = (M_{01}+M_{02})c^2-(M_{0a}+M_{0b})c^2 = \Delta M_0 c^2. \tag{4.12}$$

This formula is basic for nuclear reactions since it enables us to calculate release or absorption of energy (in the above example of kinetic energy) from the differences of the sum of rest masses before and after collision. It represents one of the main results which follow from the special theory of relativity.

Reactions needing special mention are those in which one or more of the participating particles are photons (or any other particle of zero rest mass). We have already seen that the total electromagnetic momentum and energy of a photon represents a four-vector, so that a photon can be treated as a particle. However, since it travels with the velocity of light it must have zero rest mass, because any finite mass will yield for light velocity an infinite energy. From (2.13) it follows further that the length of the energy-momentum vector of a photon is zero (a null-vector which points in the direction of the light cone). Let the energy W of a photon be denoted by E_{ph} and the momentum by G_a ($P_x = G_x$, $P_y = G_y$, $P_z = G_z$), then from (2.13)

$$G_a^2 - \frac{E_{\text{ph}}^2}{c^2} = 0, \qquad |G| = \frac{E_{\text{ph}}}{c}. \tag{4.13}$$

Since the photons have no rest mass of their own, equations like (4.12) must be modified as follows when photons take part in the reaction:

$$\Delta(\text{K.E.}) + \Delta(E_{\text{ph}}) = \Delta M_0 c^2, \tag{4.14}$$

where $\Delta(E_{\text{ph}})$ is the difference of the total photon energy before and after collision. It must be emphasized that in (4.12) or (4.14) K.E. is to be understood in its relativistic sense as defined in (2.15).

Let us consider a nucleus of rest mass N_0 which has been formed from n nucleons of rest masses M_{0i}, and let the total rest mass of the free constituents be $M_0 = \sum_{i=1}^{n} M_{0i}$. The difference in energy $(M_0 - N_0)c^2$ is called the binding energy of the nucleus. It represents the energy liberated when n nucleons at rest combine and form a nucleus of rest mass N_0.

As mentioned, the amounts of energy liberated in chemical reactions are too small to give rise to any measurable differences in rest mass. For this reason Lavoisier's law was accepted as holding strictly as long as the experimental evidence rested on chemical reactions. The situation is, however, different when we consider nuclear reactions. Here the binding energies give rise to appreciable mass defects which have long been known to exist but were not understood satisfactorily in pre-relativistic physics. (Deviations from integral values are also due in part to mixing of isotopes.)

One of the first nuclear reactions to be studied quantitatively in this respect is given by the formation of two α-particles when protons collide with Li^7 nuclei. This reaction was studied by Cockcroft and Walton and is the following:

$$Li^7 + H^1 = 2He^4.$$

The α-particles emerge with great kinetic energy which is measured through their range in air. It much exceeds the kinetic energy of the protons used to produce the reaction. In the Cockcroft–Walton experiment the energy of the protons amounted to about 0·25 MeV and the range of the α-particles in air came to about 8 cm, corresponding to an energy of 8·63 MeV. Apart from the kinetic energies involved, the rest masses of all the particles taking part in the reaction are also known with great accuracy from mass-spectrographic measurements. Because in the reaction the total number of electrons is not affected, we can deal in the following with neutral atoms rather than bare nuclei. We may thus assume that it takes place between atoms and neglect the slight differences in (chemical) energy arising from rearrangements of electrons as these are too small to be considered in this context. The initial masses (referred to O = 16·00000) are as follows: $M_{\mathrm{Li}} = 7{\cdot}01822$, $M_{\mathrm{H}} = 1{\cdot}00812$. Hence the total initial rest mass is 8·02634. The rest-mass of a helium atom is 4·00390, so that the total final mass is 8·00780. The mass difference is thus 0·01854. Since one milli-mass unit corresponds to 0·931 MeV the energy release amounts to 17·26 MeV. It appears as kinetic energy of the α-particles produced, from which we must, however, deduct the kinetic energy of the proton used (say 0·25 MeV). We thus get $2 \times 8{\cdot}63 - 0{\cdot}25 = 17{\cdot}01$, in good agreement with the value obtained from the mass difference.

All this confirms the view that we must look upon matter at rest as a form of energy, whereby if the rest mass is M_0 the energy is $M_0 c^2$. Since there exists no law of conservation of energy in one form there can exist no law of conservation of matter, i.e. of rest mass. Whenever particles combine together and new particles are formed, energy in the form of rest mass will in general undergo a change. Rest mass must be distinguished from relative mass, which is defined as the total energy of a system divided by c^2. Hence it differs from the total energy just by a factor. Since total energy is conserved, relative mass is also conserved. In terms of relative mass (4.3) becomes

$$\frac{W_1}{c^2} + \frac{W_2}{c^2} = \frac{W_a}{c^2} + \frac{W_b}{c^2}. \tag{4.15}$$

The relation between rest mass and relative mass is given by

$$m_r = m_0 \gamma = \frac{m_0}{(1-\beta^2)^{\frac{1}{2}}}. \tag{4.16}$$

In the case of photons (or any other particle of zero mass), we cannot use (4.16), because the quotient becomes indefinite. We must revert to the original definition when we get

$$m_{r\mathrm{ph}} = \frac{E_{\mathrm{ph}}}{c^2}. \tag{4.17}$$

5. The Energy-Momentum Vector of an Electron

In the preceding sections we were dealing with particles in a primitive sense. They were specified by their mass and kinematic properties. Tacitly, they were also assumed to be very small. In addition no fields were assumed to surround them and contribute to their properties. Such an idealization is useful in many applications and such particles are often referred to as bare particles. However, in general particles are surrounded by some field, and in particular, electrons are surrounded by an electromagnetic field which obeys Maxwell's equations. In this section we shall deal with the question of whether this additional feature interferes with the particle concept. In other words, the question which arises is whether an electron together with the electromagnetic field which accompanies it and which undergoes changes as the state of motion of the electron is altered, can still be treated as a particle obeying relativistic dynamics. This turns out to be the case, although in one respect the dynamics has to be extended. When the electron is accelerated a certain amount of energy is radiated and extra work has to be done to account for this loss of energy.

To begin with, consider an electron at rest. The field is then the electrostatic Coulomb field. Let W^0 be the total energy of the field and $\mathbf{P}^0$ its total momentum. For an electron at rest the field has spherical symmetry and consequently the total momentum should be zero. Thus we have

$$W^0 = \tfrac{1}{2}\iiint \mathbf{E}^{02}\, d^3x^0, \qquad \mathbf{P}^0 = 0. \tag{5.1}$$

In accordance with the theory of relativity

$$m_{0e} = \frac{W^0}{c^2} \tag{5.2}$$

will have to be looked upon as the rest mass of the electron or at least as an electromagnetic contribution to the mass. Denoting any other contribution by m_{0m} (the 'mechanical' mass) and the total rest mass by m_0, we get

$$m_0 = m_{0e} + m_{0m}. \tag{5.3}$$

m_0 would thus be the observed mass of the electron. The problem now amounts to finding out whether the energy W and momentum $\mathbf{P}$ of the electromagnetic field when the electron is moving with a velocity $\mathbf{u}$, turns out to be equal to

$$W = \frac{m_{0e}\,c^2}{(1-u^2/c^2)^{\frac{1}{2}}}, \qquad \mathbf{P} = \frac{m_{0e}\,\mathbf{u}}{(1-u^2/c^2)^{\frac{1}{2}}}, \tag{5.4}$$

or, to use a more relativistic notation, to find out whether

$$P_\mu = \left(\mathbf{P}, i\frac{W}{c}\right) = m_{0e}\,U_\mu. \tag{5.5}$$

What this really amounts to, is to determine whether a unique and relativistically invariant definition of the total energy and momentum of the field can be given so that (5.5) will hold.

Let $T_{\mu\nu}$ be the energy-momentum tensor of the electromagnetic field of an electron. Then, as was already shown on p. 113, it is possible to form a four-vector $P_\mu(\sigma)$ as follows:

$$P_\mu(\sigma) = -\frac{1}{c}\iiint\limits_\sigma T_{\mu\nu}(\sigma)\,d\sigma_\nu, \tag{5.6}$$

where σ is a space-like surface to which all observers must refer when transformations are envisaged. The difficulty with this four-vector is, as indicated, its dependence on the hypersurface σ. $P_\mu(\sigma)$ will be independent of σ only in the absence of charges, when $\partial T_{\mu\nu}/\partial x_\nu = 0$. In the presence of charges this is no longer the case and various choices of σ will lead to different energy-momentum vectors. Essentially only one choice could satisfy (5.4). This difficulty was resolved independently by Kwal and Rohrlich. They pointed out that when considering an electron in motion it is not sufficient to introduce a relativistic four-vector in accordance with (5.6), but it is also necessary to select a hypersurface σ which is linked to the motion of the electron in a relativistically invariant manner. Since σ is a space-like surface and its normal is a time-like vector, the only possible choice which results from the above condition is to have the hypersurface orthogonal to the four-velocity at the intersection of the world line with the surface. Since the charge is assumed to be concentrated in a small region or else in a point, the surface can be chosen arbitrarily outside the region of intersection (where $\partial T_{\mu\nu}/\partial x_\nu = 0$). For this reason the hyperplane may be assumed as plane. Let

$$d\sigma_\nu = d\sigma\,.\,n_\nu, \tag{5.7}$$

where n_ν is a unit vector perpendicular to the hyperplane and $d\sigma$ (a scalar) is the magnitude of the surface element $d\sigma_\nu$. Since n_ν is a time-like unit vector, we have

$$n_\nu^2 = -1. \tag{5.8}$$

$d\sigma_\nu$ and n_ν must be parallel to U_ν. Consequently, on using $U_\nu^2 = -c^2$ we get the following relations:

$$n_\nu = \frac{U_\nu}{c}, \tag{5.9}$$

$$d\sigma_\nu = d\sigma . n_\nu = d\sigma \frac{U_\nu}{c}. \tag{5.10}$$

(5.10) expresses the invariant link between the four-velocity of the electron and the hyperplane which must be chosen to define the energy-momentum vector of the electromagnetic field which accompanies the electron.

$d\sigma$ is a scalar and may be evaluated in any frame of reference. The simplest is the rest system of the electron (x_ν^0). In this case $U_k = 0$ $(k = 1, 2, 3)$, $U_4 = ic$, $n_k = 0$, $n_4 = i$ and it follows that

$$d\sigma = dx^0 dy^0 dz^0 = d^3x^0. \tag{5.11}$$

Moreover, it can now be seen that the hyperplane which has been selected in accordance with (5.10) is given by

$$t^0 = \text{const.} \tag{5.12}$$

This is in agreement with (5.1), where the surface over which the energy density is integrated has already been chosen as $t^0 = \text{const.}$ The world line in the rest system consists of a parallel to the t^0-axis and the plane $t^0 = \text{const.}$ is perpendicular to it.

The situation is now as follows: $d\sigma_\nu$ has relative to the rest system the components $d\sigma_1^0 = d\sigma_2^0 = d\sigma_3^0 = 0$, $d\sigma_4^0 = id^3x^0$. On using (5.10), for a system of coordinates relative to which the electron moves along the x-axis with a velocity $u_x = u$, we get the following result:

$$d\sigma_1 = \gamma\, d^3x^0 \frac{u}{c}, \quad d\sigma_2 = 0, \quad d\sigma_3 = 0, \quad d\sigma_4 = i\gamma\, d^3x^0. \tag{5.13}$$

This is also obtainable through a Lorentz transformation of $(0, 0, 0, id^3x^0)$, using (9.17), p. 35. Let $T_{\mu\nu}^0$ be the energy-momentum tensor in the rest system. From (4.13), p. 100, this tensor is given by

$$T_{\mu\nu}^0 = \begin{bmatrix} W^0 - E_x^{02} & -E_x^0 E_y^0 & -E_x^0 E_z^0 & 0 \\ -E_x^0 E_y^0 & W^0 - E_y^{02} & -E_y^0 E_z^0 & 0 \\ -E_x^0 E_z^0 & -E_y^0 E_z^0 & W^0 - E_z^{02} & 0 \\ 0 & 0 & 0 & -W^0 \end{bmatrix} = \begin{bmatrix} T_{11}^0 & T_{12}^0 & T_{13}^0 & 0 \\ T_{12}^0 & T_{22}^0 & T_{23}^0 & 0 \\ T_{13}^0 & T_{23}^0 & T_{33}^0 & 0 \\ 0 & 0 & 0 & T_{44}^0 \end{bmatrix}. \tag{5.14}$$

On transforming this tensor to a system which moves relative to the rest system with a velocity $-u$, we get (in the same way as in (3.6), p. 98):

$$T_{\mu\nu} = \begin{bmatrix} \gamma^2(T^0_{11}-\beta^2 T^0_{44}) & \gamma T^0_{12} & \gamma T^0_{13} & i\gamma^2\beta(T^0_{11}-T^0_{44}) \\ \gamma T^0_{12} & T^0_{22} & T^0_{23} & i\gamma\beta T^0_{12} \\ \gamma T^0_{13} & T^0_{23} & T^0_{33} & i\gamma\beta T^0_{13} \\ i\gamma^2\beta(T^0_{11}-T^0_{44}) & i\gamma\beta T^0_{12} & i\gamma\beta T^0_{13} & -\gamma^2\beta^2 T^0_{11}+\gamma^2 T^0_{44} \end{bmatrix}, \tag{5.15}$$

where $\beta = uc^{-1}$. (As a check one can verify that the trace of the matrix $T^0_{\mu\nu}$ is equal to that of $T_{\mu\nu}$, since $\sum T_{\mu\mu}$ is an invariant.) From (5.6) and (5.13) follows

$$P_\mu(\sigma) = -\frac{1}{c}\Big[\iiint \gamma\Big(T_{\mu 1}\frac{u}{c}+iT_{\mu 4}\Big)\,d^3x^0\Big]. \tag{5.16}$$

From the spherical symmetry in the rest system it follows that integrals of T^0_{12}, T^0_{13}, T^0_{23} over the whole space vanish. Also, since T^0_{4k} is zero, on inserting from (5.15) in (5.16) we get:

$$\begin{aligned} P_1(\sigma) &= -\gamma\frac{u}{c^2}\iiint T^0_{44}\,d^3x^0 = \gamma m_{0e}u_x = m_{0e}U_1, \\ P_2(\sigma) &= 0, \\ P_3(\sigma) &= 0, \\ P_4(\sigma) &= -\frac{1}{c}\iiint i\gamma^3 T^0_{44}(1-\beta^2)\,d^3x^0 = i\gamma m_{0e}c = m_{0e}U_4, \end{aligned} \tag{5.17}$$

and more generally

$$P_\mu(\sigma) = m_{0e}U_\mu. \tag{5.18}$$

In this way (5.5) is proved and the four-vector character of the energy and momentum of the field of an electron is verified.

It must be added that for a long time a great deal of confusion prevailed in connexion with the energy and momentum of the field of an electron. This was caused by not holding to the same hyperplane when transforming from the rest system to the laboratory system in which the electron moves. Thus, instead of using the surface $t^0 = \text{const.}$ in both systems, the surface $t = \text{const.}$ was used in the laboratory system. Naturally, the four-vector character of the energy and momentum of the field did not come out. According to Abraham and Lorentz, the total energy and momentum of the field were to be calculated from

$$P^{AL}_\mu = -\frac{i}{c}\iiint_{t=\text{const.}} T_{\mu 4}\,d^3x. \tag{5.19}$$

If one commits the further mistake (as also pointed out by Kwal) of using on the one hand the covariant tensor transformation (5.15) to express $T_{\mu\nu}$ in terms of $T^0_{\mu\nu}$, and on the other hand the Lorentz contraction $d^3x^0(1-\beta^2)^{\frac{1}{2}} = d^3x$ which does not represent a covariant transformation at all (under such a relation d^3x is neither a scalar nor a component of vector), one gets the following result:

$$P^{\mathrm{AL}}_{\mu} = \left(\frac{\gamma}{c}\beta\left[\iiint(-T^0_{44})\,d^3x^0 + \iiint T^0_{11}\,d^3x^0\right],\ 0,\right.$$
$$\left. 0,\ i\frac{\gamma}{c}\iiint(-T^0_{44}+\beta^2T^0_{11})\,d^3x^0\right). \tag{5.20}$$

Now
$$\iiint T^0_{11}\,d^3x^0 = \tfrac{1}{3}W^0, \tag{5.21}$$
consequently,
$$P^{\mathrm{AL}}_{\mu} = \left(\frac{4}{3}\gamma u\frac{W^0}{c^2},\ 0,\ 0,\ i\gamma\frac{W^0}{c}(1+\tfrac{1}{3}\beta^2)\right). \tag{5.22}$$

In the rest system this yields
$$P^{0\mathrm{AL}}_{\mu} = \left(0,\ 0,\ 0,\ i\frac{W^0}{c}\right). \tag{5.23}$$

However (5.22) is not obtainable from (5.23) by a four-vector transformation as is the case with (5.17).

The result (5.22) was linked with the instability of a purely electric charge. Because of electrostatic repulsion an electron cannot be stable unless it is held together by some other forces of a non-electromagnetic nature (Poincaré stresses). In the present context, the fact that without binding forces an electron cannot be stable comes out if we consider the integrated energy momentum tensor. It will be sufficient to consider an electron at rest. We get

$$\iiint T^0_{\mu\nu}\,d^3x^0 = \begin{bmatrix} \frac{1}{3}W^0 & 0 & 0 & 0 \\ 0 & \frac{1}{3}W^0 & 0 & 0 \\ 0 & 0 & \frac{1}{3}W^0 & 0 \\ 0 & 0 & 0 & -W^0 \end{bmatrix}. \tag{5.24}$$

The fact that the first three diagonal elements of the stress tensor part do not vanish indicates that a stress exists in the nature of a pressure which is uncompensated. Poincaré introduced an additional non-electromagnetic tensor which would just compensate these three terms and as can be seen from (5.20) this will also have the additional effect of abolishing the terms which cause the factors $\frac{4}{3}$ and $(1+\frac{1}{3}\beta^2)$ to appear in (5.22), instead of 1 in each term. In this way Poincaré not only accounts for the stability of the electron but also re-establishes the four-vector character of the total energy and momentum (i.e. electromagnetic+non-electromagnetic). This state of affairs has greatly contributed to lending credence to the theory of Abraham, Lorentz, and Poincaré. However, nice as the result may be, it is based on a wrong argument and is therefore unacceptable.

It is clear that the stability problem has nothing to do with the covariance property of the energy and momentum. If we follow Poincaré and introduce a 'matter tensor' to compensate for the electrostatic pressure, the required field will have to transform like the electromagnetic field, for otherwise the compensation will not hold for every velocity of the electron. This field will contribute in exactly the same way to the energy and momentum as the electromagnetic field. In other words the contribution from the matter field will by itself also form a four-vector. Experimentally we cannot distinguish between the two contributions.

6. The Equation of Motion of an Electron in the Quasi-Stationary Case

Having established the four-vector character of the energy and momentum of the field of an electron, we will now proceed with a discussion of its equation of motion. Electrons interact with electromagnetic fields and the strength of interaction is proportional to the electric charge and the field. This brings us to the first difficulty encountered when attempting to derive the equation of motion. The field is the total field, that is the external field which originates from other charges and currents and which is assumed known, and the self-field. This latter depends on the motion of the electron and its internal structure (the model). The motion is not known in advance: it is yet to be determined from the equation of motion. Consequently, it might appear that an equation of motion in the strict and conventional sense may not exist. On the other hand, experimentally, we know that a given external force will produce a definite motion so that the theory is faced with the task of reducing the self-field into an effect on certain parameters and terms which enter in the equation of motion, leaving the external field only to act as the force. We shall see that in the Lorentz theory the self-field contributes to the mass of the electron and also accounts for terms which have some similarity with the impedance which is encountered in circuit theory, particularly in aerial-theory. The self-field gives rise to radiation loss and to a term which causes a certain amount of energy to leave the electron and to return to it after a complete cycle. The first is an irreversible process, the second is reversible. To reduce the self-field into these effects it is necessary to make some assumptions about the shape and size of the electron. It is assumed that the electron is sufficiently small for the external field to be practically constant over the extension of the electron. Often the passage is also made to a point electron. This passage produces a divergence but it is possible to circumvent this difficulty by a renormalization procedure which leaves all the observables of motion and radiation unaltered without destroying relativistic covariance. It must be mentioned at this stage that in certain theories which treat the electron as a point charge, it is assumed that the electron does not interact with itself at all and that the field of the electron only produces the interaction with other charges. In particular, as already mentioned, Wheeler and Feynman have elaborated on a theory in which radiation from an electron can take place only if other charges are present to absorb the radiation. On the basis of this theory an equation of motion for an electron can be derived only if an adequate absorber is also introduced (the absorber theory). We shall deal with this aspect briefly later on.

The first elaborate theory of an electron of finite size was produced by Abraham who considered the electron to be a rigid spherical particle. Abraham attempted to explain the whole mass of the electron as due to its self-field (the reaction of the field of the electron) and accepted the rigidity and the fact that the electron holds together in spite of the electrostatic repulsion between its parts, in the same way as we deal in ordinary dynamics of rigid bodies with their motion, without inquiring how the rigid body is held together. This separation of the dynamics of the electron from the problem of its internal stability is forced upon us to this day. However, Abraham's theory could not be relativistic because the concept of a rigid body with its six degrees of freedom cannot be reconciled with the theory of relativity. For one reason, a rigid body would enable us to transmit instant signals from one locality to another. Another non-relativistic feature in Abraham's theory was the assumption that the electron remained spherical while in motion. An improvement on Abraham's theory in this respect was achieved by Lorentz who took into account the Lorentz–Fitzgerald contraction, so that a spherical electron became an oblate ellipsoid while in motion. However, this does not reconcile the theory of relativity with the concept of a rigid body which would enable us to attach an invariant significance to simultaneity and hence to a hyperplane $t =$ const., in direct contradiction to the Lorentz transformation. As a consequence of this, in Abraham's theory as well as in that of Lorentz, the calculation led to a result which is not acceptable in a relativistic theory. It led to an electromagnetic mass which is given by $\frac{4}{3}W^0/c^2$, where W^0 is the electrostatic energy, instead of W^0/c^2. This error is similar to the one which was discussed in the previous section in connexion with the energy-momentum four-vector. Hence, a relativistic theory of an electron of finite extension cannot be achieved by considering it as a rigid body, and a Lorentz contraction cannot remedy the situation. The only possible relativistic treatment of an electron of finite size can be achieved by considering it as a region of charge of a given shape, not at time $t =$ const. but over a hypersurface $t(x, y, z)$ which bears an invariant relationship to the world-tube of the electron. This relationship must also be maintained in the limit when the electron is taken to be a point charge and the world tube becomes a world line. The first to realize this was Fermi, as far back as 1922, but Fermi's paper went somehow into oblivion. Also it derived only the mass term and not the complete equation of motion including the radiative reaction. Fermi obtained for the electromagnetic mass the value W^0/c^2.

A relativistic equation of motion for a point electron was first derived by Dirac in 1938. It involved the calculation of a self-field in the vicinity of the electron. According to Dirac's theory, the electrostatic energy contributes to the mass the amount W^0/c^2 without a factor $\frac{4}{3}$, as in Fermi's theory. But somehow this result and its close relation to the energy-momentum vector did not get sufficient attention. Perhaps this was due to the fact that one of Dirac's main objects was to devise a method of subtracting the infinite electrostatic self- mass without violating relativistic covariance and without interference with the observed radiation. In Dirac's theory the remaining mass then is not electromagnetic and is not deduced but taken as observed.

The simplest approach to the problem is to postulate both the required interaction between the field and the electron as well as Maxwell's equations (as was done so far for the latter) without seeking a variational principle. The interaction will be assumed to result in a force which is given as follows:

$$F_\kappa = \frac{e}{c} F_{\kappa\lambda} U_\lambda. \tag{6.1}$$

$F_{\kappa\lambda}$ is the sum of the external field $F^{\text{in}}_{\kappa\lambda}$ and a certain average of the self-field $F^s_{\kappa\lambda}$. For an electron at rest, or when in uniform motion, this average field is zero, as a result of the symmetry of the self-field. When the electron is accelerated the average field is no longer zero.

In (6.1) it is already assumed that the electron is held together by some non-electric force which counteracts the electrostatic repulsion between the various charge elements. We thus assume that the world tube is of constant width.

A more fundamental equation than (6.1) which can be derived from a variational principle (see H. Weyl, *Space, Time, and Matter*) is the following:

$$f_\kappa = \frac{1}{c}\rho_0 F_{\kappa\lambda} U_\lambda, \tag{6.2}$$

where f_κ is a force density, ρ_0 is the rest density of the charge, and $F_{\kappa\lambda}$ the actual total field. f_κ is obtained from a force by dividing by the rest volume. Similarly, to obtain ρ_0 we divide total charge by rest volume. ρ_0 is a scalar and f_κ a four-vector.

If the electron possesses a 'mechanical mass' m_{0m}, its acceleration will give rise to an additional inertial force $-m_{0m}\dot{U}$ and (6.1) will give

$$\frac{e}{c}F_{\kappa\lambda}U_\lambda - m_{0m}\dot{U}_\kappa = 0, \tag{6.3}$$

where the dot indicates differentiation with respect to proper time. On the other hand, if one adopts the Thomson–Abraham point of view that the mass of the electron is purely of electromagnetic origin ($m_{0m} = 0$), we have instead of (6.3)

$$\frac{e}{c}F_{\kappa\lambda}U_\lambda = 0. \tag{6.4}$$

Until recently, it was thought that on relativistic grounds Abraham's idea of a purely electromagnetic mass is untenable. However, the situation has altered since the correction of the $\frac{4}{3}$ factor was achieved and the four-vector character of the energy and momentum of the field proved. It only remains to see whether any future solution of the stability problem of the electron will entail an additional mass term. The idea of a point electron and the procedure of renormalization which it requires (such as Dirac's subtraction of the electromagnetic self-energy) is also, in all probability, provisional.

The first step for deriving the equation of motion consists in getting a relativistic expression for the resultant force which arises from the self-field or the average self-field. Each element of charge and current interacts with all the other elements, and because of retardation we must consider the electron not at a given time t but over a space-time extension, so that the averaging will involve a four-dimensional integral over a region in the Minkowski space. Each element of the charge traces a world line and an extended electron will thus produce a world tube (see Fig. 27). The region of integration will be bounded by the invariant wall of the tube and by two end surfaces, σ_1 and σ_2. It is essential not only that these surfaces be adopted by all observers, but also, as pointed out first by Fermi, that they bear an invariant relationship to the world tube. This can be achieved by taking the end pieces σ_1 and σ_2 to be orthogonal to the world tube. The mistake committed by Abraham, Lorentz, and many others can be traced to their choosing surfaces $t_1 =$ const. and $t_2 =$ const. (as indicated in Fig. 27) and taking the different surfaces $t'_1 =$ const. and $t'_2 =$ const. when a change to another frame of reference is performed. Such a procedure would be justified if simultaneity was absolute and a rigid body could exist. As pointed out by Fermi, the nearest approach to a rigid body can be achieved if at each moment we transform the electron to rest and assume that in the surfaces $t_0 =$ const. the electron has the same shape and size. This leads to the surfaces which are perpendicular to the tube and have the same cross-section.

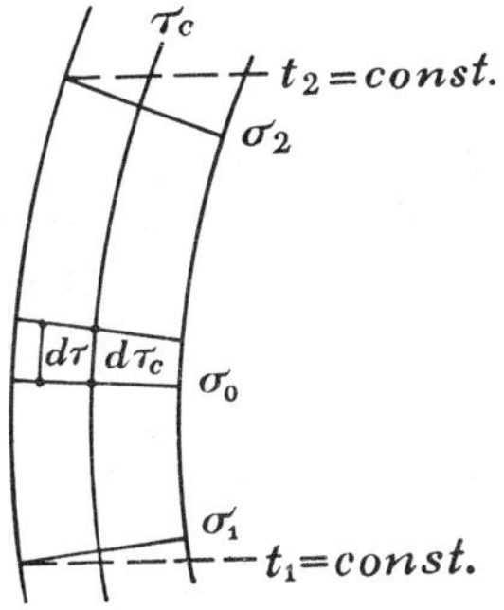

FIG. 27

When dealing with an electron of small size, it is convenient to introduce a central world line τ_c relative to which all the other lines will be referred. With the help of this central line the resultant self-force may be postulated as follows:

$$F^s_\kappa = \frac{d}{d\tau_c} \iiint\limits_{\sigma_1}^{\sigma_2}\!\!\int c^{-1} J_\lambda F^s_{\kappa\lambda}\, d\sigma\, d\tau, \tag{6.5}$$

where $d\sigma$ is the invariant magnitude of a surface element on a surface σ and $c\,d\tau$ an invariant line element perpendicular to σ. We have

$$c\, d\sigma . d\tau = d^4 x_\kappa. \tag{6.6}$$

A great simplification in the problem of deriving an equation of motion is achieved if we consider a quasi-stationary motion, in which the acceleration is small and the radiation may be neglected. The situation is parallel to the one encountered in the theory of quasi-stationary currents. To get to the quasi-stationary limit, two ways are open to us. The first, much in line with Fermi, is to work out the integral in (6.5) with the permissible approximations under quasi-stationary conditions. The other way is to make use of the fact that at each moment and corresponding velocity, the field of the electron yields an energy-momentum vector. Under quasi-stationary conditions we may assume that all the external force is doing is to alter the energy-momentum vector at a rate which is proportional to the external force. This method was not available to Fermi in 1922 because at that time the four-vector character of the energy and momentum of the field of an electron was not yet recognized. With this knowledge at our disposal, we get under quasi-stationary conditions:

$$\frac{dP_\kappa}{d\tau} = (m_{0e}+m_{0m})\frac{dU_\kappa}{d\tau} = \frac{e}{c}F^{\text{in}}_{\kappa\lambda}U_\lambda, \tag{6.7}$$

where m_{0e} depends on the electrostatic energy W^0 as follows:

$$m_{0e} = \frac{W^0}{c^2} \tag{6.8}$$

and m_{0m} is any mechanical mass which the electron may possess.

Following Fermi,† we evaluate the integral (6.5) to the relevant approximation as follows. We must first express the dependance of $d\tau$ on $d\tau_c$ in terms of the separation between the world lines τ and τ_c. Let R be the radius of curvature of a world line τ and let R_c be this radius for the central world line. This radius is closely related to the acceleration

$$\frac{1}{R_c^2} = \frac{(dx_\mu/d\tau)^2(d^2x_\mu/d\tau^2)^2-(dx_\mu/d\tau\;\,d^2x_\mu/d\tau^2)^2}{(\dot{x}_\mu^2)^3} = \frac{U_\mu^2\,\dot{U}_\mu^2-(U_\mu\,\dot{U}_\mu)^2}{(U_\mu^2)^3}. \tag{6.9}$$

On using

$$U_\mu^2 = -c^2, \qquad U_\mu\,\dot{U}_\mu = 0, \tag{6.10}$$

we get

$$\frac{1}{R_c^2} = \frac{\dot{U}_\mu^2}{c^4}, \tag{6.11}$$

$$R_c = \frac{c^2}{A}, \tag{6.12}$$

where

$$A = +(\dot{U}_\mu^2)^{\frac{1}{2}} \tag{6.13}$$

is the magnitude of the acceleration. Let κ_μ be a space-like four-vector along the radius of curvature with magnitude A/c^2. Then

$$\kappa_\mu = -\frac{\dot{U}_\mu}{c^2} \tag{6.14}$$

† The average self-force as defined in (6.5) is due to the author and not to Fermi. Fermi based his calculation on an action principle for a continuously distributed charge which retains its shape in the rest system, and argued that variations which consist of shifts of the space xyz along planes t = const. are not consistent with the theory of relativity under the conditions of the problem, and that instead the variations should consist of shifts of σ-surfaces which leave these perpendicular to the world lines. One consequence of this condition is that the region over which the action is integrated must be bounded by surfaces σ_1 and σ_2 as indicated in Fig. 27. If the action is integrated over a region bounded by the surfaces t_1 = const. and t_2 = const. and the variations are along surfaces t = const., one gets the factor $\frac{4}{3}$. In the correct method the factor is 1. (See also Chapter 4, section 1, the variation principle.) In a field theory the choice of σ_1 and σ_2 is arbitrary as long as they are space surfaces. In our present case we are dealing with an electron which retains its shape in its rest system and the perpendicularity of the surfaces to the world lines is essential.

is the curvature of the world line. Let π_μ be a space-like vector joining the intersections of τ_c and τ with σ_0. $c^2\kappa_\mu/A$ is a unit vector and as can be seen from Fig. 28,

$$R = R_c - \frac{\kappa_\mu \pi_\mu c^2}{A} \tag{6.15}$$

and to a sufficient approximation

$$\frac{d\tau}{d\tau_c} = \frac{R}{R_c} = 1 - \frac{\kappa_\mu \pi_\mu c^2}{AR_c} = 1 - \kappa_\mu \pi_\mu \tag{6.16}$$

so that
$$d\tau = d\tau_c(1-\kappa_\mu \pi_\mu) = d\tau_c\left(1+\frac{\dot{U}_\mu \pi_\mu}{c^2}\right). \tag{6.17}$$

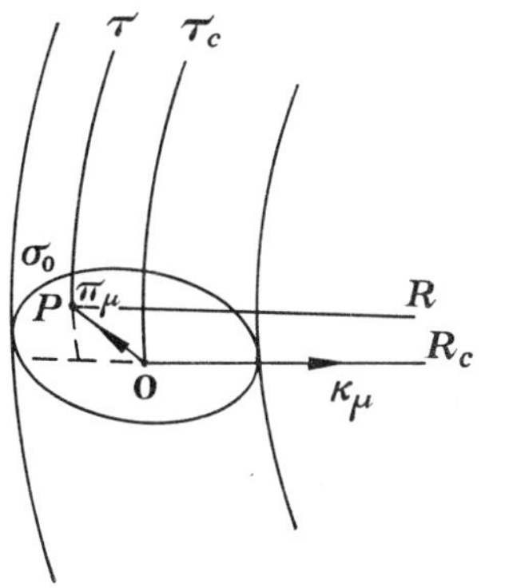

FIG. 28

Consequently, (6.5) becomes

$$F^s_\kappa = \frac{d}{d\tau} \int\int\int_{\sigma_0}^{\sigma}\int c^{-1} J_\lambda F^s_{\kappa\lambda} \left(1+\frac{\dot{U}_\mu \pi_\mu}{c^2}\right) d\sigma d\tau_c, \tag{6.18}$$

$$F^s_\kappa = \int\int\int_\sigma c^{-1} J_\lambda F^s_{\kappa\lambda} \left(1+\frac{\dot{U}_\mu \pi_\mu}{c^2}\right) d\sigma. \tag{6.19}$$

This expression becomes simpler if the rest system of the electron is introduced ($d\tau \to dt$, $\dot{U}_\mu \pi_\mu \to \dot{\mathbf{u}}\boldsymbol{\pi}$, $d\sigma \to dx^0 dy^0 dz^0$, $J_1 = J_2 = J_3 = 0$, $J_4 = ic\rho$, $F_{k\lambda} \to \mathbf{E}$, $\rho^0 dx^0 dy^0 dz^0 = de$):

$$(\mathbf{F}^s)_{\text{rest system}} = \int \mathbf{E}^s \left(1+\frac{\dot{\mathbf{u}}\boldsymbol{\pi}}{c^2}\right) de. \tag{6.20}$$

Fermi was the first to derive this formula and pointed out that it corrects the result obtained by Lorentz and repeated by many others with regard to the factor $\frac{4}{3}$ for the electromagnetic mass. If instead of the surfaces σ_0 and σ chosen by Fermi one takes $t_0 =$ const. and $t =$ const., we have $d\tau = d\tau_c$ (instead of (6.17)) and the self-force comes out to be

$$(\mathbf{F}^s)_{\text{rest system}} = \int \mathbf{E}^s \, de \quad \text{(according to Lorentz).} \tag{6.21}$$

It is this integral which yields

$$(\mathbf{F}^s)_{\text{rest system}} = \mathbf{F} = -\frac{4}{3}\frac{W^0}{c^2}\dot{\mathbf{u}}. \tag{6.22}$$

Fermi showed that if we restrict ourselves to terms linear in $\dot{\mathbf{u}}$ the additional term appearing in (6.20) is given by

$$\int \mathbf{E}^s \frac{\dot{\mathbf{u}}\boldsymbol{\pi}}{c^2} \, de = \frac{W^0}{3c^2}\dot{\mathbf{u}}, \tag{6.23}$$

so that

$$\mathbf{F}^s = -\frac{W^0}{c^2}\dot{\mathbf{u}} = -m_{0e}\dot{\mathbf{u}}. \tag{6.24}$$

From this result in the rest system it follows further that in an arbitrary inertial frame we shall have

$$F^s_\kappa = -m_{0e}\dot{U}_\kappa \tag{6.25}$$

and (6.8) will hold when an external force is present. The proof of (6.23) proceeds as follows. $\mathbf{E}^s$, which appears in the integral, is the sum of two forces: (1) the Coulomb force

$$\frac{\mathbf{r}}{r^3} \, de de',$$

where r is the distance between the charge elements de and de', and (2) a force which is proportional to the acceleration $\dot{\mathbf{u}}$ (see next section, (7.49)). Hence, when this part is multiplied by $\dot{\mathbf{u}}$ it will give a contribution which is proportional to $\dot{\mathbf{u}}^2$. To obtain the mass term we are only interested in terms with $\dot{\mathbf{u}}$ as a factor, hence we disregard the second part of $\mathbf{E}^s$ and we are left with the evaluation of

$$\mathbf{I} = \frac{1}{c^2}\iint \frac{\mathbf{r}}{r^3}(\dot{\mathbf{u}}\boldsymbol{\pi}) \, de de'. \tag{6.26}$$

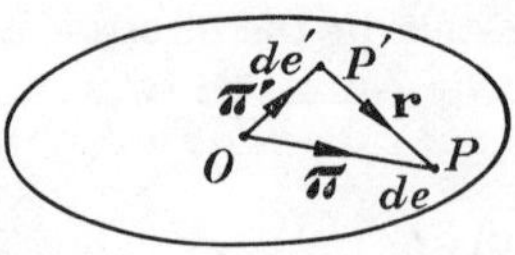

Fig. 29

The value of this integral is not changed if de and de' are exchanged. $\mathbf{r}$ becomes $-\mathbf{r}$ and $\boldsymbol{\pi}$ which points from O to P (see Fig. 29) becomes a vector $\boldsymbol{\pi}'$ which points from O to P'. Hence,

$$\mathbf{I} = \frac{1}{c^2}\iint \frac{\mathbf{r}}{r^3}(\dot{\mathbf{u}}\boldsymbol{\pi}) \, de de' = -\frac{1}{c^2}\iint \frac{\mathbf{r}}{r^3}(\dot{\mathbf{u}}\boldsymbol{\pi}') \, de de'. \tag{6.27}$$

Taking the arithmetic mean of these two expressions and noting that $\boldsymbol{\pi}-\boldsymbol{\pi}' = \mathbf{r}$, we get

$$\mathbf{I} = \frac{1}{2c^2}\iint \frac{\mathbf{r}}{r^3}[\dot{\mathbf{u}}(\boldsymbol{\pi}-\boldsymbol{\pi}')]\,dede' = \frac{1}{2c^2}\iint \frac{\mathbf{r}}{r^3}(\mathbf{r}\dot{\mathbf{u}})\,dede'. \tag{6.28}$$

Consider the x-component

$$I_x = \frac{1}{2c^2}\iint \frac{(x-x')}{r^3}[\dot{u}_x(x-x')+\dot{u}_y(y-y')+\dot{u}_z(z-z')]\,dede'. \tag{6.29}$$

Here we may replace $(x-x')^2$, $(x-x')(y-y')$, $(x-x')(z-z')$ by their average in all directions for any given r. These averages are $\frac{1}{3}r^2$, 0, 0 respectively. Hence

$$I_x = \frac{1}{6c^2}\iint \frac{dede'}{r} = \frac{W^0}{3c^2}u_x \tag{6.30}$$

and (6.23) follows.

Eqn. (6.7) for the quasi-stationary case, which may be expressed as

$$\frac{dP_\kappa}{d\tau} = m_0\frac{dU_\kappa}{d\tau} = \frac{e}{c}F^{\text{in}}_{\kappa\lambda}U_\lambda, \tag{6.31}$$

where m_0 is the total rest mass of the electron, is used whenever radiation is negligible, and has many practical applications. In particular it is used in all methods for determining the observable rest mass m_0, in which $F_{k\lambda}$ is a static or quasi-static electromagnetic field.

Expressed in terms of a Newton force (6.31) becomes

$$m_0\frac{dU_\kappa}{dt} = \left(1-\frac{u^2}{c^2}\right)^{\frac{1}{2}}\frac{e}{c}F_{\kappa\lambda}U_\lambda, \tag{6.32}$$

where $F_{\kappa\lambda} = F^{\text{in}}_{\kappa\lambda}$. In the remainder of this section the superfix 'in' is dropped.

For the three space components $\kappa = x, y, z$ we thus get

$$m_0\frac{dU_a}{dt} = \frac{e}{c}F_{a\lambda}u_\lambda, \tag{6.33}$$

where $$u_\lambda = \left(\frac{dx}{dt},\frac{dy}{dt},\frac{dz}{dt}, ic\right) = (\mathbf{u}, ic). \tag{6.34}$$

Inserting for $F_{a\lambda}$ the corresponding expressions in terms of $\mathbf{E}$ and $\mathbf{H}$ we get

$$m_0\frac{dU_a}{dt} = \frac{e}{c}[\mathbf{u},\mathbf{H}]+e\mathbf{E}. \tag{6.35}$$

The fourth component of (6.32) yields

$$\frac{dW}{dt} = e\mathbf{E}.\mathbf{u} = e\left(E_x\frac{dx}{dt}+E_y\frac{dy}{dt}+E_z\frac{dz}{dt}\right). \tag{6.36}$$

In (6.35) the right-hand side is known as the Lorentz force. Eqn. (6.36) expresses the change in energy of the electron as the work done by the electric force on the charge. The magnetic force, according to (6.35), is always perpendicular to **u** and makes no contribution to the energy of the electron.

(6.35) may be expressed entirely in terms of the ordinary velocity

$$m_0\frac{d}{dt}\left(\frac{\mathbf{u}}{(1-u^2/c^2)^{\frac{1}{2}}}\right) = \frac{d}{dt}\left(\frac{m_0\,\mathbf{u}}{(1-u^2/c^2)^{\frac{1}{2}}}\right) = \frac{e}{c}[\mathbf{u},\mathbf{H}]+e\mathbf{E}. \tag{6.37}$$

In this form the equation is readily used in most applications where it is necessary to determine the path of an electron in a given electromagnetic field and when the radiation is negligible.

In (6.31) the electromagnetic field can be expressed in terms of the four-potential

$$A_\kappa = (A_x, A_y, A_z, i\Phi), \tag{6.38}$$

$$F_{\kappa\lambda} = \partial_\kappa A_\lambda - \partial_\lambda A_\kappa.\dagger \tag{6.39}$$

We get

$$\frac{dP_\kappa}{d\tau} = \frac{e}{m_0\,c}(\partial_\kappa A_\lambda - \partial_\lambda A_\kappa)P_\lambda. \tag{6.40}$$

Considering the first component of this equation:

$$\frac{dP_1}{d\tau} = \frac{e}{m_0\,c}\left(\frac{\partial A_1}{\partial x}P_1+\frac{\partial A_2}{\partial x}P_2+\frac{\partial A_3}{\partial x}P_3+\frac{\partial i\Phi}{\partial x}\frac{iW}{c}-\right.$$

$$\left.-\frac{\partial A_1}{\partial x}P_1-\frac{\partial A_1}{\partial y}P_2-\frac{\partial A_1}{\partial z}P_3-\frac{\partial A_1}{ic\partial t}\frac{iW}{c}\right). \tag{6.41}$$

Now

$$\frac{e}{m_0\,c}\left(\frac{\partial A_1}{\partial x}P_1+\frac{\partial A_1}{\partial y}P_2+\frac{\partial A_1}{\partial z}P_3+\frac{\partial A_1}{ic\partial t}\frac{iW}{c}\right)$$

$$= \frac{e}{c}\left(\frac{\partial A_1}{\partial x}U_1+\frac{\partial A_1}{\partial y}U_2+\frac{\partial A_1}{\partial z}U_3+\frac{\partial A_1}{\partial t}\frac{1}{(1-u^2/c^2)^{\frac{1}{2}}}\right)$$

$$= \frac{e}{c}\left(\frac{\partial A_1}{\partial x}U_1+\frac{\partial A_1}{\partial y}U_2+\frac{\partial A_1}{\partial z}U_3+\frac{\partial A_1}{\partial \tau}\right). \tag{6.42}$$

This quantity represents the change of A_1 per unit of proper time as seen by the particle when it moves in the field. (However, A_1 and its change are measured in the frame of reference relative to which the motion of the particle is described.) The expression in brackets is the

† In usual vector notation they are

$$\mathbf{E} = -(\text{grad}\,\phi+\dot{\mathbf{A}}/c), \qquad \mathbf{H} = \text{curl}\,\mathbf{A}.$$

'substantial' change of A_1 in terms of the Minkowski four-velocity. Denoting this change by $dA_1/d\tau$ we get

$$\frac{dA_1}{d\tau} = \frac{\partial A_1}{\partial x}U_1 + \frac{\partial A_1}{\partial y}U_2 + \frac{\partial A_1}{\partial z}U_3 + \frac{\partial A_1}{\partial \tau}. \tag{6.43}$$

Substituting this expression in (6.41) and dealing with the other components in the same way we get

$$\frac{d}{d\tau}\left(P_a + \frac{e}{c}A_a\right) = \frac{e}{m_0 c}\partial_a A_\lambda . P_\lambda \quad (a = 1, 2, 3) \tag{6.44}$$

$$\frac{d}{d\tau}\left(\frac{iW}{c} + \frac{ie\Phi}{c}\right) = \frac{e}{m_0 c}\partial_4 A_\lambda . P_\lambda. \tag{6.45}$$

The last of these equations we can express as

$$\frac{d}{d\tau}\left(\frac{iW}{c} + \frac{ie\Phi}{c}\right) = \frac{e}{ic^2}\left(U_a \frac{\partial A_a}{\partial t} - \frac{W}{m_0 c}\frac{\partial \Phi}{\partial t}\right). \tag{6.46}$$

This equation shows that if the field is stationary so that neither A nor Φ depends explicitly on time,

$$W + e\Phi = \text{const.} \tag{6.47}$$

$e\Phi$ is the potential energy while W is the self-energy of the particle when in motion. The sum of these energies remains constant and represents the total energy of the particle in the field. If the field is not stationary the total energy changes according to (6.46). Similar statements apply to the equation dealing with the momentum. If A_a and Φ are constant in space all the derivatives with respect to x, y, z vanish. In this case $P_a + e(A_a/c)$ will not change in the course of time and we call it the total momentum of the electron. It is composed of a kinematical self-momentum P_1, P_2, P_3 and a potential momentum $(e/c)(A_1, A_2, A_3)$. It is true that if A_a and Φ are constant in space no electromagnetic field exists and we may as well omit A_a and Φ. In general, when A_a and Φ vary in space the generalized momentum $P_a + e(A_a/c)$ will vary as the particle changes position.

Eqns. (6.44) and (6.45) can be condensed to

$$\frac{d}{d\tau}\left(P_\kappa + \frac{e}{c}A_\kappa\right) = \frac{e}{m_0 c}\partial_\kappa A_\lambda . P_\lambda, \tag{6.48}$$

which may be taken as the alternative to (6.31).

Let the generalized momentum $P_\kappa + (e/c)A_\kappa$ be denoted by P_κ^*. Then we have†

$$P_\kappa + \frac{e}{c}A_\kappa = P_\kappa^*, \tag{6.49}$$

† $e = |e|$ for a positron and $-|e|$ for an electron.

therefore
$$U_\kappa \equiv \frac{dx_\kappa}{d\tau} = \frac{P^*_\kappa}{m_0} - \frac{e}{m_0 c} A_\kappa. \tag{6.50}$$
Eqn. (6.48) becomes
$$\frac{dP^*_\kappa}{d\tau} = \frac{e}{m_0 c}\left(P^*_\lambda - \frac{e}{c} A_\lambda\right)\partial_\kappa A_\lambda. \tag{6.51}$$
Eqns. (6.50) and (6.51) can be regarded as the Hamiltonian equations of motion
$$\frac{dx_\kappa}{d\tau} = \frac{\partial H}{\partial P^*_\kappa}, \tag{6.52}$$
$$\frac{dP^*_\kappa}{d\tau} = -\frac{\partial H}{\partial x_\kappa}, \tag{6.53}$$
with the Hamiltonian H given by
$$H = \frac{P^2_\lambda}{2m_0} \equiv \frac{(P^*_\lambda - eA_\lambda/c)^2}{2m_0}, \tag{6.54}$$
and in which only A_λ depends explicitly on x_κ.

The Hamiltonian equations are Lorentz invariant as a result of the invariance of H. H is not the total energy of the electron since the latter is not a scalar.

The above equations satisfy a variational principle
$$\delta I = \delta \int_{\tau_1}^{\tau_2} L(x_\kappa, U_\kappa)\, d\tau = 0, \tag{6.55}$$
where
$$L = \tfrac{1}{2} m_0 U^2_\kappa + \frac{e}{c} U_\kappa A_\kappa \tag{6.56}$$
and the dependence of L on x_κ is through A_κ.

The Euler–Lagrange equations are
$$\frac{d}{d\tau}\frac{\partial L}{\partial U_\kappa} - \frac{\partial L}{\partial x_\kappa} = 0. \tag{6.57}$$
The generalized momenta are
$$P^*_\kappa \equiv \frac{\partial L}{\partial U_\kappa} = m_0 U_\kappa + \frac{e}{c} A_\kappa = P_\kappa + \frac{e}{c} A_\kappa \tag{6.58}$$
and (6.57) yields (6.48).

From
$$P^2_\lambda = -(m_0 c)^2 \tag{6.59}$$
we get
$$\left(P^*_\lambda - \frac{eA_\lambda}{c}\right)^2 = -(m_0 c)^2. \tag{6.60}$$
This yields for the total energy of the particle
$$W \equiv \frac{cP^*_4}{i} = c\left[(m_0 c)^2 + \left(\mathbf{P}^* - \frac{e\mathbf{A}}{c}\right)^2\right]^{\frac{1}{2}} + e\phi, \tag{6.61}$$

where $\mathbf{P}$ is the space part of P_λ and $A_\lambda = (\mathbf{A}, i\phi)$. For small velocities, $(\mathbf{P}^* - e\mathbf{A}/c)^2 \ll m_0^2 c^2$, $\mathbf{P} \to \mathbf{p}$, we get

$$W = \text{K.E.} + \text{P.E.} + m_0 c^2 = \frac{1}{2m_0}\left(\mathbf{P}^* - \frac{e\mathbf{A}}{c}\right)^2 + e\phi + m_0 c^2$$

$$= \frac{\mathbf{p}^2}{2m_0} + e\phi + m_0 c^2. \tag{6.62}$$

It is possible to get a set of canonical equations also by using W as a Hamiltonian instead of (6.54). The resulting equations are relativistic but do not display this property at a glance like (6.52)–(6.54). Using W as a Hamiltonian and letting P_a and x_a be the components of $\mathbf{P}$ and $\mathbf{x}$ respectively, we have

$$\frac{\partial W}{\partial P_a} = \frac{dx_a}{dt}, \tag{6.63}$$

$$\frac{\partial W}{\partial x_a} = -\frac{dP_a^*}{dt}. \tag{6.64}$$

In fact, using $\mathbf{P} = m_0 \mathbf{u}(1 - u^2/c^2)^{-\frac{1}{2}}$, we confirm from (6.63) the relation $u_a = dx_a/dt$:

$$\frac{dx_a}{dt} = \frac{(P_a^* - eA_a/c)c}{[(m_0 c)^2 + (\mathbf{P}^* - e\mathbf{A}/c)^2]^{\frac{1}{2}}}$$

$$= \frac{m_0 u_a c}{(1 - u^2/c^2)^{\frac{1}{2}}[(m_0 c)^2 + (m_0 u)^2/(1 - u^2/c^2)]^{\frac{1}{2}}} = u_a. \tag{6.65}$$

Similarly, (6.64) gives, on using (6.65),

$$\frac{\partial W}{\partial x_a} = \frac{c(P_b - eA_b/c)}{[(m_0 c)^2 + (\mathbf{P}^* - e\mathbf{A}/c)^2]^{\frac{1}{2}}}\left(-\frac{e}{c}\frac{\partial A_b}{\partial x_a}\right) + e\frac{\partial \phi}{\partial x_a} = -e\left(u_b \frac{\partial A_b}{\partial x_a} - \frac{\partial \phi}{\partial x_a}\right). \tag{6.66}$$

Consequently, (6.64) is equivalent to

$$\frac{dP_a^*}{dt} = e\left(u_b \frac{\partial A_b}{\partial x_a} - \frac{\partial \phi}{\partial x_a}\right). \tag{6.67}$$

On using

$$\frac{dA_a}{dt} = \frac{\partial A_a}{\partial t} + u_c \frac{\partial A_a}{\partial x_c} \tag{6.68}$$

and subtracting (after multiplication by e/c) from (6.67), we get

$$\frac{d}{dt}\left(P_a^* - \frac{eA_a}{c}\right) = -\left(e\frac{\partial \phi}{\partial x_a} + \frac{e}{c}\frac{\partial A_a}{\partial t}\right) + \frac{e}{c}\left(u_b \frac{\partial A_b}{\partial x_a} - u_c \frac{\partial A_a}{\partial x_c}\right). \tag{6.69}$$

Hence

$$\frac{dP_a}{dt} = \frac{e}{c}(\partial_a A_\lambda - \partial_\lambda A_a)u_\lambda \quad (u_\lambda) = (u_a, ic) \tag{6.70}$$

and in agreement with (6.33)

$$\frac{dP_a}{dt} = \frac{e}{c} F_{a\lambda} u_\lambda. \tag{6.71}$$

The next objective is to derive an equation of motion for a charged particle without neglecting the radiation. For this purpose it will be necessary to deal at first with the field of a moving point charge.

7. The Field of a Moving Point Charge

In theoretical investigations the concept of a point charge holds an important place and in this section the Liénard–Wiechert potentials of a moving point charge and the corresponding electromagnetic field will be introduced. This will prove useful at a later stage, when dealing with the equation of motion of a charged particle without neglecting the radiation.

Let a charge e be moving with a velocity $\dot{\xi}_\mu$ and assume ρ_0 to be the rest-density of the charge. Then $\rho_0 \dot{\xi}_\mu$ is the four-current density J_μ. For mathematical reasons it is convenient to express this current density with the help of a scalar density function $g(x^s_\kappa - \xi_\kappa)$ of dimension L^{-4} in the four-dimensional world as follows:

$$J_\mu(x^s_\kappa) = ec \int_{-\infty}^{+\infty} \dot{\xi}_\mu g(x^s_\kappa - \xi_\kappa)\, d\tau, \tag{7.1}$$

where τ is the proper time, x^s_κ is a space-time point in the region of the current (the source), and

$$\iiiint g(x^s_\kappa - \xi_\kappa)\, d^3x\, cdt = 1. \tag{7.2}$$

The kernel $g(x^s_\kappa - \xi_\kappa)$ acts like a Green's function for the current density, the source being the central world line of the moving electron. The coordinates along this line are denoted by ξ and the four-velocity is $U_\mu = \dot{\xi}_\mu$. Each element $d\xi_\mu$ is responsible for a current density

$$dJ_\mu(x^s_\kappa) = ceg(x^s_\kappa - \xi_\kappa)\, d\xi_\mu,$$

so that

$$J_\mu(x^s_\kappa) = \int dJ_\mu(x^s_\kappa) = ec \int g(x^s_\kappa - \xi_\kappa)\, d\xi_\mu = ec \int g(x^s_\kappa - \xi_\kappa)\dot{\xi}_\mu\, d\tau. \tag{7.3}$$

For a current which is not strictly filamentary, $g(x^s_\kappa - \xi_\kappa)$ would account for a spreading of the world line into a tube. For a point charge and a filamentary current the scalar g in (7.1) must be a δ^4-function:

$$g(x^s_\kappa - \xi_\kappa) = \delta^4(x^s_\kappa - \xi_\kappa) = \delta(x^s_1 - \xi_1)\delta(x^s_2 - \xi_2)\delta(x^s_3 - \xi_3)\delta(i^{-1}(x^s_4 - \xi_4)), \tag{7.4}$$

so that

$$J_\mu(x^s_\kappa) = ec \int \delta^4(x^s_\kappa - \xi_\kappa)\dot{\xi}_\mu\, d\tau. \tag{7.5}$$

The weight function δ^4 has the effect of producing a current along the world line only where the electron happens to be at time τ. The function $\delta^4(x^s_\kappa - \xi_\kappa)$ is a relativistic invariant since it is zero everywhere except at the world point $x^s_\kappa = \xi_\kappa$ and a Lorentz transformation merely changes the coordinates of the point without changing the definition of the δ-function.

The next task is to evaluate the retarded and advanced potential of the current as given by (7.5) at an arbitrary field point x_κ (see p. 124):

$$A_\mu(x_\kappa) = \frac{1}{c} \iiiint J_\mu(x^s_\kappa) D(x^s_\kappa - x_\kappa)\, d^4x^s$$

$$= e \iiiiint \dot{\xi}_\mu \delta^4(x^s_\kappa - \xi_\kappa) D(x^s_\kappa - x_\kappa)\, d\tau\, d^4x^s. \tag{7.6}$$

Hence
$$A_\mu(x_\kappa) = e \int_{-\infty}^{+\infty} \dot{\xi}_\mu D(x_\kappa - \xi_\kappa)\, d\tau. \tag{7.7}$$

For D we shall use D^r and D^a respectively, as given in (9.61) and (9.62) (p. 124) where for x^s we insert $\xi_\mu(\tau)$. On applying the formula

$$\int \delta[g(\tau)] f(\tau)\, d\tau = \sum_{\tau_0} \int \frac{\delta(\tau - \tau_0)}{|\dot{g}(\tau_0)|} f(\tau)\, d\tau = \sum \frac{f(\tau_0)}{|\dot{g}(\tau_0)|}, \tag{7.8}$$

where τ_0 is a root of $g(\tau) = 0$, to

$$A^r_\mu(x_\kappa) = \frac{e}{2\pi} \int \dot{\xi}_\mu \delta[(x_\kappa - \xi_\kappa)^2] \Theta(x_\kappa - \xi_\kappa)\, d\tau, \tag{7.9}$$

$$A^a_\mu(x_\kappa) = \frac{e}{2\pi} \int \dot{\xi}_\mu \delta[(x_\kappa - \xi_\kappa)^2] \Theta(\xi_\kappa - x_\kappa)\, d\tau, \tag{7.10}$$

and taking into account that τ^r and τ^a make $(x_\kappa - \xi_\kappa)^2 = 0$ and that $\Theta(\pm(x_\kappa - \xi_\kappa))$ blots out anything on the forward and backward cone respectively, we get the following result:

$$A^r_\mu(x_\kappa) = \frac{e}{4\pi} \sum_{\tau=\tau^r,\tau^a} \frac{\dot{\xi}_\mu(\tau) \Theta(x_\kappa - \xi_\kappa)}{|(x_\kappa - \xi_\kappa)\dot{\xi}_\kappa|} = \frac{e}{4\pi} \frac{\dot{\xi}_\mu}{|(x_\kappa - \xi_\kappa)\dot{\xi}_\kappa|}\bigg|_{\text{at } \tau=\tau^r}, \tag{7.11}$$

$$A^a_\mu(x_\kappa) = \frac{e}{4\pi} \sum_{\tau=\tau^r,\tau^a} \frac{\dot{\xi}_\mu(\tau) \Theta(\xi_\kappa - x_\kappa)}{|(x_\kappa - \xi_\kappa)\dot{\xi}_\kappa|} = \frac{e}{4\pi} \frac{\dot{\xi}_\mu}{|(x_\kappa - \xi_\kappa)\dot{\xi}_\kappa|}\bigg|_{\text{at } \tau=\tau^a}. \tag{7.12}$$

The scalar product $(x_\kappa - \xi_\kappa)\dot{\xi}_\kappa$ in (7.11) is negative, while the corresponding product in (7.12) is positive. Hence

$$A^r_\mu(x_\kappa) = -\frac{e}{4\pi} \frac{U_\mu}{R_\kappa U_\kappa}\bigg|_{\text{at } \tau=\tau^r}, \tag{7.13}$$

$$A^a_\mu(x_\kappa) = \frac{e}{4\pi} \frac{U_\mu}{R_\kappa U_\kappa}\bigg|_{\text{at } \tau=\tau^a}, \tag{7.14}$$

where
$$R_\kappa = x_\kappa - \xi_\kappa(\tau^r) \quad \text{in (7.13)}, \tag{7.15}$$
$$R_\kappa = x_\kappa - \xi_\kappa(\tau^a) \quad \text{in (7.14)}. \tag{7.16}$$

Eqns. (7.13) and (7.14) are the Liénard–Wiechert potentials in covariant form. Their expression in terms of **u** and **r** (the radius vector from source point to field point) is obtained as follows:

$$R_\kappa = \{\mathbf{r}, ic(t-t^s)\} = (\mathbf{r}, ir), \tag{7.17}$$

$$\begin{aligned} U_\kappa R_\kappa = \gamma\mathbf{ur} + ic\gamma(x_4-\xi_4) &= \gamma\{\mathbf{ur}+ic(ict-ict^s)\} \\ &= \gamma\{\mathbf{ur}-c^2t+c^2(t\mp r/c)\} \\ &= \gamma(\mathbf{ur}\mp cr). \end{aligned} \tag{7.18}$$

Using again $U_\kappa = (\gamma\mathbf{u}, \gamma ic)$, we get from (7.13) and (7.14)

$$\mathbf{A}(x_\kappa) = \mp\frac{e\mathbf{u}}{4\pi(\mathbf{ur}\pm cr)} = \frac{e\mathbf{u}}{4\pi c(r\mp\mathbf{ur}/c)}, \tag{7.19}$$

$$\phi(x_\kappa) = \mp\frac{ec}{4\pi(\mathbf{ur}\pm cr)} = \frac{e}{4\pi(r\mp\mathbf{ur}/c)}. \tag{7.20}$$

In all these formulae, the upper sign applies to the retarded potentials and the lower sign to the advanced potentials.

The retarded and advanced potentials can be expressed in yet another form if the null vector R_κ is resolved into the sum of a time-like vector and a space-like vector as follows:

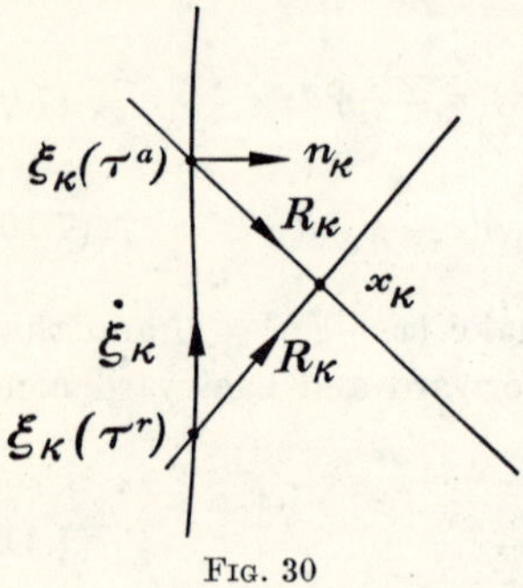

FIG. 30

$$R^r_\kappa = \rho\left(n_\kappa + \frac{1}{c}U_\kappa\right), \tag{7.21}$$

where n_κ is a unit vector ($n_\kappa^2 = 1$, $U_\kappa^2/c^2 = -1$) and ρ is a scalar. It is also assumed that n_κ and U_κ are orthogonal to each other and that n_κ points in the direction of the radius-vector **r**. In the advanced case (see Fig. 30) we have

$$R^a_\kappa = \rho\left(n_\kappa - \frac{1}{c}U_\kappa\right). \tag{7.22}$$

Also
$$R_\kappa U_\kappa = \rho\left(n_\kappa + \frac{1}{c}U_\kappa\right)U_\kappa = -c\rho \quad \text{(ret. case)}, \tag{7.23}$$

$$R_\kappa U_\kappa = \rho\left(n_\kappa - \frac{1}{c}U_\kappa\right)U_\kappa = c\rho \quad \text{(adv. case)}, \tag{7.24}$$

$$R_\kappa n_\kappa = \rho\left(n_\kappa \pm \frac{1}{c}U_\kappa\right)n_\kappa = \rho. \tag{7.25}$$

The scalar ρ may be evaluated in the rest system of the electron, where $U_\kappa^0 = (0, 0, 0, ic)$, $n_\kappa^0 = (\mathbf{r}^0/r^0, 0)$. On using (7.17), it follows that

$$\rho = R_\kappa^0 n_\kappa^0 = (\mathbf{r}^0, ir^0)\left(\frac{\mathbf{r}^0}{r^0}, 0\right) = r^0. \tag{7.26}$$

Hence, ρ is the space distance between the field point and the source point at retarded or advanced time in the rest system of the particle. From (7.13) and (7.14), (7.23) and (7.24) it follows that

$$A_\mu = \frac{1}{4\pi c}\frac{eU_\mu}{\rho}. \tag{7.27}$$

This formula applies to both retarded and advanced potentials but U_μ and ρ must be taken at the retarded and advanced times respectively.

In deriving the field from the Liénard–Wiechert potentials, it must be realized that they are functions of both x_κ and U_κ. Hence,

$$\frac{dA_\mu(x)}{dx_\kappa} = \frac{\partial A}{\partial U_\lambda}\frac{\partial U_\lambda}{\partial x_\kappa}+\frac{\partial A_\mu}{\partial x_\kappa} = \frac{\partial A_\mu}{\partial U_\lambda}\frac{\partial U_\lambda}{\partial \tau}\frac{\partial \tau}{\partial x_\kappa}+\frac{\partial A_\mu}{\partial x_\kappa}. \tag{7.28}$$

Moreover, the field point x_κ and the source point $\xi_\kappa(\tau)$ must always lie on a null vector. This imposes a condition on $\partial\tau/\partial x_\kappa$:

$$\frac{dR_\kappa^2}{d\tau} = 2R_\kappa\frac{dR_\kappa}{d\tau} = 2R_\kappa(\dot{x}_\kappa - \dot{\xi}_\kappa) = 2R_\kappa(\dot{x}_\kappa - U_\kappa) = 0. \tag{7.29}$$

Consequently,
$$R_\kappa \dot{x}_\kappa - \rho\left(n_\kappa \pm \frac{1}{c}U_\kappa\right)U_\kappa = 0. \tag{7.30}$$

Hence,
$$\frac{d\tau}{dx_\kappa} = \mp\frac{R_\kappa}{c\rho}. \tag{7.31}$$

From (7.28) and (7.31) it follows that

$$\frac{dA_\mu}{dx_\nu} = \frac{\partial A_\mu}{\partial x_\nu}+\frac{\partial A_\mu}{\partial U_\lambda}\frac{dA_\lambda}{d\tau}\left(\mp\frac{R_\nu}{c\rho}\right). \tag{7.32}$$

The first term on the right-hand side gives

$$\frac{\partial A_\mu}{\partial x_\nu} = \frac{e}{4\pi c}\frac{\partial(U_\mu/\rho)}{\partial x_\nu} = -\frac{e}{4\pi c}\frac{U_\mu}{\rho^2}\frac{\partial\rho}{\partial x_\nu} = -\frac{e}{4\pi c}\frac{U_\mu}{\rho^2}n_\nu, \tag{7.33}$$

where use is made of†
$$\frac{\partial\rho}{\partial x_\nu} = n_\nu. \tag{7.34}$$

† $\rho = (x_\nu - \xi_\nu)n_\nu$, and (7.34) follows.

The second term in (7.32) gives, apart from the factor $(e/4\pi c)$,

$$\mp\left[\frac{R_\nu}{c\rho^2}\frac{dU_\mu}{d\tau}\pm\frac{R_\nu U_\mu}{c^2\rho^3}\frac{\partial(R_\kappa U_\kappa)}{\partial U_\lambda}\frac{dU_\lambda}{d\tau}\right] = \mp\frac{R_\nu}{c\rho^2}\frac{dU_\mu}{d\tau}-\frac{R_\nu U_\mu}{c^2\rho^3}R_\kappa\frac{dU_\kappa}{d\tau}$$

$$= \mp\frac{R_\nu}{c\rho^2}\frac{dU_\mu}{d\tau}-\frac{R_\nu U_\mu}{c^2\rho^2}\left(n_\kappa\pm\frac{U_\kappa}{c}\right)\frac{dU_\kappa}{d\tau}$$

$$= \mp\frac{R_\nu}{c\rho^2}\frac{dU_\mu}{d\tau}-\frac{R_\nu U_\mu}{c^2\rho^2}n_\kappa\frac{dU_\kappa}{d\tau}. \tag{7.35}$$

Altogether we get

$$\frac{dA_\mu}{dx_\nu} = \frac{e}{4\pi c}\left[-\frac{U_\mu}{\rho^2}n_\nu+\frac{R_\nu}{c\rho^2}\left(\mp\frac{dU_\mu}{d\tau}-\frac{1}{c}U_\mu n_\kappa\frac{dU_\kappa}{d\tau}\right)\right]. \tag{7.36}$$

For the electromagnetic field the result is

$$F_{\mu\nu} = \frac{dA_\nu}{dx_\mu}-\frac{dA_\mu}{dx_\nu}$$

$$= \frac{e}{4\pi c}\left[\frac{1}{\rho^2}(U_\mu n_\nu-U_\nu n_\mu)-\frac{1}{c\rho}n_\mu\left(\frac{1}{c}U_\nu n_\kappa\frac{dU_\kappa}{d\tau}\pm\frac{dU_\nu}{d\tau}\right)+\right.$$

$$\left.+\frac{1}{c\rho}n_\nu\left(\frac{1}{c}U_\mu n_\kappa\frac{dU_\kappa}{d\tau}\pm\frac{dU_\mu}{d\tau}\right)+\frac{1}{c^2\rho}\left(U_\nu\frac{dU_\mu}{d\tau}-U_\mu\frac{dU_\nu}{d\tau}\right)\right]. \tag{7.37}$$

In (7.37) the upper sign applies to F^{ret} and the lower sign to F^{adv}.

Another way of deriving the electromagnetic field was given by Dirac. Instead of using the retarded and advanced potentials in the form (7.27) they are left in integral form (7.9) and (7.10). From these we get

$$\frac{\partial A_\mu(x_\nu)}{\partial x_\nu} = \frac{e}{2\pi}\int\dot{\xi}_\mu\frac{\partial[\delta\{(x_\kappa-\xi_\kappa)^2\}\Theta\{\pm(x_\nu-\xi_\nu)\}]}{\partial(x_\nu-\xi_\nu)}\,d\tau, \tag{7.38}$$

where ∂x_ν has been replaced by $\partial(x_\nu-\xi_\nu)$. In the present calculation the condition for x_ν and ξ_ν lying on the light cone is automatically satisfied through the δ-function so that x_ν and ξ_ν are independent of each other. Also only ξ_ν is dependent on τ but not x_ν. Eqn. (7.38) gives

$$\frac{\partial A_\mu(x_\nu)}{\partial x_\nu} = \frac{e}{\pi}\int\dot{\xi}_\mu\frac{\partial[\delta\{(x_\kappa-\xi_\kappa)^2\}\Theta\{\pm(x_\nu-\xi_\nu)\}](x_\nu-\xi_\nu)}{\partial(x_\nu-\xi_\nu)^2}\,d\tau. \tag{7.39}$$

On using

$$\frac{d(x_\nu-\xi_\nu)}{d\tau} = -\frac{d\xi_\nu}{d\tau} = -\dot{\xi}_\nu \tag{7.40}$$

we get

$$\frac{\partial A_\mu(x_\nu)}{\partial x_\nu} = -\frac{e}{2\pi}\int\frac{\dot{\xi}_\mu(x_\nu-\xi_\nu)}{\dot{\xi}_\kappa(x_\kappa-\xi_\kappa)}\frac{d}{d\tau}[\delta\{(x_\kappa-\xi_\kappa)^2\}\Theta\{\pm(x_\nu-\xi_\nu)\}]\,d\tau. \tag{7.41}$$

Partial integration leads to†

$$\frac{\partial A_\mu(x_\nu)}{\partial x_\nu} = \frac{e}{2\pi}\int \frac{d}{d\tau}\frac{\dot\xi_\mu(x_\nu-\xi_\nu)}{\dot\xi_\kappa(x_\kappa-\xi_\kappa)}\delta\{(x_\kappa-\xi_\kappa)^2\}\Theta\{\pm(x_\nu-\xi_\nu)\}\,d\tau. \tag{7.42}$$

Using (7.8) to evaluate the integral, it follows that

$$\frac{\partial A^r_\mu(x_\nu)}{\partial x_\nu} = \frac{e}{4\pi}\frac{1}{(x_\kappa-\xi_\kappa)\dot\xi_\kappa}\frac{d}{d\tau}\left[\frac{\dot\xi_\mu(x_\nu-\xi_\nu)}{\dot\xi_\kappa(x_\kappa-\xi_\kappa)}\right]_{\tau^r} = \frac{e}{4\pi c\rho}\frac{d}{d\tau}\left(\frac{U_\mu R_\nu}{c\rho}\right)_{\tau^r}, \tag{7.43}$$

$$\frac{\partial A^a_\mu(x_\nu)}{\partial x_\nu} = \frac{e}{4\pi}\frac{1}{(x_\kappa-\xi_\kappa)\dot\xi_\kappa}\frac{d}{d\tau}\left[\frac{\dot\xi_\mu(x_\nu-\xi_\nu)}{\dot\xi_\kappa(x_\kappa-\xi_\kappa)}\right]_{\tau^a} = \frac{e}{4\pi c\rho}\frac{d}{d\tau}\left(\frac{U_\mu R_\nu}{c\rho}\right)_{\tau^a}. \tag{7.44}$$

The retarded and advanced times are the roots of

$$(x_\nu-\xi_\nu)^2 = \mathbf{r}^2-c^2(t-t^s)^2 = 0 \tag{7.45}$$

after transforming t^s into the proper time.

For the field we get

$$F_{\mu\nu}(x_\nu) = \frac{\partial A_\nu}{\partial x_\mu}-\frac{\partial A_\mu}{\partial x_\nu} = \frac{e}{4\pi(x_\kappa-\xi_\kappa)\dot\xi_\kappa}\frac{d}{d\tau}\left[\frac{\dot\xi_\nu(x_\mu-\xi_\mu)-\dot\xi_\mu(x_\nu-\xi_\nu)}{(x_\kappa-\xi_\kappa)\dot\xi_\kappa}\right] \tag{7.46}$$

$$= \frac{e}{4\rho c^2}\frac{d}{d\tau}\left[\frac{U_\nu R_\mu-U_\mu R_\nu}{\rho}\right].$$

This formula applies both for retarded and advanced fields. However, in the first case it is necessary to take the values at τ^r and in the second case at τ^a. On carrying out the differentiation in (7.46), it should be remembered that

$$\frac{dR_\mu}{d\tau} = \frac{d}{d\tau}(x_\mu-\xi_\mu) = -\dot\xi_\mu = -U_\mu, \tag{7.47}$$

and use must be made of (7.22)–(7.25) and of $U_\kappa \dot U_\kappa = 0$. Then (7.37) follows.

Of special interest is the field of an electron in the rest system, when $U_\nu = (0,0,0,ic)$ and $\dot U_\nu$ is not necessarily zero. In (7.37) we must use $\rho = r^0$, $n_\nu = (n_k, 0) = (\mathbf{r}^0/r^0, 0)$, $\dot U_\nu = (\dot U_k, 0) = (\dot{\mathbf{u}}, 0)$, because $\gamma = 1$. We then get

$$i^{-1}F_{4k} = E_k = \frac{e}{4\pi}\left[\frac{n_k}{r^{0^2}}+\frac{1}{c^2 r^0}\left(n_k n_l\frac{dU_l}{d\tau}-\frac{dU_k}{d\tau}\right)\right], \tag{7.48}$$

$$\mathbf{E} = \frac{e}{4\pi}\left[\frac{\mathbf{n}}{r^{0^2}}+\frac{1}{c^2 r^0}[\mathbf{n},[\mathbf{n},\dot{\mathbf{u}}]]\right]. \tag{7.49}$$

Similarly

$$H_m = F_{kl} = \mp\frac{e}{4\pi c^2}\frac{1}{r^0}\left(n_k\frac{dU_l}{d\tau}-n_l\frac{dU_k}{d\tau}\right), \tag{7.50}$$

† The limits of integration are $-\infty$, ξ_ν with $\xi_4 < x_4$ in the retarded case and $\xi_4 > x_4$ in the advanced case.

$$\mathbf{H} = \mp \frac{e}{4\pi c^2} \frac{1}{r^0} [\mathbf{n}, \dot{\mathbf{u}}]. \tag{7.51}$$

In the case of $\mathbf{E}$ the same formula applies in the retarded and advanced field. However, all the quantities must be taken at the retarded and advanced times respectively. For the magnetic field we must also reverse the sign. From (7.49) and (7.51) it follows for Poynting's vector

$$\mathbf{P}^r = c[\mathbf{E}, \mathbf{H}] = \frac{e^2}{16\pi^2 r^{0^2} c^3} [[\mathbf{n}^0, [\mathbf{n}^0, \dot{\mathbf{u}}]],\ [\mathbf{n}^0, \dot{\mathbf{u}}]], \tag{7.52}$$

$$\mathbf{P}^r = \frac{e^2}{16\pi^2 r^{0^2} c^3}\, \mathbf{n}^0 [\mathbf{n}^0, \dot{\mathbf{u}}]^2. \tag{7.53}$$

Similarly,

$$\mathbf{P}^a = -\mathbf{P}^r. \tag{7.54}$$

The components of $\mathbf{P}^r$ and $\mathbf{P}^a$ in the direction of the radius-vector are given by $\mathbf{P}^r \mathbf{n}^0$ and $\mathbf{P}^a \mathbf{n}^0$ and consequently we get for the total flow of energy

$$\mathscr{R} = \frac{e^2}{16\pi^2 r^{0^2} c^3} \int [\mathbf{n}^0, \dot{\mathbf{u}}] r^{0^2}\, d\Omega = \frac{e^2}{16\pi^2 c^3} \int \dot{\mathbf{u}}^2 \sin^2\theta\, d\Omega = \frac{2}{3} \frac{e^2}{4\pi c^3} \dot{\mathbf{u}}^2. \tag{7.55}$$

From the scalar nature of this expression, we must expect it to be an invariant and the only relativistic invariant which will give the same expression in the rest system is

$$\mathscr{R} = \frac{2}{3} \frac{e^2}{4\pi c^3} \dot{U}_\mu^2. \tag{7.56}$$

We shall come across this expression for the energy loss again, when dealing with the equation of motion of the electron. In the advanced case we get the same value for the radiation but it is directed towards the electron and not away from it.

8. The Equation of Motion of an Electron (Radiation Included)

We shall now proceed to deal with the problem of deriving an equation of motion of an electron without neglecting the radiation. It is necessary to distinguish between two types of theories, those attributing to the electron a finite size and those treating it as a point charge. In the following we shall limit ourselves to the much simpler case of a point electron. The first to derive a relativistic equation of motion for a point electron which can claim to be exact, was Dirac (1938). His theory will be presented here with some modifications.

In trying to derive an equation of motion it seems natural to seek an expression for the force exerted by an electron on itself. A similar procedure is common, for instance, in antenna theory where the evaluation of the counter e.m.f. developed along the aerial when a current is maintained yields the total impedance, i.e. both the reactance as well as the radiation resistance (in contrast with the evaluation of Poynting's vector at a great distance from the aerial which gives only the radiation resistance). The reactance contains, apart from the self-inductance, a radiative reactance, which is closely connected with the retardation between a current element and the corresponding field along another element. Whereas the radiation involves a loss of energy which has to be supplied by the source of power, the reactance only causes a certain amount of power to flow out of and into the aerial. However, just as in the case of an infinitely thin aerial where the self-inductance is infinite, so also in the case of a point electron, the self-force contains a diverging term (leading to an infinite mass). Moreover, when dealing strictly with a point electron, it is impossible to carry out an integration over it. For this reason, Dirac did not proceed by directly evaluating the self-force. Instead he made use of the law of conservation of energy and momentum of the electron and the field.

Whether we wish to calculate the self-force or the flow of energy and momentum at the electron, we need an expression for the electromagnetic field in the immediate neighbourhood of the world line of the electron. Now, although (7.37) is an exact expression for the field, both near and far from the world line, it is not suitable for our purpose. The reason is best seen if we think in terms of a self-force, and regard the point charge as a limiting case of a distributed charge. The charge is spread over a three-dimensional region and this region must be selected in a relativistic manner. To consider the whole region at a time t, in a frame in which the electron moves, in other words to select a hyperplane $t = \text{const.}$, is not justifiable. This would amount to attributing absolute meaning to simultaneity. The correct method is to consider the charge over a hyperplane which is orthogonal to the world tube formed by the extended charge. Consequently, what is required is an expression for the field at a point x_κ which forms the tip of a space-like vector which is perpendicular, say, to the central world line of the electron at a given time τ_0. Equation (7.37) does not yield this information directly. It correlates the field with times τ^r and τ^a. As seen from Fig. 30, the nearer we get to the central world line, the more will the point τ_0 lie in the middle between τ^r and τ^a. This raises the question whether it would not be more appropriate to consider as the field due to the electron, the average of the retarded and advanced fields, i.e.

$$F^{+}_{\mu\nu} = \tfrac{1}{2}(F^{r}_{\mu\nu}+F^{a}_{\mu\nu}).$$

Maxwell's equations are invariant with respect to time reversal and the corresponding transformation in $F^{+}_{\mu\nu}$ would be to exchange r and a and this would leave $F^{+}_{\mu\nu}$ invariant. Hence, by using $F^{+}_{\mu\nu}$ we would preserve the symmetry in time which is inherent in Maxwell's equations. It may be argued that by attaching physical significance to the advanced field, the law of causality is being violated. However, this is not the case. In Maxwell's theory fields and currents are inseparable from each other, and to signal one as a cause and the other as an effect is not possible on purely electromagnetic grounds. Consequently, any general relation we are trying to derive from Maxwell's theory should as far as possible preserve the inherent symmetries of the Maxwell theory. Just as there is no question of Maxwell's equations violating the law of causality, the same also applies to any relation derived from them, whether symmetric or not. A causal relationship can be considered only if cause and effect can be localized in space-time, and arises only when considering solutions of the relevant differential equations. However, another difficulty arises. The equation of motion which would result on using $F^{+}_{\mu\nu}$ under explicit use of the law of conservation of energy and momentum, would give no account of the observed radiation and would contain no damping term. $F^{+}_{\mu\nu}$ being the sum of retarded and advanced fields yields no radiation either into the source or away from it and so no gain or loss of energy would be incurred by the electron. $F^{r}_{\mu\nu}$ is a diverging wave and $F^{a}_{\mu\nu}$ a converging wave, both of equal strength, and the total flow of power is zero. We can account for an outflow or an inflow of power, by adding to $F^{+}_{\mu\nu}$ a field $\pm F^{-}_{\mu\nu}$, where $F^{-}_{\mu\nu}$ is a solution of the homogeneous wave equation already introduced in (9.34), Chapter 3 and given by

$$F^{-}_{\mu\nu} = \tfrac{1}{2}(F^{r}_{\mu\nu} - F^{a}_{\mu\nu}).$$

The radiation which goes with $F^{-}_{\mu\nu}$ is equal to that of $F^{r}_{\mu\nu}$ (or $F^{a}_{\mu\nu}$). Experience indicates radiation away from the source, so that we have to add $F^{-}_{\mu\nu}$ to $F^{+}_{\mu\nu}$ and this gives for the total field

$$F_{\mu\nu} = \tfrac{1}{2}(F^{r}_{\mu\nu} + F^{a}_{\mu\nu}) + \tfrac{1}{2}(F^{r}_{\mu\nu} - F^{a}_{\mu\nu}) = F^{r}_{\mu\nu}. \tag{8.1}$$

So we are led to the retarded field, which is often explained as due to the law of causality. A radiation into the world line would necessitate consideration of the field $(F^{+}_{\mu\nu} - F^{-}_{\mu\nu})$ and this would lead to field $F^{a}_{\mu\nu}$. The radiation, whether into the source or away from it, introduces an asymmetry in time, or a time arrow, and this cannot be explained in electromagnetic terms. An explanation of the radiation term was first offered by Wheeler and Feynman and with this we shall deal later.

In obtaining the near field, it will be sufficient to get an expression in the form of a power series in the distance of the field point from the world-line or else in the proper time difference $\tau_0-\tau^r = \tau^a-\tau_0 = \sigma$.

Let x_κ be a field point in the neighbourhood of $\xi_\kappa(\tau_0)$, so that

$$x_\kappa = \xi_\kappa(\tau_0)+\pi_\kappa, \tag{8.2}$$

where π_κ is orthogonal to the world line of the electron at $\xi_\kappa(\tau_0)$,

$$\dot{\xi}_\kappa(\tau_0)\pi_\kappa = U_\kappa(\tau_0)\pi_\kappa = 0. \tag{8.3}$$

To each point along the direction of π_κ belongs a certain retarded proper

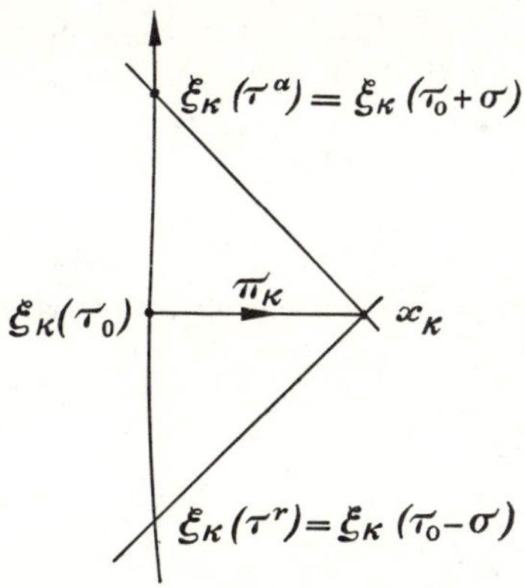

FIG. 31

time $\tau_0-\sigma$ and an advanced time $\tau_0+\sigma$. π_κ and $c\sigma$ are of the same order and as σ tends to zero so does π_κ. The following relation holds:

$$[\xi_\kappa(\tau_0)+\pi_\kappa-\xi_\kappa(\tau_0\mp\sigma)]^2 = 0, \tag{8.4}$$

which expresses that $\xi_\kappa(\tau_0\mp\sigma)$ are on the light cone through x_κ. Expanding $\xi_\kappa(\tau_0\mp\sigma)$, we get

$$x_\kappa-\xi_\kappa(\tau_0\mp\sigma) = x_\kappa-\left(\xi_\kappa\mp\sigma\dot{\xi}_\kappa+\frac{\sigma^2}{2}\ddot{\xi}_\kappa\mp\frac{\sigma^3}{6}\dddot{\xi}_\kappa+\ldots\right), \tag{8.5}$$

where to simplify the writing, it is not indicated explicitly that ξ_κ, U_κ, $\dot{U}_\kappa$, etc., are to be taken at τ_0. On the other hand, any quantity at $\tau_0\mp\sigma$ will be indicated explicitly. For the expansion of the four-velocity we get:

$$\dot{\xi}_\kappa(\tau_0\mp\sigma) = U_\kappa\mp\sigma\dot{U}_\kappa+\tfrac{1}{2}\sigma^2\ddot{U}_\kappa\mp\tfrac{1}{6}\sigma^3\dddot{U}_\kappa+\ldots. \tag{8.6}$$

Since τ and σ are real and it would be inconvenient to have dt and $d\tau$ (or $d\sigma$) of opposite sign, we must use for the invariant metrical form the expression $c^2\tau^2 = c^2t^2-x^2-y^2-z^2$ and not $x^2+y^2+z^2-c^2t^2$. In a pseudo-euclidean space we must then use ix, iy, iz, ct as coordinates. With these, the scalar product x_μ^2 (or of any other four-vector) would have the opposite sign from the one obtained on using x, y, z, ict. It follows that the square of a time-like vector is now positive and that of a space-like vector, negative. Hence

$$U_\kappa^2 = c^2, \qquad \pi_\kappa^2 = -\epsilon^2 \quad (\epsilon > 0). \tag{8.7}$$

From (8.5) and (8.6) we get

$$(x_\kappa - \xi_\kappa)\dot{\xi}_\kappa|_{\tau_0 \mp \sigma} = \pm\sigma(c^2 - \pi_\kappa \dot{U}_\kappa) + \tfrac{1}{2}\sigma^2 \pi_\kappa \ddot{U}_\kappa \mp \tfrac{1}{6}\sigma^3 \dot{U}_\kappa^2. \tag{8.8}$$

Using (7.46) for the field, it is necessary to have the following expansion:

$$[(x_\kappa - \xi_\kappa)\dot{\xi}_\kappa]^{-1}_{\tau_0 \mp \sigma} = \frac{1}{\mp\sigma(c^2 - \pi_\kappa \dot{U}_\kappa)}\left(1 \mp \frac{1}{2c^2}\sigma\pi_\kappa \ddot{U}_\kappa + \frac{1}{6c^2}\sigma^2 \dot{U}_\kappa^2 \ldots\right). \tag{8.9}$$

From the above follows

$$\dot{\xi}_\mu(x_\nu - \xi_\nu) - \dot{\xi}_\nu(x_\mu - \xi_\mu)$$
$$= U_\mu \pi_\nu \mp \sigma \dot{U}_\mu \pi_\nu - \tfrac{1}{2}\sigma^2 \dot{U}_\mu U_\nu + \tfrac{1}{2}\sigma^2 \ddot{U}_\mu \pi_\nu \pm \tfrac{1}{3}\sigma^3 \ddot{U}_\mu U_\nu - (\nu, \mu), \tag{8.10}$$

where (ν, μ) indicates the same terms again but with μ and ν exchanged. On using (8.9):

$$\frac{\dot{\xi}_\mu(x_\nu - \xi_\nu) - \dot{\xi}_\nu(x_\mu - \xi_\mu)}{\dot{\xi}_\kappa(x_\kappa - \xi_\kappa)}$$
$$= \frac{1}{c^2 - \pi_\kappa \dot{U}_\kappa}\left(\sigma^{-1} U_\mu \pi_\nu - U_\mu \pi_\nu \mp \tfrac{1}{2}\sigma \dot{U}_\mu U_\nu \pm \tfrac{1}{2}\sigma \ddot{U}_\mu \pi_\nu + \tfrac{1}{3}\sigma^2 \ddot{U}_\mu U_\nu - \right.$$
$$\left. - \frac{1}{2c^2}\pi_\kappa \ddot{U}_\kappa U_\mu \pi_\nu + \frac{1}{6c^2}\sigma \dot{U}_\kappa^2 U_\mu \pi_\nu\right) - (\nu, \mu). \tag{8.11}$$

This expression must now be differentiated with respect to $\tau = \tau_0 \mp \sigma$. This can be replaced by $-d/d\sigma$ in the retarded case and $d/d\sigma$ in the advanced case. For the retarded case:

$$\frac{d}{d\tau}\frac{\dot{\xi}_\mu(x_\nu - \xi_\nu) - \dot{\xi}_\nu(x_\mu - \xi_\mu)}{\dot{\xi}_\kappa(x_\kappa - \xi_\kappa)}$$
$$= \frac{1}{c^2 - \pi_\kappa \dot{U}_\kappa}\left(-\sigma^{-2} U_\mu \pi_\nu - \tfrac{1}{2}\dot{U}_\mu U_\nu + \tfrac{1}{2}\ddot{U}_\mu \pi_\nu + \tfrac{2}{3}\sigma \ddot{U}_\mu U_\nu + \right.$$
$$\left. + \frac{1}{6c^2}\dot{U}_\kappa^2 U_\mu \pi_\nu\right) - (\nu, \mu). \tag{8.12}$$

To obtain the electromagnetic field, this expression must be multiplied again by (8.8) and by $e/4\pi$. This gives, in terms of order σ^{-2}, σ^{-1}, σ^0, where for instance, $\sigma^{-3}\pi_\nu$ is of order σ^{-2}:

$$F_{\mu\nu}^{\text{ret}} = \frac{e}{4\pi(c^2 - \pi_\kappa \dot{U}_\kappa)^2}\left(-\sigma^{-3} U_\mu \pi_\nu - \tfrac{1}{2}\sigma^{-1}\dot{U}_\mu U_\nu + \tfrac{1}{2}\sigma^{-1}\ddot{U}_\mu \pi_\nu + \tfrac{2}{3}\ddot{U}_\mu U_\nu + \right.$$
$$\left. + \frac{\sigma^{-2}}{2c^2}\pi_\kappa \ddot{U}_\kappa U_\mu \pi_\nu\right) - (\nu, \mu). \tag{8.13}$$

So far, σ has been treated as a free variable, but in fact it is to be determined for any π_κ from (8.4). In terms of the power series in σ and in view of $\pi_\kappa \dot{U}_\kappa = 0$, $U_\kappa^2 = c^2$, $U_\kappa \dot{U}_\kappa = 0$, $U_\kappa \ddot{U}_\kappa = -\dot{U}_\kappa^2$, we get from (8.4):

$$\pi_\kappa^2 + \sigma^2(c^2 - \pi_\kappa \dot{U}_\kappa) + \tfrac{1}{3}\sigma^3 \pi_\kappa \ddot{U}_\kappa - \frac{\sigma^4}{12}\dot{U}_\kappa^2 = 0. \tag{8.14}$$

To first order, $\epsilon = c\sigma$. This would be exactly the case if the motion was not accelerated, as can be seen in the rest system, where time and proper time are equal. Now, the accuracy of (8.14) will not be affected if σ^3 and σ^4 are replaced by $\epsilon^3 c^{-3}$ and $\epsilon^4 c^{-4}$ respectively. This gives

$$-\epsilon^2 + \sigma^2(c^2 - \pi_\kappa \dot{U}_\kappa) + \frac{\epsilon^3}{3c^3}\pi_\kappa \ddot{U}_\kappa - \frac{\epsilon^4}{12c^4}\dot{U}_\kappa^2 = 0. \tag{8.15}$$

Hence,

$$\sigma = \epsilon(c^2 - \pi_\kappa \dot{U}_\kappa)^{-\frac{1}{2}}\left(1 - \frac{\epsilon}{6c^3}\pi_\kappa \ddot{U}_\kappa + \frac{\epsilon^2}{24c^4}\dot{U}_\kappa^2\right), \tag{8.16}$$

$$\sigma^{-1} = \epsilon^{-1}(c^2 - \pi_\kappa \dot{U}_\kappa)^{\frac{1}{2}}\left(1 + \frac{\epsilon}{6c^3}\pi_\kappa \ddot{U}_\kappa - \frac{\epsilon^2}{24c^4}\dot{U}_\kappa^2\right). \tag{8.17}$$

Inserting (8.17) in (8.13), and retaining only powers of order ϵ^{-2}, ϵ^{-1}, ϵ^0, we get:

$$F_{\mu\nu}^{\text{ret}} = \frac{e}{4\pi}(c^2 - \pi_\kappa \dot{U}_\kappa)^{-\frac{1}{2}}\Big(-\epsilon^{-3}U_\mu \pi_\nu - \tfrac{1}{2}\epsilon^{-1}(c^2 + \pi_\kappa \ddot{U}_\kappa)\dot{U}_\mu U_\nu + \\ + \frac{1}{8c^4}\epsilon^{-1}\dot{U}_\kappa^2 U_\mu \pi_\nu + \tfrac{1}{2}\epsilon^{-1}\ddot{U}_\mu \pi_\nu + \tfrac{2}{3}\dddot{U}_\mu U_\nu\Big) - (\nu, \mu), \tag{8.18}$$

where the last two terms have been brought under the same brackets with the other terms for convenience, since this has no effect on the accuracy to the required order. In the case of these two terms, the square root of $c^2 - \pi_\kappa \dot{U}_k$ could be replaced by c^2.

To obtain the advanced field, we must change the sign of the whole expression and also replace ϵ by $-\epsilon$. This gives

$$F_{\mu\nu}^{\text{adv}} = \frac{-e}{4\pi}(c^2 - \pi_\kappa \dot{U}_\kappa)^{-\frac{1}{2}}\Big(\epsilon^{-3}U_\mu \pi_\nu + \tfrac{1}{2}\epsilon^{-1}(c^2 + \pi_\kappa \ddot{U}_\kappa)\dot{U}_\mu U_\nu - \\ - \frac{1}{8c^4}\epsilon^{-1}\dot{U}_\kappa^2 U_\mu \pi_\nu + \tfrac{1}{2}\epsilon^{-1}\ddot{U}_\mu \pi_\nu + \tfrac{2}{3}\dddot{U}_\mu U_\nu\Big) - (\nu, \mu). \tag{8.19}$$

For half the sum and half the difference of these fields we get

$$\tfrac{1}{2}(F_{\mu\nu}^{\text{ret}} + F_{\mu\nu}^{\text{adv}}) = \frac{e}{4\pi}(c^2 - \pi_\kappa \dot{U}_\kappa)^{-\frac{1}{2}}\Big[-\Big(\epsilon^{-3} - \frac{1}{8c^4}\epsilon^{-1}\dot{U}_\kappa^2\Big)(U_\mu \pi_\nu - U_\nu \pi_\mu) + \\ + \frac{1}{2c^4}\epsilon^{-1}(c^2 + \pi_\kappa \ddot{U}_\kappa)(\dot{U}_\mu U_\nu - \dot{U}_\nu U_\mu) + \frac{1}{2c^2}\epsilon^{-1}(\ddot{U}_\mu \pi_\nu - \ddot{U}_\nu \pi_\mu)\Big] \equiv F_{\mu\nu}^{+}, \tag{8.20}$$

$$\tfrac{1}{2}(F_{\mu\nu}^{\text{ret}} - F_{\mu\nu}^{\text{adv}}) = \frac{e}{4\pi c^4}\tfrac{2}{3}(\dddot{U}_\mu U_\nu - \dddot{U}_\nu U_\mu) \equiv F_{\mu\nu}^{-}. \tag{8.21}$$

In the last expression, the factor $(c^2 - \pi_\kappa \dot{U}_\kappa)^{-\frac{1}{2}}$ has been replaced by c^{-1} as this is permissible in the present degree of approximation. At the world line (8.21) holds exactly. $F^-_{\mu\nu}$ has no singularity, corresponding with the fact that it is a solution of the homogeneous wave equation.

We must now turn our attention to the question of the flow of energy and momentum in the neighbourhood of the electron. For this purpose let us consider a section of the world line between τ_1 and τ_2 and imagine it enveloped by a time-like surface Σ which forms a narrow tube around the world line. At τ_1 and τ_2 we close the tube by two space-like surfaces, σ_1 and σ_2, perpendicular to the world line. Because of the presence of the source inside the tube, the situation is now different from the one considered in section 8 of Chapter 3. There we considered a region in which $\partial T_{\kappa\lambda}/\partial x_\lambda = 0$ applies, and by means of Gauss's theorem in a four-dimensional space, we calculated the amount of energy and momentum over the space-like surfaces σ_1 and σ_2 (in other words the amount of energy and momentum within three-dimensional space regions) and found these quantities to be equal. The argument was based on the assumption that the field on the boundaries of the regions σ_1 and σ_2 vanishes at the whole considered time and consequently this applied also over the whole time-like hyper-surface Σ. As a result of this assumption we had

$$\frac{1}{c}\iiint T_{\kappa\lambda}\, d\Sigma_\kappa = 0. \tag{8.22}$$

In our present case the field is not negligible at the edges of σ_1 and σ_2 and (8.22) does not hold, also the divergence of $T_{\kappa\lambda}$ is different from zero. Gauss's theorem now gives

$$P_\lambda(\sigma_2) - P_\lambda(\sigma_1) + \frac{1}{c}\iiint T_{k\lambda}\, d\Sigma_\kappa = \frac{1}{c}\iiiint \frac{\partial T_{\kappa\lambda}}{\partial x_\kappa}\, d^4x$$
$$= -\iiiint K_\lambda\, d^4x. \tag{8.23}$$

The last term is the integral of the density of force (see p. 111). Now, the total force, arising from the incident field and the reaction of the self-field, must vanish so that if Σ surrounds the world line, the integral of K_λ must vanish and we get for the law of conservation of energy and momentum

$$P_\lambda(\sigma_2) - P_\lambda(\sigma_1) = -\frac{1}{c}\iiint T_{\kappa\lambda}\, d\Sigma_\kappa. \tag{8.24}$$

In (8.24) $P_\lambda(\sigma)$ does not represent the total energy and momentum of the field but only the part contained in σ. The relation shows that the integral over Σ is a perfect differential, irrespective of any particular choice of Σ. For let us consider two tubes Σ_1 and Σ_2 based on the same edges of σ_1 and σ_2. In the region between them the divergence of $T_{\kappa\lambda}$ vanishes and consequently the integrals over the tubes have the same value, which depends only on the conditions at σ_1 and σ_2. We may therefore use the following relation for a surface which is infinitely near to the world line:

$$c^{-1}\iiint T_{\kappa\lambda}\,d\Sigma_\kappa = -\int\limits_{\tau_1}^{\tau_2}\dot{P}_\lambda(\tau)\,d\tau \qquad \left(d\Sigma_\kappa = d\Sigma\frac{\pi_\kappa}{\epsilon}\right). \tag{8.25}$$

To determine the flow of energy and momentum through the surface of the tube, we need an expression for $d\Sigma$. It may be expressed as $dS.ds$, where $ds = c\,d\tau$ and dS is a surface element of a sphere of radius ϵ surrounding the electron. (In a three-dimensional Minkowski world, as in Fig. 28, dS is given by $\epsilon.d\phi$.) For a given $d\tau$, ds can be obtained from the curvature of the world line, as was done by Fermi and described in section 6, or else from the following reasoning. The surface of the tube is defined by means of the equation

$$(x_\kappa - \xi_\kappa)^2 \equiv \pi_\kappa^2 = -\epsilon^2, \tag{8.26}$$

$$(x_\kappa - \xi_\kappa)U_\kappa \equiv \pi_\kappa U_\kappa = 0. \tag{8.27}$$

Consider now the point $\xi_\kappa + (d\xi_\kappa/d\tau)d\tau = \xi_\kappa + U_\kappa\,d\tau$ and a point $x_\kappa + dx_\kappa$: which lies on the surface. For the variation of this point we get from (8.26) and (8.27)

$$\pi_\kappa(dx_\kappa - U_\kappa\,d\tau) = 0, \tag{8.28}$$

$$(dx_\kappa - U_\kappa\,d\tau)U_\kappa + \pi_\kappa\dot{U}_\kappa\,d\tau = 0. \tag{8.29}$$

Eqns. (8.27) and (8.28) yield

$$\pi_\kappa\,dx_\kappa = 0, \tag{8.30}$$

while (8.29) yields (with $U_\kappa^2 = c^2$)

$$dx_\kappa\,U_\kappa = (c^2 - \pi_\kappa\dot{U}_\kappa)\,d\tau. \tag{8.31}$$

Eqn. (8.30) tells us that any shift dx_κ along the surface of the tube is orthogonal to π_κ, so that the surface element of the tube is orthogonal to the cross-section of the tube. Eqn. (8.31) tells us that we can decompose dx_κ into two vectors d_1x_κ and d_2x_κ, where the first is orthogonal to U_κ and the second parallel to it and where the magnitude of d_2x is given by

$$|d_2x_\kappa| \equiv ds = (1/c)(c^2 - \pi_\kappa\dot{U}_\kappa)\,d\tau. \tag{8.32}$$

Consequently, we have to evaluate the integral

$$-\frac{1}{c}\iiint T_{\mu\nu}\frac{\pi_\nu}{\epsilon}\epsilon^2\,d\Omega ds = -\frac{1}{c^2}\iiint T_{\mu\nu}\frac{\pi_\nu}{\epsilon}\epsilon^2\,d\Omega(c^2 - \pi_\kappa\dot{U}_\kappa)\,d\tau, \tag{8.33}$$

where $d\Omega$ is a solid angle extended from the centre of the electron ($\epsilon^2 d\Omega = dS$). The tensor $T_{\mu\nu}$ depends on the total field and not only on the retarded field. The total field is given by

$$F_{\mu\nu} = F^{\text{ret}}_{\mu\nu} + F^{\text{in}}_{\mu\nu} = \tfrac{1}{2}(F^{r}_{\mu\nu} + F^{a}_{\mu\nu}) + f_{\mu\nu}, \tag{8.34}$$

where

$$f_{\mu\nu} = F^{\text{in}} + \tfrac{1}{2}(F^{r}_{\mu\nu} - F^{a}_{\mu\nu}). \tag{8.35}$$

The essential fact about $f_{\mu\nu}$ is that, like $F^{-}_{\mu\nu}$, it is finite over the world line and from (8.21) we have

$$\frac{e}{c} f_{\mu\nu} U_{\nu} = \frac{e}{c} F^{\text{in}}_{\mu\nu} U_{\nu} + \frac{1}{4\pi} \frac{2e^{2}}{3c^{3}} \left(\ddot{U} + \frac{1}{c^{2}} \dot{U}^{2}_{\kappa} U_{\mu}\right). \tag{8.36}$$

Using the expressions for F^{r} and F^{a} and (8.34) we get $F_{\mu\nu}$. The tensor $T_{\mu\nu}$ required is given by (see (4.12), p. 99)

$$T_{\mu\rho} = F_{\mu\nu} F_{\rho\nu} - \tfrac{1}{4}\delta_{\mu\rho} F_{\kappa\lambda} F_{\kappa\lambda}. \tag{8.37}$$

The evaluation of this tensor is not required to terms after ϵ^{-2}. Moreover we only need

$$T_{\mu\rho}(\pi_{\rho}/\epsilon).$$

The evaluation of this quantity is straightforward and leads to

$$T_{\mu\rho}\pi_{\rho} = \frac{e^{2}}{16\pi^{2}} (c^{2} - \pi_{\kappa} \dot{U}_{\kappa})^{-1} \left\{ \left(\frac{c^{2}}{2} \epsilon^{-4} + \frac{1}{2c^{2}} \epsilon^{-2} \dot{U}^{2}_{\kappa} \right) \pi_{\mu} - \right.$$

$$\left. - \frac{1}{2c^{2}} \epsilon^{-2} (c^{2} + \tfrac{3}{2}\pi_{\kappa} \dot{U}_{\kappa}) \dot{U}_{\mu} \right\} + \frac{e}{4\pi c} \epsilon^{-1} f_{\mu\nu} U_{\nu}. \tag{8.38}$$

Inserting this in (8.33) and taking into account that $(c^{2} - \pi_{\kappa} \dot{U}_{\kappa})^{-1}$ cancels, also that when integrating over all directions of $d\Omega$, terms which are linear in π_{κ} can make no contribution to the integral, we get, after neglecting terms which vanish with ϵ (changing units and dropping powers of 4π):

$$-\int \dot{P}_{\mu}(\tau)\, d\tau = \frac{1}{c} \iiint T_{\mu\nu} \frac{\pi_{\nu}}{\epsilon}\, d\Sigma = -\int\limits_{\epsilon \to 0} \left(\frac{e^{2}}{2c^{2}} \epsilon^{-1} \dot{U}_{\mu} - \frac{e}{c} U_{\nu} f_{\mu\nu} \right) d\tau. \tag{8.39}$$

Hence, the integrand is given by

$$(1/2c^{2}) e^{2} \epsilon^{-1} \dot{U}_{\mu} - (e/c) f_{\mu\nu} U_{\nu} = \dot{P}_{\mu}. \tag{8.40}$$

P_{μ} is the energy-momentum residing inside a cross-section of the world tube, for the limit $\epsilon \to 0$. The first term in (8.40) is the electrostatic energy residing outside the cross-section, which is a sphere of radius ϵ. As this decreases so does P_{μ}, while the energy residing outside increases. Let us then assume that as $\epsilon \to 0$, $P_{\mu} \to 0$. The first term then gives the total electromagnetic mass and we get

$$(1/c^{2}) W_{\text{el.st.}} \dot{U}_{\mu} = (e/c) f_{\mu\nu} U_{\nu}. \tag{8.41}$$

The difficulty here lies in the divergence of $W_{\text{el.st.}}$. This is due to treating the electron as a point charge. There is, therefore, some justification for replacing this term by the experimental mass, which may even be only partly of electromagnetic nature. So we get

$$m_{0} \dot{U}_{\mu} = (e/c) f_{\mu\nu} U_{\nu}. \tag{8.41a}$$

That this procedure is consistent with a purely electromagnetic theory was shown by Dirac. From Maxwell's equations we can only conclude that the flow of energy-momentum through the walls of the tube is a perfect differential. So far it has been assumed as $\dot{P}_\mu$. But it is possible to introduce other quantities, say $\dot{B}_\mu$. The choice is not unique and all we can say about any $\dot{B}_\mu$ is that it must be orthogonal to the four-velocity. Since $f_{\mu\nu}$ is anti-symmetrical,

$$\dot{B}_\mu U_\mu = (1/2c^2)e^2\epsilon^{-1}\dot{U}_\mu U_\mu - (e/c)f_{\mu\nu} U_\mu U_\nu \equiv 0. \tag{8.42}$$

The simplest B_μ which is consistent with this orthogonality is

$$B_\mu = kU_\mu, \tag{8.43}$$

where k is a constant which is independent of U_μ, but may depend on ϵ. This last property makes it possible to select k in such a way that the infinite electrostatic term cancels out or is subtracted. Let

$$k = (1/2c^2)e^2\epsilon^{-1} - m_0. \tag{8.43 a}$$

Then, replacing $\dot{P}_\mu$ in (8.40) by $\dot{B}_\mu$ and using (8.43) and (8.43 a), we get (8.41 a). The procedure of subtracting the infinite term or renormalizing the mass seems the only way at present to deal with the divergence encountered when the electron is assumed to be a point charge.

It is interesting to note that the electrostatic energy appearing in (8.40) and subsequent formulae is without the factor $\frac{4}{3}$. This is due to the fact that Dirac, like Fermi, uses surfaces which are orthogonal to the world line. Somehow this contrast with the Lorentz theory did not attract notice.

On using (8.36) we get

$$m_0 \dot{U}_\mu - (2e^2/3c^3)\{\ddot{U}_\mu + (1/c^2)\dot{U}_\kappa^2 U_\mu\} = (e/c)F_{\mu\nu}^{\text{in}} U_\nu = F_\mu^{\text{in}}, \tag{8.44}$$

where F_μ^{in} is the external electromagnetic force acting on the electron.

In (8.44) the term with $\ddot{U}_\mu$ is a perfect differential. Consequently, after some cycle in the motion of the electron with respect to $\dot{U}_\mu$ the effect of this term on the balance of energy and momentum will vanish. On the other hand, the term with $\dot{U}_\kappa^2 U_\mu$ is not a perfect differential and produces a constant loss of energy due to its negative sign. This result originates from our using a field which represents a diverging wave at a great distance from the source. Let us consider the fourth component of the radiation term $-(2e^2/3c^5)\dot{U}_\kappa^2 U_\mu$. It is given by $-(2e^2/3c^5)\dot{U}_\kappa^2 \gamma c$. (We are using a space ix_k, ct.) Because $\dot{U}_\mu$ is orthogonal to U_μ, it is space like and its square is negative. Hence the term we are now considering is positive and the incident force must perform a positive work to account for it. On multiplying $-(2e^2/3c^5)\dot{U}_\kappa^2 U_\mu$ by U_μ we get the relativistic generalization of power. In this case it is the radiation power which was mentioned earlier,

$$R = -\frac{2e^2}{3c^5}\,\dot{U}_\kappa^2\, U_\mu\, U_\mu = -\frac{2e^2}{3c^3}\,\dot{U}_\kappa^2. \tag{8.45}$$

In the rest system of the electron this reduces to the Larmor formula $R = +(2e^2/3c^5)\dot{u}_k^2$ (since $U_\mu^0 = (i\gamma^0 u_k, \gamma^0 c) = (iu_k, c)$). If we consider in (8.44) the first and second terms for $\mu = 4$, they become

$$\frac{1}{c}\frac{d}{d\tau}(m_0\,\gamma c^2) \quad \text{and} \quad \frac{1}{c}\frac{d}{d\tau}\left(\frac{2e^2}{3c^3}\frac{d\gamma}{d\tau}\right).$$

The first contains the self-energy and the second was called by Schott the acceleration energy. Both these energies are reversible over any cycle of the motion, after which the electron returns to the same value of U_μ. In terms of circuit theory the acceleration energy gives rise to a radiation reactance. Although the power concerned with this term is not radiated away and is only emitted and reabsorbed by the electron, it is an essential part of the radiation process. If (8.44) is multiplied by U_μ, the right hand side vanishes on account of the antisymmetry of $F_{\mu\nu}^{\text{in}}$. The term with the mass also vanishes. As to the radiative terms, they do not vanish individually but they cancel each other:

$$\ddot{U}_\mu\, U_\mu + (1/c^2)\dot{U}_\kappa^2\, U_\mu^2 = -\dot{U}_\mu^2 + \dot{U}_\kappa^2 = 0.$$

As to the real self-inductance, this is either zero owing to the subtraction of the self-energy or else included in m_0. When using this analogy with circuit theory we must bear in mind that the circuit equations, even when radiation is included, are linear, whereas the equation of motion is not linear as a result of the occurrence of $\dot{U}_\kappa^2$.

Dirac's theory does not supply a specific model through which the radiative reaction could be explained but it makes use of the experimental fact that the field of an electron represents a divergent wave at great distance from source.

A model which would explain the radiative feature of the equation of motion would in fact explain why the retarded solution only occurs in practice and the irreversible character of the motion in spite of the fact that Maxwell's equations are symmetric in time. A theory to this effect was advanced by Wheeler and Feynman. The basic idea of their theory is that the presence of many other charged particles in the universe is essential for the understanding of the radiative reaction, and statistical considerations must also be taken into account. To bring in the effect of the universe recalls Mach's principle, according to which the inertial term $m_0 \dot{U}_\mu$ can be understood only if we consider the gravitational effect on m_0 of all the masses in the universe. More specifically, the theory of Wheeler and Feynman makes the following assumptions: (1) As regards the electromagnetic aspects, the theory must be symmetrical in time. (2) An irreversible feature is introduced from statistical considerations involving all the charged particles of the universe. These interact with the considered electron and are called for the purpose of this theory the absorber. (3) An electron cannot interact with itself, and a force acting on it results from other electrons. In accordance with (1), it is assumed that field of each electron is given by

$$F_{\mu\nu} = \tfrac{1}{2}(F^r_{\mu\nu}+F^a_{\mu\nu}). \tag{8.46}$$

As regards (3), an electron could not radiate electromagnetic energy if other electrons did not exist also. This idea was proposed by Tetrode in 1922.

Let us denote the field of the considered electron by $F_{\mu\nu}$ and the sum of the fields of all the other electrons by $G_{\mu\nu}$. We have

$$F_{\mu\nu} = \tfrac{1}{2}(F^r_{\mu\nu}+F^a_{\mu\nu}), \tag{8.47}$$

$$G_{\mu\nu} = \tfrac{1}{2}(G^r_{\mu\nu}+G^a_{\mu\nu}). \tag{8.48}$$

Assuming that no radiation escapes from the whole system, so that we have a complete absorber, we get

$$\tfrac{1}{2}(G^r_{\mu\nu}+G^a_{\mu\nu}+F^r_{\mu\nu}+F^a_{\mu\nu}) = 0 \quad \text{(outside the whole system).} \tag{8.49}$$

Since advanced and retarded fields cannot cancel each other, it follows from (8.49) that

$$G^r_{\mu\nu}+F^r_{\mu\nu} = 0 \quad \text{(outside the whole system),} \tag{8.50}$$

$$G^a_{\mu\nu}+F^a_{\mu\nu} = 0 \quad (\quad ,, \qquad ,, \qquad ,, \quad). \tag{8.51}$$

Consequently,

$$\tfrac{1}{2}(G^r_{\mu\nu}+F^r_{\mu\nu})-\tfrac{1}{2}(G^a_{\mu\nu}+F^a_{\mu\nu}) = 0 \quad \text{(outside the whole system).} \tag{8.52}$$

Now, the field in (8.52) has no singularities anywhere, and since it vanishes in the whole outer region it must vanish everywhere, and so we get

$$G^r_{\mu\nu}+F^r_{\mu\nu} = G^a_{\mu\nu}+F^a_{\mu\nu}. \tag{8.53}$$

By assumption, the force acting on the considered electron is determined by $\tfrac{1}{2}(G^r_{\mu\nu}+G^a_{\mu\nu})$, which arises from all the other electrons. This force may be resolved as follows:

$$\tfrac{1}{2}(G^r_{\mu\nu}+G^a_{\mu\nu}) = G^r_{\mu\nu}+\tfrac{1}{2}(F^r_{\mu\nu}-F^a_{\mu\nu})-\tfrac{1}{2}[(F^r_{\mu\nu}+G^r_{\mu\nu})-(F^a_{\mu\nu}+G^a_{\mu\nu})]. \tag{8.54}$$

According to (8.53), the last term in (8.54) vanishes and we get

$$\tfrac{1}{2}(G^r_{\mu\nu}+G^a_{\mu\nu}) = G^r_{\mu\nu}+\tfrac{1}{2}(F^r_{\mu\nu}-F^a_{\mu\nu}). \tag{8.55}$$

Nothing has been done so far to upset the invariance with respect to time reversal. Although the individual terms on the right-hand side of (8.55) are not invariant against time reversal, their sum is. In fact by bringing $G^r_{\mu\nu}$ to the left side we get

$$\tfrac{1}{2}(G^a_{\mu\nu}-G^r_{\mu\nu}) = \tfrac{1}{2}(F^r_{\mu\nu}-F^a_{\mu\nu}) \tag{8.56}$$

and here it is evident that the relation is invariant against time reversal.

Equating the force $\frac{1}{2}(G^r_{\mu\nu}+G^a_{\mu\nu})(e/c)U_\nu$, which acts on the electron to $m_0\dot{U}_\mu$, we get

$$m_0\dot{U}_\mu = \frac{e}{c}G^r_{\mu\nu}U_\nu+\frac{e}{2c}(F^r_{\mu\nu}-F^a_{\mu\nu})U_\nu. \tag{8.57}$$

On using Dirac's formula for $\frac{1}{2}(F^r_{\mu\nu}-F^a_{\mu\nu})$, we get

$$m_0\dot{U}_\mu = \frac{e}{c}G^r_{\mu\nu}U_\nu+\frac{2e^2}{3c^3}\left(\ddot{U}_\mu+\frac{1}{c^2}\dot{U}^2_\kappa U_\mu\right). \tag{8.58}$$

It is important to realize that resolving the field of the absorber according to (8.54) is not the only way. With equal right we can use

$$\begin{aligned}\tfrac{1}{2}(G^r_{\mu\nu}+G^a_{\mu\nu}) &= G^a_{\mu\nu}-\tfrac{1}{2}(F^r_{\mu\nu}-F^a_{\mu\nu})-\tfrac{1}{2}[(F^a_{\mu\nu}+G^a_{\mu\nu})-(F^r_{\mu\nu}+G^r_{\mu\nu})]\\ &= G^a_{\mu\nu}-\tfrac{1}{2}(F^r_{\mu\nu}-F^a_{\mu\nu}).\end{aligned} \tag{8.59}$$

This leads to

$$m_0\dot{U}_\mu = \frac{e}{c}G^a_{\mu\nu}U_\nu-\frac{2e^2}{3c^3}\left(\ddot{U}_\mu+\frac{1}{c^2}\dot{U}^2_\kappa U_\mu\right). \tag{8.60}$$

There is nothing in the theory to suggest any preference between (8.58) and (8.60), where in the first of these equations the radiative term appears with a positive sign while in the second it appears with a negative sign. Both equations are equally legitimate and valid. However, to get corresponding results, it must be realized that the initial conditions of the absorber play an important part in the solution. These conditions are entirely different for G^r and G^a which appear in the equations of motion. The initial conditions determine whether the electron will lose energy to the absorber or whether the latter will lose energy to the electron. In the view of Wheeler and Feynman, the required initial conditions for the electron to gain energy are highly improbable and this explains the fact that an accelerated electron always loses energy through radiation. This does not introduce any preference between (8.58) and (8.60). Provided the initial conditions are used appropriately in both equations, both will lead to the observed loss of energy by an accelerated electron. Suppose we use the equation of motion in the form (8.58) and let us assume that for a certain (short) time $t_0 \to t$ a given force (the nature of which is immaterial) is acting on the electron producing a certain motion. In the form (8.58) we must now seek the retarded effects which the motion of the electron is producing via the absorber. These will appear after the time t_0, so that initially we may look upon the force $(e/c)G^r_{\mu\nu}\,U_\nu$ as due only to the chaotic motion of the absorber which existed prior to setting the electron in motion. For a completely chaotic motion the resulting force on the electron will be negligible and under such circumstances the equation of motion may be written as

$$m_0\dot{U}_\mu = \frac{2e^2}{3c^3}\left(\ddot{U}_\mu+\frac{1}{c^2}\dot{U}^2_\kappa\,U_\mu\right)+F^{\text{ext}}_\mu. \tag{8.61}$$

Thus we get the usual equation of motion which is obtained through the use of the retarded field when it is assumed that the electron exerts a self-force and no reference to an absorber is present. On the other hand, if we use the equation of motion in the form (8.60), we cannot proceed in the same way as in the case of eqn. (8.58) by arguing that $(e/c)G^a_{\mu\nu}\,U_\nu$ is negligible, because, apart from the field resulting from the chaotic motion, we have in addition a field created in the past and related to the future motion of the electron (between t_0 and t) arriving at the electron just at the time interval $t_0 \to t$. This field is not negligible and we can get its value from (8.53). Initially, $G^r_{\mu\nu}$ is resulting from the chaotic motion only and so we may approximate this equation by

$$G^a_{\mu\nu} = F^r_{\ \mu\nu}-F^a_{\ \mu\nu}. \tag{8.62}$$

Since this is twice as big as the radiation reactance and with opposite sign to the one which appears in (8.59) and (8.60), we arrive at the same equation as in (8.61).

Another aspect dealt with at great length in the paper by Wheeler and Feynman is that of incomplete absorption. One conclusion they reached is that if the left-hand side of (8.52) does not vanish, this expression should be identified with Dirac's incident field. Apart from this they also mentioned that an expanding universe might bear on the problem of the one-sidedness of the radiation. This was further elaborated upon by Hogarth and later by Hoyle and Narlikar. The expansion brings a continuous modification in the absorber so that the absorber of the past is different from that of the future. In Hogarth's view the directedness in time cannot be explained by statistical considerations but must depend on the expansion of the universe.

The equation of motion is of an unusual type and beset with difficulties because it is of third order in the time-derivative of the position-variable. Consequently, as an initial value problem it is necessary to specify not only the initial position and velocity but also an initial acceleration, against experimental evidence. Even without an external force the electron can move with acceleration if this is initially different from zero (runaway solutions). Dirac proposed to alter the nature of the problem by prescribing initial position and velocity and final (asymptotic) acceleration. Rohrlich has shown that the equation of motion can be replaced by an integro-differential equation which takes into account the restriction with regard to the asymptotic behaviour to avoid runaway solutions. This equation requires only the usual initial conditions. It is typical of the new situation that the law of causality in its strict classical form does not hold. Already a short time before a force is applied the electron begins accelerating and the acceleration begins to diminish a short time before the applied force is stopped (assuming that such a step force can be produced). The time in question is of the order r_0/c, where r_0 is the classical radius of the electron. This radius is given by $e^2/2m_0c^2$ (as obtained by equating the electrostatic energy $\frac{1}{2}(e^2/r_0)$ to m_0c^2). Thus, the short period in question is of the order of $e^2/2m_0c^3$ which is about 10^{-23} sec. It is the time it would take an electromagnetic wave to cross an electron. It is possible to interpret this result by saying that the electron has finite size and that at the instant the fringe of the electron experiences a force, it is transmitted with infinite velocity through the electron (breakdown of ordinary space-time laws inside the electron).

We are on less speculative ground and in agreement with experimental evidence when the radiation may be treated as a small perturbation of the motion which would have taken place without radiation, i.e. if the acceleration is sufficiently small.

As an example let us consider an electron moving in a magnetic field where as a result of $F_{4\lambda} = (iE_x, iE_y, iE_z) = 0$, no work is done by the external field on the electron. Any radiation of electromagnetic energy is therefore at the cost of the energy which the electron possesses. Assuming that the orbit is in a plane perpendicular to the magnetic field, the motion must end with the electron coming to rest with its rest energy, m_0c^2. The amount of radiation at any time per unit of proper time is equal to

$$\frac{dW}{d\tau} = \frac{d}{d\tau}\left(\frac{m_0c^2}{(1-u^2/c^2)^{\frac{1}{2}}}\right) = \frac{d}{d\tau}\left(m_0c^2\frac{dt}{d\tau}\right). \tag{8.63}$$

Using (8.44) with the right-hand side equal to zero, we get after multiplication of the fourth component by $-ic$

$$-m_0 ic\frac{dU_4}{d\tau} \equiv \frac{d}{d\tau}\left(\frac{m_0 c^2}{(1-u^2/c^2)^{\frac{1}{2}}}\right) = \frac{dW}{d\tau} = -\frac{2e^2 i}{3c^2}\left\{\frac{d^2U_4}{d\tau^2}-\frac{1}{c^2}\left(\frac{dU_\lambda}{d\tau}\right)^2 U_4\right\}$$

$$= \frac{2}{3}\frac{e^2}{c^3}\left\{\frac{d^2}{d\tau^2}\left(\frac{c^2\,dt}{d\tau}\right)-\left(\frac{dU_\lambda}{d\tau}\right)^2\frac{dt}{d\tau}\right\}$$

$$= \frac{2}{3}\frac{e^2}{c^3}\left\{c^2\frac{d^3t}{d\tau^3}-\left(\frac{dU_\lambda}{d\tau}\right)^2\frac{dt}{d\tau}\right\}. \tag{8.64}$$

At any given instant the electron will move in a circular orbit which corresponds to some angular frequency ω. To simplify matters a little let us assume that the radius of the orbit does not change appreciably in a full period. We may then look upon the velocity u as a constant, so that

$$\frac{d^2t}{d\tau^2} = \frac{d}{d\tau}\left(\frac{1}{(1-u^2/c^2)^{\frac{1}{2}}}\right) \approx 0, \qquad \frac{d^3t}{d\tau^3} \approx 0. \tag{8.65}$$

Eqn. (8.64) then yields for the loss of energy

$$-\frac{dW}{d\tau} = \frac{2e^2}{3c^3}\left(\frac{dU_\lambda}{d\tau}\right)^2\frac{dt}{d\tau}. \tag{8.66}$$

For the square of the four-acceleration we may get another expression in terms of the ordinary velocity and the ordinary acceleration. Thus

$$\left(\frac{dU_1}{d\tau}\right)^2 = \left(\frac{d(u_x\,dt/d\tau)}{d\tau}\right)^2 = \left(\frac{du_x}{d\tau}\frac{dt}{d\tau}+u_x\frac{d^2t}{d\tau^2}\right)^2 = \left\{\frac{du_x}{dt}\left(\frac{dt}{d\tau}\right)^2+u_x\frac{d^2t}{d\tau^2}\right\}^2. \tag{8.67}$$

Similar expressions hold for U_2, U_3. As to U_4, we have

$$\left(\frac{dU_4}{d\tau}\right)^2 = \left(ic\frac{d^2t}{d\tau^2}\right)^2 = -c^2\left(\frac{d^2t}{d\tau^2}\right)^2. \tag{8.68}$$

Using the approximations expressed in (8.65) we thus get

$$\left(\frac{dU_\lambda}{d\tau}\right)^2 = \left[\left(\frac{du_x}{dt}\right)^2+\left(\frac{du_y}{dt}\right)^2+\left(\frac{du_z}{dt}\right)^2\right]\left(\frac{dt}{d\tau}\right)^4 = w^2\left(1-\frac{u^2}{c^2}\right)^{-2}, \tag{8.69}$$

where w is the ordinary acceleration of the electron, which satisfies the relation

$$w = u^2/r = r\omega^2 \tag{8.70}$$

(r is the radius of the orbit). Eqn. (8.66) becomes

$$-\frac{dW}{d\tau} = \frac{2e^2}{3c^3}\frac{w^2}{(1-\beta^2)^2}\frac{dt}{d\tau} = \frac{2}{3}\frac{e^2}{c^3}\frac{r^2\omega^4}{(1-\beta^2)^2}\frac{dt}{d\tau} \quad \left(\beta = \frac{u}{c}\right), \tag{8.71}$$

therefore

$$-\frac{dW}{dt} = \frac{2e^2r^2\omega^4}{3c^3(1-\beta^2)^2} = \frac{2e^2\beta^4c}{3r^2(1-\beta^2)^2}. \tag{8.72}$$

Of greater practical value is a formula which yields the loss of energy per unit of time in terms of the magnetic field and β rather than in terms of r and β. Using

$$\frac{m_0 u^2}{r(1-\beta^2)^{\frac{1}{2}}} = \frac{euH}{c}, \tag{8.73}$$

which follows in first approximation from the equation of motion when radiation is neglected, we get

$$-\frac{dW}{dt} = \frac{2}{3}\frac{e^4H^2\beta^2}{c^3m_0^2(1-\beta^2)}. \tag{8.74}$$

Using the Liénard–Wiechert potential for a given motion of an electron, it is possible to calculate the radiation in its spectral and directional properties. We shall not, however, pursue this question further, beyond mentioning that under certain circumstances the maximum of intensity lies in the visible region and that the effect has been observed in betatrons.

We have considered a magnetostatic field which does no work on the electron. In this case the radiation is at the expense of the energy of the electron, and the motion ends with the electron coming to rest (or, more generally, with a motion of constant velocity along the lines of the magnetic field). If we consider a quasi-circular orbit in an electric field, the right-hand side in (8.44) yields $(ie/c)(E_x U_1+E_y U_2+E_z U_3)$ and we get instead of (8.74)

$$-\frac{dW}{dt} = \frac{2e^2\beta^4c}{3r^2(1-\beta^2)^2}+e(u_x E_x+u_y E_y+u_z E_z) = \frac{2e^2\beta^4c}{3r^2(1-\beta^2)^2}+\frac{e\,d\Phi}{dt}, \tag{8.75}$$

where Φ is the potential from which the field is derivable.

9. Orbital and Spin Angular Momentum

In adapting the concept of angular momentum of classical dynamics to the theory of relativity we naturally arrive at the following definition of an angular momentum tensor:

$$M_{\kappa\lambda} = x_\kappa P_\lambda - x_\lambda P_\kappa. \tag{9.1}$$

It follows that

$$M_{\kappa\lambda} = -M_{\lambda\kappa}. \tag{9.2}$$

Hence, the angular momentum contains six non-vanishing components. From the law of conservation of the energy-momentum vector $(dP_\kappa/d\tau = 0)$ for a free particle it follows that the angular momentum is also conserved:

$$\begin{aligned}\frac{dM_{\kappa\lambda}}{d\tau} &= \frac{dx_\kappa}{d\tau}P_\lambda + x_\kappa\frac{dP_\lambda}{d\tau} - \frac{dx_\lambda}{d\tau}P_\kappa - x_\lambda\frac{dP_\kappa}{d\tau} \\ &= U_\kappa P_\lambda - U_\lambda P_\kappa = m_0 U_\kappa U_\lambda - m_0 U_\lambda U_\kappa \equiv 0. \end{aligned} \tag{9.3}$$

Next, let us consider a system of particles which may interact with each other. In the corresponding case in classical dynamics we have Newton's law of action and reaction at our disposal, according to which at each instant action is equal to reaction, and this enables us to derive the law of conservation of angular momentum, even when several particles are present. In the theory of relativity matters are more complicated for the following reasons. Firstly, the concept of two particles at a given time is not relativistically invariant and can therefore play no fundamental role in the dynamics of the system. Secondly, according to the theory of relativity, action of one particle on another cannot travel faster than light. As a result of this situation it is also necessary to take into account the field which carries the interaction, if the law of conservation of total angular momentum is to be derived from the laws of conservation of energy and momentum. Matters are simpler in the case of collisions in which the interaction is assumed to occur when the interacting particles are at the same world point. Here the law of conservation of angular momentum follows immediately from the law of conservation of momentum and energy. We shall not deal with this question further and simply assume that conservation of angular momentum holds as an experimental fact.

Attempts have been made to consider also particles endowed with an intrinsic angular momentum, which, unlike the orbital angular momentum, is independent of a point of reference and is called the spin (Fraenkel, Mathisson, Lubanski, Weyssenhoff and Raabe).

Like the orbital angular momentum the spin also must be determined by an anti-symmetrical tensor which will be denoted by $S_{\kappa\lambda}$. The total angular momentum $J_{\kappa\lambda}$ is then given by

$$J_{\kappa\lambda} = M_{\kappa\lambda}+S_{\kappa\lambda} = x_\kappa P_\lambda - x_\lambda P_\kappa + S_{\kappa\lambda}. \tag{9.4}$$

The orbital angular momentum has in general only three non-vanishing components, when the particle is considered in a rest system ($U_1 = U_2 = U_3 = 0$):

$$M^0_{\kappa\lambda} = \begin{bmatrix} 0 & 0 & 0 & x_1 P^0_4 \\ 0 & 0 & 0 & x_2 P^0_4 \\ 0 & 0 & 0 & x_3 P^0_4 \\ -x_1 P^0_4 & -x_2 P^0_4 & -x_3 P^0_4 & 0 \end{bmatrix}. \tag{9.5}$$

We shall assume that in the rest system the spin is similarly characterized by three components. However, instead of assuming the three 'time' components as non-vanishing we must assume that the three space components are non-vanishing:

$$S^0_{\kappa\lambda} = \begin{bmatrix} 0 & S^0_{12} & S^0_{13} & 0 \\ -S^0_{12} & 0 & S^0_{23} & 0 \\ -S^0_{13} & -S^0_{23} & 0 & 0 \\ 0 & 0 & 0 & 0 \end{bmatrix}. \tag{9.6}$$

Thus in the rest system, where the particle is at rest, the spin has three components S^0_{12}, S^0_{23}, S^0_{31} and the square of the magnitude of the spin may be defined by

$$S^2 = S_{\kappa\lambda}S_{\kappa\lambda} = (S^0_{12})^2+(S^0_{23})^2+(S^0_{31})^2. \tag{9.7}$$

S^2 and S are invariants.

Eqn. (9.6) may be expressed also by the following relation:

$$S_{\kappa\lambda}U_\lambda = 0. \tag{9.8}$$

(In the rest system $U_1 = U_2 = U_3 = 0$ and only $U_4 \neq 0$. Consequently, $S_{\kappa 4}U_4 = 0$ and hence $S_{14} = S_{24} = S_{34} = 0$, while $S_{11} = S_{22} = S_{33} = 0$ on account of the anti-symmetry of $S_{\kappa\lambda}$.)

Similarly, (9.5) may be expressed by

$$M_{[\kappa\lambda}P_{\mu]} = 0, \tag{9.9}$$

where the square brackets indicate that we must sum up over all permutations of κ, λ, μ, taking the even permutations with positive sign and the odd permutations with negative sign.

Suppose now we aim at establishing equations of motion of the following form:

$$\frac{dP_\kappa}{d\tau} = K_\kappa, \tag{9.10}$$

$$\frac{dJ_{\kappa\lambda}}{d\tau} = \frac{d(M_{\kappa\lambda}+S_{\kappa\lambda})}{d\tau} = N_{\kappa\lambda}, \tag{9.11}$$

where $N_{\kappa\lambda}$ is a torque.

$dM_{\kappa\lambda}/d\tau$ is actually already determined by (9.10):

$$\frac{dM_{\kappa\lambda}}{d\tau} = U_\kappa P_\lambda + x_\kappa \frac{dP_\lambda}{d\tau} - U_\lambda P_\kappa - x_\lambda \frac{dP_\kappa}{d\tau} = x_\kappa K_\lambda - x_\lambda K_\kappa, \tag{9.12}$$

where we have used $P_\kappa = m_0 U_\kappa$. It follows, therefore, that (9.11) must be replaced by an equation of the form

$$\frac{dS_{\kappa\lambda}}{d\tau} = N_{\kappa\lambda} - x_\kappa K_\lambda + x_\lambda K_\kappa \equiv N'_{\kappa\lambda}, \tag{9.13}$$

where $N'_{\kappa\lambda}$ is a torque which acts on the spin. Such a scheme is not, however, acceptable for two reasons. Firstly, it will result in no interaction between spin and orbit. We shall be confronted as it were with two independent dynamical properties of the particle. In particular, when dealing with several interacting particles we shall obtain separate laws of conservation of orbital angular momentum and of spin angular momentum, whereas experimentally we know that only total angular momentum is conserved. Secondly, we have to consider also eqn. (9.8). In general, solutions of (9.10) and (9.13) will not satisfy this condition.

To avoid these results, Weyssenhoff and Raabe suggested that we assume that when a particle has an intrinsic spin the energy-momentum vector is no longer parallel to the four-velocity vector. A brief account of their theory will now follow.

Let us consider a free particle. Fom the law of conservation of total angular momentum we get

$$\frac{dJ_{\kappa\lambda}}{d\tau} = \frac{d}{d\tau}(x_\kappa P_\lambda - x_\lambda P_\kappa + S_{\kappa\lambda}) = 0. \tag{9.14}$$

Making use of the law of conservation of P_κ, (9.14) yields

$$U_\kappa P_\lambda - U_\lambda P_\kappa + \frac{dS_{\kappa\lambda}}{d\tau} = 0. \tag{9.15}$$

Hence, as long as U_κ is not parallel to P_κ, $S_{\kappa\lambda}$ will not be a constant. ($U_\kappa P_\lambda - U_\lambda P_\kappa$ is zero only when the vectors are parallel and does not vanish otherwise.) The assumption introduced by Weyssenhoff and Raabe therefore amounts to assuming that even in the case of a free particle the spin is not conserved. To ensure conservation of total angular momentum, the particle will have to move on acceleratedly. (Later we shall see that it moves in a circular path.)

On multiplying (9.15) by U_λ and using $U_\lambda^2 = -c^2$ we get

$$U_\kappa U_\lambda P_\lambda + c^2 P_\kappa + \frac{dS_{\kappa\lambda}}{d\tau} U_\lambda = 0; \tag{9.16}$$

therefore
$$P_\kappa = -\frac{1}{c^2}(U_\lambda P_\lambda)U_\kappa - \frac{1}{c^2}\frac{dS_{\kappa\lambda}}{d\tau} U_\lambda. \tag{9.17}$$

Introducing the invariant

$$\mu_0 = -\frac{1}{c^2} U_\lambda P_\lambda, \tag{9.18}$$

we get the following expression for the energy-momentum vector:

$$P_\kappa = \mu_0 U_\kappa - \frac{1}{c^2}\frac{dS_{\kappa\lambda}}{d\tau} U_\lambda. \tag{9.19}$$

It is a satisfactory feature of the theory that the energy-momentum vector should depend also on the spin. The second term on the right-hand side of (9.19) gives the deviation from parallelism between P_κ

and U_κ. From the anti-symmetry of $S_{\kappa\lambda}$ and the symmetry of $U_\kappa U_\lambda$ it follows that deviation is orthogonal to U_κ:

$$\frac{dS_{\kappa\lambda}}{d\tau} U_\lambda U_\kappa \equiv 0. \tag{9.20}$$

We shall now prove that for a free particle both μ_0 as defined by (9.18) and the magnitude S of the spin as defined by (9.7) are conserved. Let us begin with μ_0. Because $dP_\kappa/d\tau = 0$, we get

$$\frac{d\mu_0}{d\tau} = -\frac{1}{c^2}\frac{dU_\kappa}{d\tau} P_\kappa - \frac{1}{c^2}\frac{dP_\kappa}{d\tau} U_\kappa = -\frac{1}{c^2}\frac{dU_\kappa}{d\tau} P_\kappa. \tag{9.21}$$

On using (9.19), we get

$$\frac{d\mu_0}{d\tau} = -\frac{1}{c^2}\frac{dU_\kappa}{d\tau}\left(\mu_0 U_\kappa - \frac{1}{c^2}\frac{dS_{\kappa\lambda}}{d\tau} U_\lambda\right) \equiv 0. \tag{9.22}$$

(U_κ is orthogonal to $dU_\kappa/d\tau$ and

$$\frac{dU_\kappa}{d\tau} U_\lambda \frac{dS_{\kappa\lambda}}{d\tau} = -\frac{dU_\kappa}{d\tau}\frac{dU_\lambda}{d\tau} S_{\kappa\lambda}$$

on account of (9.8) when differentiated with respect to τ. $S_{\kappa\lambda}$ is anti-symmetrical while $dU_\kappa/d\tau \times dU_\lambda/d\tau$ is symmetrical, hence the product of these two tensors is zero.)

Next we deal with S. Multiplying (9.15) by $S_{\kappa\lambda}$ and using (9.8), we get

$$-S_{\kappa\lambda}\frac{dS_{\kappa\lambda}}{d\tau} = S_{\kappa\lambda} U_\kappa P_\lambda - S_{\kappa\lambda} U_\lambda P_\kappa = -(S_{\lambda\kappa} U_\kappa) P_\lambda - (S_{\kappa\lambda} U_\lambda) P_\kappa = 0; \tag{9.23}$$

therefore
$$\frac{1}{2}\frac{d}{d\tau}(S_{\kappa\lambda} S_{\kappa\lambda}) = \frac{1}{2}\frac{dS^2}{d\tau} = 0, \tag{9.24}$$

so
$$\frac{dS}{d\tau} = 0. \tag{9.25}$$

As a result of (9.8), which yields

$$\frac{dS_{\kappa\lambda}}{d\tau} U_\lambda + S_{\kappa\lambda}\frac{dU_\lambda}{d\tau} = 0, \tag{9.26}$$

we can also express (9.19) as follows:

$$P_\kappa = \mu_0 U_\kappa + \frac{1}{c^2} S_{\kappa\lambda}\frac{dU_\lambda}{d\tau}. \tag{9.27}$$

Using the laws of conservation of P_κ and μ_0, (9.27) yields

$$\frac{dP_\kappa}{d\tau} = \mu_0 \frac{dU_\kappa}{d\tau} + \frac{1}{c^2} S_{\kappa\lambda}\frac{d^2 U_\lambda}{d\tau^2} + \frac{1}{c^2}\frac{dS_{\kappa\lambda}}{d\tau}\frac{dU_\lambda}{d\tau} = 0. \tag{9.28}$$

The last term on the left-hand side of this equation is zero as can be proved as follows. When we multiply (9.15) by $dU_\lambda/d\tau$ we get

$$U_\kappa P_\lambda \frac{dU_\lambda}{d\tau} = -\frac{dS_{\kappa\lambda}}{d\tau}\frac{dU_\lambda}{d\tau}. \tag{9.29}$$

On the other hand, (9.27) yields

$$P_\kappa \frac{dU_\kappa}{d\tau} = \mu_0 U_\kappa \frac{dU_\kappa}{d\tau} + \frac{1}{c^2} S_{\kappa\lambda}\frac{dU_\lambda}{d\tau}\frac{dU_\kappa}{d\tau} \equiv 0. \tag{9.30}$$

Hence, each side of (9.29) is zero. (9.28) thus simplifies to the following equation:

$$\frac{dP_\kappa}{d\tau} = \mu_0 \frac{dU_\kappa}{d\tau} + \frac{1}{c^2} S_{\kappa\lambda}\frac{d^2U_\lambda}{d\tau^2} = 0. \tag{9.31}$$

The corresponding equation of motion when a force is present will be given by

$$\mu_0 \frac{dU_\kappa}{d\tau} + \frac{1}{c_2} S_{\kappa\lambda}\frac{d^2U_\lambda}{d\tau^2} = K_\kappa. \tag{9.32}$$

Similarly, we can eliminate P_κ from (9.15) by using (9.27):

$$\frac{dS_{\kappa\lambda}}{d\tau} - \frac{1}{c^2} S_{\kappa\mu}\frac{dU_\mu}{d\tau} U_\lambda + \frac{1}{c^2} S_{\lambda\mu}\frac{dU_\mu}{d\tau} U_\kappa = 0. \tag{9.33}$$

The corresponding equation of motion when a torque $N_{\kappa\lambda}$ is present which acts on the spin, is given by

$$\frac{dS_{\kappa\lambda}}{d\tau} - \frac{1}{c^2} S_{\kappa\mu}\frac{dU_\mu}{d\tau} U_\lambda + \frac{1}{c^2} S_{\lambda\mu}\frac{dU_\mu}{d\tau} U_\kappa = N_{\kappa\lambda}. \tag{9.34}$$

Eqns. (9.32) and (9.34) must be solved by taking (9.8) into account, and this is possible because (9.32) is of second order in $d/d\tau$.

Let us consider a free particle. We have already mentioned that the lack of parallelism between P_κ and U_κ forces the particle to travel always in an accelerated motion. Weyssenhoff and Raabe themselves point out that this result contradicts common experience and shows that it is not possible to construct a classical model of a spin-particle. The spin occurs in fields possessing several components. Such fields have no complete analogy with classical particles and must be quantized. Weyssenhoff and Raabe integrated eqns. (9.31) and (9.33) and showed that in the zero-momentum frame ($P_1 = P_2 = P_3 = 0$), which must be distinguished from the rest system ($U_1 = U_2 = U_3 = 0$), the orbit of a free particle according to their theory is always circular. In this frame (which we denote by $^{(p)}$) (9.15) yields

$$\frac{dS_{ab}^{(p)}}{d\tau} = 0 \quad (a, b = x, y, z), \tag{9.35}$$

so that not only is S^2 constant (as was derived earlier) but so also is the vector $\mathbf{S}^{(p)}$ with components $S_{12}^{(p)}$, $S_{23}^{(p)}$, $S_{31}^{(p)}$. The circular orbit is in a plane perpendicular to $\mathbf{S}^{(p)}$ and proceeds with angular velocity

$$\omega = \frac{\mu_0 c^2}{S^{(p)}}\left(1-\frac{u^2}{c^2}\right)^{\frac{1}{2}} = \frac{M_0 c^2}{S^{(p)}}, \tag{9.36}$$

where u is obtained from the total energy in the p-frame through (9.18): $P_4^{(p)} = i\mu_0 c(1-u^2/c^2)^{\frac{1}{2}}$ and not through a relation $P^{(p)} = i\mu_0 c(1-u^2/c^2)^{-\frac{1}{2}}$ as would be the case for a spin-less particle of mass μ_0. In (9.36) $M_0 \equiv \mu_0(1-u^2/c^2)^{\frac{1}{2}}$. For the radius of the orbit we have

$$r = u/\omega = uS^{(p)}/M_0 c^2. \tag{9.37}$$

Relative to any other frame the orbit will be a relativistic combination of a linear motion with constant velocity and the circular motion.

CHAPTER 6

ELEMENTS OF HYDRODYNAMICS

1. The Equations of Motion

We shall consider at first a highly idealized fluid, one in which the interaction between one part and another is negligible. In other words, neither viscosity nor pressure is taken to exist. Such a fluid is more like a swarm of non-interacting particles or a dust-cloud. We know already that under such circumstances the rest mass of any portion of the fluid will be conserved, and that an external force will also have no effect on the rest mass. However, this will not be assumed in the following, but rather confirmed in a different way. The intention is to obtain a relativistic hydrodynamics which rests as closely as possible on classical hydrodynamics, independently of the results obtained in Chapter 5 where we dealt with the dynamics of a particle, although we may let ourselves be guided somewhat by them.

Let δm_0 be the rest mass of a small drop of fluid. If δV_0 is the rest volume of the drop, i.e. the volume relative to a frame which moves with the drop, we have

$$\delta m_0 = \rho_0 \delta V_0, \tag{1.1}$$

where ρ_0 is the rest density of the rest mass. Both ρ_0 and δV_0 are invariants, but neither is conserved, since both change in the course of time. Relative to some arbitrary frame K, the volume will be δV and the density of the rest mass some ρ_0^K where again $\delta m_0 = \rho_0^K \delta V$. Sometimes the notation ρ_{00} (or ρ_0^0) is used instead of ρ_0, and ρ_0^K is replaced by ρ_0. In such notation ρ_0 is then not an invariant. It yields the densities of rest mass relative to an observer who measures the volume of the drop while in motion. In our notation this applies to ρ_0^K.

From particle dynamics we know already that the theory of relativity replaces the mass by the relative mass and so we introduce a relative mass δm which is defined by

$$\delta m = \frac{\delta m_0}{(1-u^2/c^2)^{\frac{1}{2}}} = \gamma\,\delta m_0 = \rho\,\delta V, \tag{1.2}$$

where ρ is the density of relative mass. The relation between δV as measured in K, and δV_0 as measured in the rest system, follows from the Lorentz contraction

$$\delta V = \gamma^{-1}\delta V_0. \tag{1.3}$$

Hence, using

$$\delta m_0 = \rho_0 \delta V_0 = \gamma^{-1}\rho\,\delta V = \gamma^{-2}\rho\,\delta V_0 = \rho_0^K \delta V = \rho_0^K \gamma^{-1}\delta V_0, \quad (1.4)$$

we get

$$\rho = \gamma^2 \rho_0, \quad \rho_0^K = \gamma\rho_0. \quad (1.5)$$

Let us now assume that the liquid is exposed to a volume force of density F_a $(a = x, y, z)$ and let us postulate a law of motion as follows, where u_a is the velocity:

$$F_a \delta V = \frac{d(\delta m\, u_a)}{dt} = \frac{d(\rho\,\delta V\, u_a)}{dt}. \quad (1.6)$$

For the moment we say nothing about the mass δm but we shall find that to achieve relativistic invariance we shall have to identify it with the relative mass as defined by (1.2). In (1.6) d/dt must be looked upon as a substantial derivation, as distinct from the local differentiation $\partial/\partial t$, since our attention is now focused on a given liquid drop. For the substantial time-derivative of any quantity $\Phi(x_a, t)$ we have

$$\frac{d\Phi}{dt} = u_a \frac{\partial\Phi}{\partial x_a} + \frac{\partial\Phi}{\partial t}. \quad (1.7)$$

As to the volume element δV we have

$$\frac{d}{dt}\delta V = \frac{\partial u_a}{\partial x_a}\delta V. \quad (1.8)$$

Consequently,

$$\frac{d(\Phi\,\delta V)}{dt} = \frac{\partial(\Phi u_a)}{\partial x_a}\delta V + \frac{\partial\Phi}{\partial t}\delta V. \quad (1.9)$$

Applying this formula to the right-hand side of (1.6) by letting $\Phi = \rho u_b$, we get (after exchanging indices a, b)

$$F_a = \frac{\partial(\rho u_a u_b)}{\partial x_b} + \frac{\partial(\rho u_a)}{\partial t} \quad (a = x, y, z). \quad (1.10)$$

To these three equations we can add a fourth which refers to the power density:

$$(F_a u_a)\,\delta V = \frac{d(\delta W)}{dt} = \frac{d(w\,\delta V)}{dt}, \quad (1.11)$$

where δW is the energy of the drop and w the energy density. Applying (1.9) again, this time by letting $\Phi = w$, we get

$$F_a u_a = \frac{\partial(w u_a)}{\partial x_a} + \frac{\partial w}{\partial t}. \quad (1.12)$$

Let us now introduce a quantity $T_{\mu\nu}$ defined by

$$T_{ab} = \rho u_a u_b, \quad T_{a4} = ic\rho u_a, \quad T_{4b} = \frac{iw u_b}{c}, \quad T_{44} = -w, \quad (1.13)$$

and a quantity F_μ defined by

$$F_\mu = (F_a, iF_a u_a/c). \tag{1.14}$$

Then eqns. (1.10) and (1.12) can be summed up as follows:

$$F_\mu = \frac{\partial T_{\mu\nu}}{\partial x_\nu}. \tag{1.15}$$

To guarantee relativistic invariance of these equations of motion we have only to make sure that $T_{\mu\nu}$ will be a four-tensor and F_μ a four-vector. We note that if we identify δm as relative mass in accordance with (1.2) and thus make use of the relation (1.5), the following twelve of the components of $T_{\mu\nu}$ will have the form of components of a four-tensor:

$$T_{ab} = \rho u_a u_b = \rho_0 \gamma^2 u_a u_b = \rho_0 U_a U_b, \tag{1.16}$$

$$T_{a4} = ic\rho u_a = ic\rho_0 \gamma^2 u_a = \rho_0 U_a U_4, \tag{1.17}$$

where $U_\mu = (U_a, U_4)$ is the four-velocity vector, U_4 being equal to $ic\gamma$. It is now plausible to assume that if we wish $T_{\mu\nu}$ to be a four-tensor, the remaining four components will have to depend in the same way on U_μ and ρ_0 as the twelve components listed in (1.16), (1.17). We then get

$$T_{4b} = \rho_0 U_4 U_b, \qquad T_{44} = \rho_0 U_4 U_4 \equiv -\rho_0 \gamma^2 c^2 = -\rho c^2; \tag{1.18}$$

therefore

$$T_{\mu\nu} = \rho_0 U_\mu U_\nu. \tag{1.19}$$

The tensor $T_{\mu\nu}$ is thus symmetrical, and in particular

$$T_{a4} = T_{4a}. \tag{1.20}$$

Further down, we shall prove that the symmetry of $T_{\mu\nu}$ is actually a necessary consequence of the symmetry of T_{ab} (see 1.16).

From (1.12) and (1.13) we get

$$ic\rho u_a = \frac{iw}{c} u_a; \tag{1.21}$$

therefore

$$w = \rho c^2. \tag{1.22}$$

This relation agrees with the last of the relations in (1.18) since, according to (1.13), $T_{44} = -w$. In particular, in the rest system, the density of energy will be given by

$$w_0 = \rho_0 c^2. \tag{1.23}$$

In the case of particle dynamics we were led naturally to an energy-momentum vector in which the fourth component is equal (apart from a factor ic^{-1}) to the energy mc^2, which becomes $m_0 c^2$ in the rest system. We now see that by attempting to modify classical hydrodynamics and make it relativistically covariant we are led naturally to a four-tensor $T_{\mu\nu}$, the component T_{44} of which represents energy density

$w = \rho c^2$. With the significance of the other components of $T_{\mu\nu}$ we shall deal in section 3. The tensor $T_{\mu\nu}$ is called the kinetic energy-momentum tensor. For relativistic invariance F_μ in (1.15) must be a four-vector. This represents a condition which follows from the theory of relativity with regard to the nature of a force density F_a: it must form together with $iF_a u_a/c$ a four-vector. Conversely, in a covariant relation of the form (1.15), the first three components will represent the density of a Newton force, and the fourth component the power delivered by the force per unit of volume.

As mentioned already, the symmetry of $T_{\mu\nu}$ can also be proved and we do not have to use a plausibility argument. More precisely, it can be proved that because of the symmetry of the space components ($T_{ab} = T_{ba}$) it necessarily follows that $T_{\mu\nu} = T_{\nu\mu}$, if this quantity is to transform as a tensor and T_{ab} is to remain symmetrical in every frame of reference. This last property must hold because in introducing the quantity $T_{ab} = \rho u_a u_b = \rho_0 U_a U_b$ we used an arbitrary system of coordinates, and hence this relation should hold in every frame of reference. The proof of the symmetry of $T_{\mu\nu}$ proceeds as follows. If $T_{\mu\nu}$ is to be a tensor it must transform according to

$$T_{\mu'\nu'} = L_{\mu'\kappa} L_{\nu'\lambda} T_{\kappa\lambda}. \tag{1.24}$$

For the space components this yields

$$T_{a'b'} = L_{a'a} L_{b'b} T_{ab} + L_{a'4} L_{b'b} T_{4b} + L_{a'a} L_{b'4} T_{a4} + L_{a'4} L_{b'4} T_{44}. \tag{1.25}$$

Altogether the right-hand side of (1.25) contains $9+3+3+1 = 16$ terms. Now, the first of the sums (the double sum over a, b) is symmetrical in a', b', because of $T_{ab} = T_{ba}$:

$$L_{a'a} L_{b'b} T_{ab} \equiv L_{a'b} L_{b'a} T_{ba} = L_{a'b} L_{b'a} T_{ab} \equiv L_{b'a} L_{a'b} T_{ab}. \tag{1.26}$$

Also, the last term in (1.25) is symmetrical in a', b'. Therefore if $T_{a'b'}$ is to be equal to $T_{b'a'}$, the sum of the remaining two terms in (1.25) must also be symmetrical. For these terms we have

$$\begin{aligned} L_{a'4} L_{b'b} T_{4b} + L_{a'a} L_{b'4} T_{a4} &= L_{a'4} L_{b'b} T_{4b} + L_{b'4} L_{a'a} T_{a4} \\ &\equiv L_{a'4} L_{b'b} T_{4b} + L_{b'4} L_{a'b} T_{b4}. \end{aligned} \tag{1.27}$$

On exchanging a' and b' we get

$$L_{b'4} L_{a'b} T_{4b} + L_{a'4} L_{b'b} T_{b4}. \tag{1.28}$$

This can be equal to (1.27) only if

$$T_{4b} = T_{b4}. \tag{1.29}$$

In this way the symmetry of $T_{\mu\nu}$ is proved and (1.21), (1.22) follow.

2. The Conservation of Rest Mass in the Absence of Pressure

The essential step in deriving the equation of motion (1.15) consisted in introducing the relative mass and relative density which are not invariant with respect to different observers in motion relative to each other. In classical dynamics ρ is identified with the ρ_0 of the relativistic theory and a law of conservation of mass (eqn. of continuity for ρ_0) is introduced, so that the equation of motion

$$\rho_0 \delta V \frac{du_a}{dt} = F_a \delta V \tag{2.1}$$

is identical with

$$\frac{d(\rho_0 \delta V u_a)}{dt} = F_a \delta V, \tag{2.2}$$

since $\rho_0 \delta V$ is the total mass of the drop and by assumption this is conserved. This leads us to the following classical equations:

$$F_\mu = \frac{\partial t_{\mu\nu}}{\partial x_\nu}, \tag{2.3}$$

where

$$t_{ab} = \rho_0 u_a u_b, \quad t_{a4} = ic\rho_0 u_a, \quad t_{4b} = (iw/c)u_b, \quad t_{44} = -w, \tag{2.4}$$

and

$$w = \tfrac{1}{2}\rho_0 u^2 + \text{const.} \tag{2.5}$$

Eqns. (2.3) look superficially like relativistic equations, but of course they are not, since $t_{\mu\nu}$ is not a four-tensor.

In the relativistic theory ρ_0 is replaced by the relative ρ and this quantity does not satisfy an equation of continuity so that relative mass is not conserved. In fact, by virtue of $w = \rho c^2$, ρ is directly proportional to the energy density and this depends on the work done by the external force. From (1.12) we get

$$\frac{\partial(\rho c^2 u_a)}{\partial x_a} + \frac{\partial(\rho c^2)}{\partial t} = F_a u_a; \tag{2.6}$$

therefore

$$\text{div}(\rho \mathbf{u}) + \frac{\partial \rho}{\partial t} = \frac{F_a u_a}{c^2}. \tag{2.7}$$

It is true that we must not look upon the left-hand side of this equation as the expression which should occur in a covariant equation of continuity, because $(\rho u_a, ic\rho) = (\rho_0 \gamma U_a, ic\rho_0 \gamma^2) = \rho_0 \gamma(U_a, ic\gamma)$ is not a four-vector as γ is not an invariant. However, eqn. (2.7) results from our relativistic theory and expresses in a conventional (non-covariant) way the non-conservation of relative mass.

On the other hand, we shall now prove that rest mass is conserved, so that ρ_0 satisfies an equation of continuity. This equation we shall derive in covariant form. To this end let us consider the scalar $F_\mu U_\mu$. From (1.14) and $U_4 = i\gamma c$ we get

$$F_\mu U_\mu = F_a U_a + F_4 U_4 = \gamma F_a u_a - \gamma F_a u_a \equiv 0. \tag{2.8}$$

Next we can rewrite (1.15) as follows:

$$F_\mu = \frac{\partial}{\partial x_\nu}(\rho_0 U_\mu U_\nu) = U_\mu \frac{\partial}{\partial x_\nu}(\rho_0 U_\nu) + \rho_0 U_\nu \frac{\partial}{\partial x_\nu} U_\mu. \tag{2.9}$$

The last term in (2.9) has the form of a substantial derivative of U_μ with respect to proper time:

$$\rho_0 U_\nu \frac{\partial U_\mu}{\partial x_\nu} \equiv \rho_0 U_1 \frac{\partial U_\mu}{\partial x_1} + \rho_0 U_2 \frac{\partial U_\mu}{\partial x_2} + \rho_0 U_3 \frac{\partial U_\mu}{\partial x_3} + \rho_0 U_4 \frac{\partial U_\mu}{ic\,\partial t}$$

$$= \rho_0 \gamma \left(u_a \frac{\partial U_\mu}{\partial x_a} + \frac{\partial U_\mu}{\partial t} \right) = \rho_0 \gamma \frac{dU_\mu}{dt} \equiv \rho_0 \frac{dU_\mu}{d\tau}. \tag{2.10}$$

Hence, (2.9) becomes

$$F_\mu = U_\mu \frac{\partial(\rho_0 U_\nu)}{\partial x_\nu} + \rho_0 \frac{dU_\mu}{d\tau}. \tag{2.11}$$

On multiplying this equation by U_μ we know from (2.8) that we should get zero:

$$F_\mu U_\mu = U_\mu^2 \frac{\partial(\rho_0 U_\nu)}{\partial x_\nu} + \rho_0 U_\mu \frac{\partial U_\mu}{d\tau} = -c^2 \frac{\partial(\rho_0 U_\nu)}{\partial x_\nu} + \tfrac{1}{2}\rho_0 \frac{d(U_\mu^2)}{d\tau} = 0. \tag{2.12}$$

Since $U_\mu^2 = -c^2$ the last term in (2.12) is zero and this yields

$$\frac{\partial(\rho_0 U_\nu)}{\partial x_\nu} = 0. \tag{2.13}$$

In terms of ρ_0^K, the density of rest mass as measured relative to an arbitrary frame K, we have, using $\rho_0 = \gamma^{-1}\rho_0^K$,

$$\frac{\partial(\rho_0^K \gamma^{-1} U_\nu)}{\partial x_\nu} = \frac{\partial(\rho_0^K \gamma^{-1} \gamma u_a)}{\partial x_a} + \frac{\partial(\rho_0^K \gamma^{-1} i\gamma c)}{ic\partial t} = 0; \tag{2.14}$$

therefore

$$\frac{\partial(\rho_0^K u_a)}{\partial x_a} + \frac{\partial \rho_0^K}{\partial t} = 0. \tag{2.15}$$

This is an equation of continuity in terms of ρ_0^K, and it expresses in differential form the law of conservation of rest mass. We also see that the covariant form of such an equation is given by (2.13). When we dealt with the equation of continuity of electricity we used the notation ρ which now corresponds to our ρ_0^K. We did not introduce the concept of density of electricity at rest. The three current components J_1, J_2, J_3 correspond to $\rho_0^K u_a$. In the same way as we derived the law of conservation of total charge, we can now conclude that total rest mass is conserved. Let

$$m_0(\sigma) = -\frac{1}{c}\iiint \rho_0 U_\mu \, d\sigma_\mu \tag{2.16}$$

be defined as the total rest mass over a space-like surface σ, then as in (8.14) of Chapter 3 we have

$$m_0(\sigma_1) = m_0(\sigma_2) = m_0. \tag{2.17}$$

In particular, if σ is a surface $t =$ constant, we get

$$m_0(\sigma) = m_0 = -c^{-1}\iiint \rho_0 U_4 \, d\sigma_4 = \iiint \rho_0 i\gamma(ic)^{-1} \, dxdydz; \tag{2.18}$$

therefore
$$m_0 = \iiint \rho_0 \gamma \, dxdydz = \iiint \rho_0^K \, dxdydz. \tag{2.19}$$

If each volume element is transformed to rest we shall have

$$m_0 = \iiint \rho_0 \, dx_0 \, dy_0 \, dz_0. \tag{2.20}$$

3. The Energy-Momentum Tensor

Eqns. (1.15) in the absence of external forces become

$$\frac{\partial T_{\mu\nu}}{\partial x_\nu} = 0. \tag{3.1}$$

These relations express the laws of conservation of momentum and energy in differential form. Let

$$T_{ab} = \rho_0 U_a U_b = \rho_0 \gamma^2 u_a u_b = \rho u_a u_b = g_a u_b, \tag{3.2}$$

where
$$g_a = \rho u_a \tag{3.3}$$

is the momentum density. The first three equations in (3.1) yield

$$\frac{\partial(g_a u_b)}{\partial x_b} + \frac{\partial(g_a)}{\partial t} = 0 \quad (a = x, y, z). \tag{3.4}$$

For each value of a, i.e. for each component of momentum g_a, we thus get an equation of continuity. The quantity $g_a u_b$ represents the current density of the 'a'-component of momentum in the b-direction. Similarly, the fourth equation in (3.1) for $\mu = 4$ yields

$$\frac{\partial(wu_b)}{\partial x_b} + \frac{\partial(w)}{\partial t} = 0, \tag{3.5}$$

which is an equation of continuity for the energy. In our present case any flow of energy proceeds through the flow of matter and as this has the velocity u_a the current density of energy is given by wu_a. For the energy-momentum tensor $T_{\mu\nu}$ we get

$$T_{\mu\nu} \equiv \begin{pmatrix} (T_{ab}) & (T_{a4}) \\ (T_{4b}) & (T_{44}) \end{pmatrix} = \begin{pmatrix} (g_a u_b) & (icg_a) \\ (iwu_b/c) & (-w) \end{pmatrix}. \tag{3.6}$$

We know already that this tensor is symmetrical, so that

$$\frac{wu_a}{c} = cg_a, \tag{3.7}$$

and this yields

$$g_a = \frac{w}{c^2} u_a, \tag{3.8}$$

which is another way of expressing $w = \rho c^2$, since $g_a = \rho u_a$. We can sum up the situation as follows. In the case of a particle we are led to an energy-momentum vector P_μ and the laws of conservation are given by

$$\frac{dP_\mu}{d\tau} = 0. \tag{3.9}$$

In the case of a fluid we are led to an energy-momentum tensor which is symmetrical. The laws of conservation in differential form are given by (3.1).

Let σ be a space-like surface. Then, we define the total energy-momentum vector by

$$P_\mu(\sigma) = -\frac{1}{c} \iiint T_{\mu\nu}\, d\sigma_\nu \tag{3.10}$$

and, as a result of (3.1), we conclude in the same way as we did with the energy-momentum tensor of the electromagnetic field that

$$P_\mu(\sigma_1) = P_\mu(\sigma_2). \tag{3.11}$$

In particular, we may use for σ a surface $t =$ constant for which $d\sigma_1 = d\sigma_2 = d\sigma_3 = 0$, $d\sigma_4 = i\, dxdydz$. We then get

$$P_\mu = \iiint T_{\mu 4}\, d\sigma_4, \tag{3.12}$$

$$P_a = \iiint g_a\, dxdydz, \qquad P_4 = \frac{i}{c} \iiint w\, dxdydz. \tag{3.13}$$

4. The Presence of Pressure

By definition an ideal liquid is one which possesses no viscosity so that the force between two neighbouring elements is perpendicular to the surface which separates them. So far we have assumed that this force is negligibly small. This assumption will now be dropped and we shall assume that a normal pressure is present.

In the general case of a continuous medium the stresses are described in the classical theory by a three-dimensional symmetrical tensor of second rank. In the theory of relativity we shall have to enlarge it to a four-dimensional tensor. However, in our present case, when only a normal pressure $p_0(x, y, z, t)$ is assumed to exist, the classical tensor has only three components each equal to p_0. We shall therefore assume that in the relativistic theory the relevant tensor is given by $p_0 \delta_{\kappa\lambda}$.

Let us assume that no volume forces are present and that the fluid is not enclosed in any vessel with movable pistons. In this case the system is 'closed', and energy and momentum must be conserved. We can again expect that a symmetrical energy-momentum tensor exists with vanishing divergence, so that the laws of conservation in differential form will be given by

$$\frac{\partial T_{\mu\nu}}{\partial x_\nu} = 0. \tag{4.1}$$

To obtain the new tensor $T_{\mu\nu}$, let us assume that the considered drop is at rest as a whole.† It may appear at first sight that in such a region a suitable generalization of our previous tensor will be given by

$$T^0_{\mu\nu} = \begin{bmatrix} p_0 & 0 & 0 & 0 \\ 0 & p_0 & 0 & 0 \\ 0 & 0 & p_0 & 0 \\ 0 & 0 & 0 & p_0 - \rho_0 c^2 \end{bmatrix} = p_0 \delta_{\mu\nu} + \Theta^0_{\mu\nu}, \tag{4.2}$$

where $\Theta_{\mu\nu}$ now denotes our previous energy-momentum tensor when pressure was neglected. For a drop at rest $\Theta^0_{\mu\nu}$ has only one non-vanishing component: $\Theta^0_{44} = -\rho_0 c^2$. Eqn. (4.1) would yield, as it should in this case,

$$\frac{\partial p_0}{\partial x_a} = 0, \tag{4.3}$$

and also

$$\frac{\partial(p_0 - \rho_0 c^2)}{\partial t} = 0. \tag{4.4}$$

However, the tensor $T^0_{\mu\nu}$ as given by (4.2) is not acceptable for the following reason: the component $-T_{44}$, which should represent the energy of the drop, depends according to (4.2) explicitly on the pressure p_0, and not solely on ρ_0 which is a function of p_0. This is unacceptable and we must subtract the term p_0 which appears at the bottom of the diagonal terms. Such subtraction must be done in a Lorentz-invariant manner and just to strike off the term p_0 will not do because the result will no longer be a tensor. We must subtract from $\Theta_{\mu\nu}$ a tensor $S_{\mu\nu}$ which happens to have in the rest system only one non-vanishing component, $S^0_{44} = p_0$. The only way of achieving this result is obtained through

$$S_{\mu\nu} = -\frac{p_0}{c^2} U_\mu U_\nu. \tag{4.5}$$

In the rest system $S_{\mu\nu}$ has only one non-vanishing component, $S^0_{44} = p_0$. Hence, we form the tensor

$$T_{\mu\nu} = p_0 \delta_{\mu\nu} + \Theta_{\mu\nu} - S_{\mu\nu} = p_0 \delta_{\mu\nu} + \rho_0 U_\mu U_\nu + (p_0/c^2) U_\mu U_\nu \tag{4.6}$$

† The drop will tend either to expand or contract depending on the sign of the pressure.

and postulate that it represents the energy-momentum tensor in the presence of interaction through pressure. Eqn. (4.1) then yields the following equation of motion:

$$\frac{\partial p_0}{\partial x_\mu}+\frac{\partial[(\rho_0+p_0/c^2)U_\mu U_\nu]}{\partial x_\nu}=0. \tag{4.7}$$

Relative to an arbitrary frame, $T_{\mu\nu}$ will be given by

$$T_{\mu\nu}=\begin{bmatrix} p_0\,\delta_{ab}+(\rho_0+p_0/c^2)\gamma^2 u_a u_b & (\rho_0+p_0/c^2)\gamma^2 u_a ic \\ (\rho_0+p_0/c^2)\gamma^2 u_b ic & p_0-(\rho_0+p_0/c^2)\gamma^2 c^2 \end{bmatrix}$$

$$=\begin{bmatrix} p_0\,\delta_{ab}+(\rho_0+p_0/c^2)\gamma^2 u_a u_b & (\rho_0+p_0/c^2)\gamma^2 u_a ic \\ (\rho_0+p_0/c^2)\gamma^2 u_b ic & -(\rho_0 c^2+p_0 u^2/c^2)\gamma^2 \end{bmatrix}. \tag{4.8}$$

Let the energy density be denoted as before by w. We have

$$w=\left(\rho_0 c^2+\frac{p_0 u^2}{c^2}\right)\gamma^2. \tag{4.9}$$

As will be seen, the pressure enters into this equation only when the drop is in motion. Using (4.1) for $\mu = 4$, we get for the energy relation

$$\frac{\partial[(\rho_0+p_0/c^2)\gamma^2 u_b ic]}{\partial x_b}-\frac{\partial[(\rho_0 c^2+p_0 u^2/c^2)\gamma^2]}{ic\,\partial t}=0. \tag{4.10}$$

On using

$$(\rho_0 c^2+p_0)\gamma^2 \equiv (\rho_0 c^2+p_0 u^2/c^2)\gamma^2+p_0\gamma^2-(p_0 u^2/c^2)\gamma^2 = w+p_0, \tag{4.11}$$

eqn. (4.10) becomes

$$\frac{\partial(wu_b)}{\partial x_b}+\frac{\partial(p_0 u_b)}{\partial x_b}+\frac{\partial w}{\partial t}=0. \tag{4.12}$$

This equation is different from an equation of continuity for w. Hence energy is not conserved locally. The rate of work done by the pressure, $p_0 u$, acts as a source or sink of energy according to whether the divergence of $p_0\,\mathbf{u}$ is positive or negative. $p_0\,\mathbf{u}$ represents a flow of energy which results from the existence of a pressure. In spite of this effect total energy is conserved because from (4.1) we get

$$\iiint\limits_{t=\text{const}_1} w\,dxdydz = \iiint\limits_{t=\text{const}_2} w\,dxdydz. \tag{4.13}$$

Hence the sources are balanced by the sinks of energy.

The existence of the energy flow $p_0\,\mathbf{u}$ and the necessity of taking into account its divergence in (4.12) is of course not a relativistic effect and from a classical point of view we should also have to express the balance of energy according to (4.12). However, the expression for the energy density w would have been different, $\frac{1}{2}\rho u^2+\text{const}$, and not as given in (4.9). Moreover, according to classical theory $\rho\mathbf{u}$ would have yielded the total density of momentum, whereas according to the theory of relativity the density of momentum is given by

$$g_a = \frac{wu_a}{c^2} + \frac{p_0 u_a}{c^2}, \tag{4.14}$$

where the second term attributes momentum also to the energy flow which is caused by the pressure. From (4.14) and (4.9) we get

$$\begin{aligned} g_a &= \rho_0 \gamma^2 u_a + \frac{p_0 u^2}{c^4}\gamma^2 u_a + \frac{p_0}{c^2} u_a \\ &= \rho_0 \gamma^2 u_a + \frac{p_0}{c^2}\gamma^2 u_a = \left(\rho_0 + \frac{p_0}{c^2}\right)\gamma^2 u_a = \frac{T_{a4}}{ic}, \end{aligned} \tag{4.15}$$

in agreement with the corresponding term in (4.8).

Just as energy is not conserved locally, so also momentum is not conserved locally. Gradients of pressure contribute to the flow of momentum and (4.1) for $\mu = 1, 2, 3$ yields

$$\frac{\partial p_0}{\partial x_a} + \frac{\partial((\rho_0 + p_0/c^2)\gamma^2 u_a u_b)}{\partial x_b} + \frac{\partial((\rho_0 + p_0/c^2)\gamma^2 u_a)}{\partial t} = 0; \tag{4.16}$$

therefore

$$\frac{\partial p_0}{\partial x_a} + \frac{\partial(g_a u_b)}{\partial x_b} + \frac{\partial g_a}{\partial t} = 0. \tag{4.17}$$

In spite of the first term in (4.17), total momentum is conserved, since (4.1) yields

$$\iiint\limits_{t=\text{const}_1} g_a\, dxdydz = \iiint\limits_{t=\text{const}_2} g_a\, dxdydz \tag{4.18}$$

We shall now prove that when the elements of the fluid interact with each other so that $p_0 \neq 0$, rest mass is no longer conserved. Eqn. (4.7) can be rewritten as follows, by making use of (2.10):

$$\begin{aligned} &\frac{\partial p_0}{\partial x_\mu} + U_\mu \frac{\partial((\rho_0 + p_0/c^2)U_\nu)}{\partial x_\nu} + \left(\rho_0 + \frac{p_0}{c^2}\right)\frac{\partial U_\mu}{\partial x_\nu} U_\nu \\ &\qquad = \frac{\partial p_0}{\partial x_\mu} + U_\mu \frac{\partial((\rho_0 + p_0/c^2)U_\nu)}{\partial x_\nu} + \left(\rho_0 + \frac{p_0}{c^2}\right)\frac{dU_\mu}{d\tau} = 0. \end{aligned} \tag{4.19}$$

On multiplying this equation by U_μ and using

$$\frac{\partial p_0}{\partial x_\mu} U_\mu = \frac{dp_0}{d\tau}, \qquad \frac{dU_\mu}{d\tau} U_\mu = 0, \qquad U_\mu^2 = -c^2, \tag{4.20}$$

we get

$$\frac{dp_0}{d\tau} - c^2 \frac{\partial((\rho_0 + p_0/c^2)U_\nu)}{\partial x_\nu} = 0; \tag{4.21}$$

therefore

$$\frac{dp_0}{d\tau} - \frac{\partial((\rho_0 c^2 + p_0)U_\nu)}{\partial x_\nu} = 0. \tag{4.22}$$

On using
$$\frac{\partial(p_0 U_\nu)}{\partial x_\nu} = p_0 \frac{\partial U_\nu}{\partial x_\nu} + \frac{\partial p_0}{\partial x_\nu} U_\nu = p_0 \frac{\partial U_\nu}{\partial x_\nu} + \frac{dp_0}{d\tau}, \tag{4.23}$$

we get
$$p_0 \frac{\partial U_\nu}{\partial x_\nu} + \frac{\partial(\rho_0 c^2 U_\nu)}{\partial x_\nu} = 0; \tag{4.24}$$

therefore
$$\frac{\partial(\rho_0 U_\nu)}{\partial x_\nu} = -\frac{p_0}{c^2} \frac{\partial U_\nu}{\partial x_\nu}. \tag{4.25}$$

Hence ρ_0 does not satisfy an equation of continuity. Let the total rest mass over a space-like surface be defined, as in (2.16), by

$$m_0(\sigma) = -\frac{1}{c} \iiint \rho_0 U_\nu \, \delta\sigma_\nu. \tag{4.26}$$

If we let the space-like surface be $t =$ constant, and if we divide the whole region of integration into small elements and take each at rest, we shall have $U_a = 0$, $U_4 = i\gamma c \to ic$ and

$$m_0(t = \text{const}) = \iiint \rho_0 \, \delta x_0 \, \delta y_0 \, \delta z_0. \tag{4.27}$$

On using Gauss's theorem† we get from (4.25), as on previous occasions,

$$\frac{1}{ic} \iiiint \frac{\partial}{\partial x_\nu} (\rho_0 U_\nu) \, \delta^4 x = m_0(\sigma_2) - m_0(\sigma_1), \tag{4.28}$$

so that
$$m_0(\sigma_2) - m_0(\sigma_1) = -\frac{1}{ic} \iiiint \frac{p_0}{c^2} \frac{\partial U_\nu}{\partial x_\nu} \delta^4 x. \tag{4.29}$$

The expression on the right-hand side can be given a simple meaning. Since $\delta^4 x = \delta x \delta y \delta z . ic\, \delta t = \delta x_0 \, \delta y_0 \, \delta z_0 . ic\, \delta t = ic\, \delta V_0 \, \delta\tau$, we have

$$\frac{\partial U_\nu}{\partial x_\nu} \delta^4 x = \frac{\partial U_\nu}{\partial x_\nu} ic\, \delta V_0 \, \delta\tau. \tag{4.30}$$

This invariant simply yields the change in rest volume produced in a time $d\tau$ which results from any non-uniformity in motion through the considered fluid drop, measured relative to an arbitrary frame in which the four-velocity is U_μ. We have only to convince ourselves that we get the correct value in one frame of reference and our statement is proved, because the expression in (4.30) is an invariant. For the particular frame we take the rest frame where $U_\nu = (U_a, i\gamma c) \to (u_a, ic)$, $\delta V_0 \equiv \delta V$, $\tau \equiv t$. We then get

$$\frac{\partial U_\nu}{\partial x_\nu} \delta V_0 \to \left(\frac{\partial u_a}{\partial x_a} + \frac{\partial(ic)}{ic\, \partial t}\right) \delta V = \frac{\partial u_a}{\partial x_a} \delta V \equiv \frac{d(\delta V)}{dt}, \tag{4.31}$$

where we have used $dc/dt \equiv 0$. This is the correct value in view of (1.8).

† The use of Gauss's theorem requires multiplying the right-hand side of (4.26) by i and making $d\sigma_4 = dxdydz$, without a factor i.

Thus, we have quite generally

$$d(\delta V_0) = \frac{1}{ic}\frac{\partial U_\nu}{\partial x_\nu}\delta^4 x \tag{4.32}$$

and

$$\frac{d(\delta V_0)}{d\tau} = \frac{\partial U_\nu}{\partial x_\nu}\delta V_0. \tag{4.33}$$

If we let σ_1 and σ_2 correspond to an infinitesimal difference $d\tau$ in proper time we get from (4.29):

$$dm_0(\sigma) = -\iiint_\sigma \frac{p_0}{c^2}\frac{d(\delta V_0)}{d\tau}\,d\tau. \tag{4.34}$$

In the integrand we have, apart from the factor c^{-2}, the whole work of compression done per unit of proper time through a rate of change of each element δV_0 which is given by $d(\delta V_0)/d\tau$. This work divided by c^2 gives the equivalent effect on rest mass. A compression will be accompanied by an increase in proper mass and an expansion by a decrease of proper mass. In (4.29) we have the total change which occurs when the mass over σ_1 is compared with the mass over σ_2

$$m_0(\sigma_2) - m_0(\sigma_1) = -\iiiint \frac{p_0}{c^2}\,d(\delta V_0). \tag{4.35}$$

The expression on the right-hand side of (4.35) corresponds to the usual $\int p\,dv$. In (4.35) we have a detailed space-time description of this quantity, which is necessitated in the general case when the pressure is not constant within the fluid.

It may be added that in the general case, when the stresses are not of the simple type of a pressure and when they are described by a stress tensor $p_{ab} = p_{ba}$, the energy-momentum tensor is given by

$$T_{\mu\nu} = \begin{bmatrix} p_{ab} + \rho_0 U_a U_b + p_{ac} U_c U_b/c^2 & \rho_0 U_a U_4 + p_{ac} U_c U_4/c^2 \\ \rho_0 U_b U_4 + p_{bc} U_c U_4/c^2 & -\rho_0 c^2 - p_{ab} U_a U_b/c^2 \end{bmatrix}. \tag{4.36}$$

The relation (4.12) is now replaced by

$$\frac{\partial(wu_a)}{\partial x_a} + \frac{\partial(p_{ab} u_b)}{\partial x_a} + \frac{\partial w}{\partial t} = 0. \tag{4.37}$$

CHAPTER 7

SPINORS

1. Introduction

In Chapter 2, when dealing with three-dimensional tensors, it was stated that the group of rotations has irreducible representations of odd dimensions only. However, if the conditions for a representation are somewhat relaxed there exist also irreducible representations of even dimensions. Let R_a, R_b, R_c be rotations such that

$$R_a R_b = R_c. \tag{1.1}$$

Let $S(R_a)$, $S(R_b)$, $S(R_c)$ be matrices representing these rotations. Then according to the modified concept, we do not insist upon

$$S(R_a)S(R_b) = S(R_c)$$

but the following relation suffices:

$$S(R_a)S(R_b) = \pm S(R_c). \tag{1.2}$$

In other words, suppose we have assigned to each rotation R a matrix $S(R)$ and assume that we then find that the multiplication table of the group is not strictly obeyed, but only to a factor ± 1. Then, in accordance with the new point of view, we accept the set of matrices $S(R)$ as a representation. It is said to be two-valued, as distinct from the true or faithful representations which we have encountered earlier. In every true representation we can alter the sign of some of the representing matrices. For instance, if each rotation is represented by 1 we can get a new representation by changing some (or all) of the 'ones' to -1. However, we are not going to look upon such representations as genuinely two-valued. On the contrary we shall always assume that we have done our best to remove discrepancy as far as sign is concerned between the multiplication table of the group elements and the corresponding table of the representing matrices. A representation is genuinely two-valued only if the above discrepancy is unavoidable, so that if we try to improve matters in one relation of the type (1.2), say by changing the sign of one of the matrices, we encounter a new discrepancy in some other relation. In our example, the representation $R_a = \pm 1$, $R_b = \pm 1$, etc., can always be changed into a true representation by changing every -1 into $+1$, but whenever we speak of a two-valued representation it is to be understood that such an improvement in the representation cannot be achieved.

The matrices $S(R)$ which may have been finally selected to yield a two-valued representation cannot possibly form a group, but they may be part of a group with a greater number of matrices. Indeed, if we take our matrices $S(R)$ and add to this set all the matrices $-S(R)$ we get a group, because now the product of any two matrices is certainly contained in the set. The group obtained in this way has twice as many elements as the group of rotations. The latter may now be looked upon as a true representation of the groups of matrices $S(R)$, $-S(R)$. In this representation each pair of matrices $S(R)$ and $-S(R)$ is represented by the same rotation R. The group $S(R)$, $-S(R)$ is said to be homomorphic to the group of rotations, but the reverse is not the case.

In classical physics, two-valued representations cannot be admitted, because each physical entity enters into physical relations with a definite magnitude and an ambiguity in sign cannot be tolerated. For this reason, the two-valued representations do not figure in classical physics. Matters are different in quantum theory. Here we deal with complex state vectors **A**, **B**, etc., in a space which has, as will be explained later, an infinite number of dimensions. As far as state vectors are concerned only such quantities as $|\mathbf{A}^*\mathbf{B}|^2$ have absolute significance, representing a transition probability from the state **A** to the state **B** (or vice versa). It follows that all physical results which may be derived from the theory are invariant against so-called gauge transformations of the first kind, which are given by

$$\mathbf{A}' = e^{i\alpha}\mathbf{A}, \qquad \mathbf{B}' = e^{i\beta}\mathbf{B}, \quad \text{etc.,} \tag{1.3}$$

where α, β are arbitrary real constants.† The transformation (1.3) does not affect the probability $|\mathbf{A}^*\mathbf{B}|^2$. From this it follows that in quantum theory we can tolerate representations which are not faithful, not only with respect to sign ± 1, but also to any phase factor $\exp(i\alpha)$. Thus, instead of (1.2) we may also admit

$$S(R_a)S(R_b) = e^{i\alpha(R_a,R_b)}S(R_c), \tag{1.4}$$

where the phase factor α may depend, as indicated, on R_a, R_b. Suppose we start with a state vector **A** in K. After transformation from K to K_a through a rotation R_a we get a new state vector $\mathbf{A}_a = S(R_a)\mathbf{A}$, and by transforming from K_a to K_b a vector $\mathbf{A}_{ba} = S(R_b)S(R_a)\mathbf{A}$. On the

† All vectors $e^{i\alpha}\mathbf{A}$ with α going through the whole range of its values represent the same state. Such a set of vectors is called a ray. Hence in quantum mechanics states correspond to rays.

other hand, suppose the direct change from K to K_b, involving a matrix $R_c = R_b R_a$, leads to a result which differs from the former by a factor $\exp(i\alpha)$, i.e. to $\exp(i\alpha)\mathbf{A}_{ba}$. In classical physics such divergence in results cannot be tolerated, but in quantum theory $\mathbf{A}_{ba}$ and $\exp(i\alpha)\mathbf{A}_{ba}$ represent the same state and yield the same values for transition probabilities. Consequently a representation up to a factor of unit modulus is permissible.

Although according to (1.4) we could tolerate in quantum theory a more general type of representation than suggested in (1.2), there is no need to go beyond the latter, because it can be shown that any representation of the group of rotations is either a true representation or one which is two-valued.

All that was said above with regard to the group of rotations applies also to the Lorentz group.

Let us assume that a Lorentz transformation L involves a transformation U_L in the state vectors, where U_L are operators or matrices which transform vectors:

$$L\colon \mathbf{A} \to \mathbf{A}' = U_L \mathbf{A}, \qquad \mathbf{B} \to \mathbf{B}' = U_L \mathbf{B}, \quad \text{etc.} \tag{1.5}$$

Then, from the foregoing we know already that the U_L need not constitute a faithful representation and we may have

$$U_{L_1} U_{L_2} = \pm U_{L_1 L_2}. \tag{1.6}$$

In quantum mechanics we are interested only in unitary and anti-unitary operators U_L. In the unitary case U_L satisfies the defining conditions

$$(U_L \mathbf{A})^* U_L \mathbf{B} = \mathbf{A}^*\mathbf{B}, \qquad U_L(a\mathbf{A}+b\mathbf{B}) = aU_L \mathbf{A}+bU_L \mathbf{B}, \tag{1.7}$$

and in the anti-unitary case U_L satisfies the relations

$$(U_L \mathbf{A})^* U_L \mathbf{B} = (\mathbf{A}^*\mathbf{B})^* = \mathbf{A}\mathbf{B}^*, \qquad U_L(a\mathbf{A}+b\mathbf{B}) = a^*U_L \mathbf{A}+b^*U_L \mathbf{B}. \tag{1.8}$$

In both these cases, transition probabilities will remain invariant for Lorentz transformations:

$$|(U_L \mathbf{A})^* U_L \mathbf{B}|^2 = |\mathbf{A}^*\mathbf{B}|^2. \tag{1.9}$$

Representations which do not satisfy this relation cannot have physical significance.

It might be thought that operators U_L which satisfy such relations as

$$(U_L \mathbf{A})^* U_L \mathbf{B} = e^{i\phi_L}\mathbf{A}^*\mathbf{B}, \tag{1.10}$$

$$(U_L \mathbf{A})^* U_L \mathbf{B} = e^{i\phi_L}\mathbf{A}\mathbf{B}^*, \tag{1.11}$$

where ϕ_L is different from zero, could also be admitted. Such U_L would satisfy (1.9). However, it would follow that $\phi_{L_1L_2} = \phi_{L_2L_1} = \phi_{L_1} + \phi_{L_2}$. But as the Lorentz group is not commutative, this can only mean that ϕ_L is independent of L and leads to a constant phase factor. As such it is of no further interest and can be set equal to 1.

Although (1.7) and (1.8) are consistent with (1.9), it does not follow that for a given group of transformations which displays a symmetry through an invariance of the type (1.9) (i.e. $|\mathbf{A}'^*\mathbf{B}'|^2 = |\mathbf{A}^*\mathbf{B}|^2$, where $\mathbf{A}'$ and $\mathbf{B}'$ are the transformed states under the symmetry group) there should exist unitary or anti-unitary operators (matrices) to represent the elements of the group. A theorem by Wigner assures us that this is the case for every symmetry group and not only for the Lorentz group. However, the representation is generally only up to a factor of unit modulus (a phase factor). For the Lorentz group, the representation is only up to a factor ± 1 and no faithful representation exists. For the sub-group of space rotations both faithful and two-valued representations exist. In (1.9) we can multiply $\mathbf{A}$ and $\mathbf{B}$ by arbitrary phase factors. They will still represent the same states and yield the same transition probability. On the other hand the scalar products as they appear in (1.7) and (1.8) change their value if $\mathbf{A}$ is replaced by $e^{i\alpha}\mathbf{A}$ and $\mathbf{B}$ by $e^{i\beta}\mathbf{B}$ (by a factor $e^{-i(\alpha-\beta)}$). The situation which follows from Wigner's theorem that we can attach invariant meaning to these products (either one or the other, depending on the elements of the group), signifies that we can dispose of the arbitrary phases of the rays in such a way as to replace the latter by vectors, with only one common phase factor remaining arbitrary ($\alpha = \beta = ...$) which can be normalized to 1. This involves a great simplification since it is easier to deal with vectors. So in every quantum mechanical theory, we can assume that the arbitrary phases of the rays have already been selected in such a way as to yield a space of vectors, and we also know that the operators representing (up to a factor) the elements of a group of symmetry, will be unitary or anti-unitary.

From (1.7) and (1.8) we see that if U_L conserves the value of $\mathbf{A}^*\mathbf{B}$, so will also $e^{i\phi_L} U_L$, where ϕ_L is an arbitrary phase,

$$(e^{i\phi_L} U_L \mathbf{A})^* e^{i\phi_L} U_L \mathbf{B} = (U_L \mathbf{A})^* U_L \mathbf{B}.$$

However, as already pointed out, the arbitrariness in the representation of L can be reduced to ± 1 in the case of the Lorentz group.

The product of two unitary transformations is again unitary and the product of two anti-unitary transformations is also unitary and not anti-unitary. Consequently, the retricted Lorentz group can be represented only by unitary operators, and the anti-unitary transformations will belong to the improper Lorentz transformations which contain either space or time inversions. This correlates with the fact that the product of two improper Lorentz transformations is a proper Lorentz transformation. It can be shown further that either the space inversion or the time reversal must be represented by an anti-unitary transformation but not both. A decision in this matter can only be reached on the basis of the physical requirement that a Lorentz transformation shall not change the sign of the energy of a state. If a state has positive energy for one observer, this should be the case also for all other equivalent observers. On the basis of this requirement it follows that the time inversion and not the space inversion must be connected with an anti-unitary transformation.†

When U_L can be expressed in matrix form, it can be shown that the condition of unitarity (1.7) leads to

$$\tilde{U}_L^* = U_L^{-1}. \tag{1.12}$$

Hence, a unitary transformation is the analogue of a rotation in a space in which the scalar product is defined by $\mathbf{A}^*\mathbf{B}$ instead of $\mathbf{AB}$.

It turns out that the Lorentz group does not possess unitary transformations of finite dimensions. All the unitary representations of the Lorentz group consist of operators or matrices of infinite dimensions. Only the subgroup of rotations can be represented by unitary transformations of finite dimensions.

We shall not deal with the unitary (and anti-unitary) representations of the Lorentz group and merely confine ourselves to the following remark as regards their role in quantum mechanics.

Since the unitary representations of the Lorentz group are of infinite dimensions, it follows that if quantum theory is to be in accord with the theory of relativity the state vectors, **A**, **B**,... necessarily belong to a space of infinite dimensions. A space of vectors in which a scalar product $\mathbf{A}^*\mathbf{B}$ exists for any pair of vectors is called a Hilbert space. Hence, the state vectors of quantum theory form a Hilbert space of infinite dimensions. More specifically, the Hilbert space is a space of functions in which every vector is associated with one particular function (and more generally with a functional). Let **A** be associated with

† Suppose we represent the time inversion by the transformation

$$T\colon \mathbf{A} \to \mathbf{A}' = \mathbf{A}^*, \qquad \mathbf{B} \to \mathbf{B}' = \mathbf{B}^*.$$

This is an anti-unitary transformation, since $\mathbf{A}^*\mathbf{B} \to \mathbf{A}\mathbf{B}^*$.

Now the transformation $t' = -t$ will change the energy factor e^{iEt} into e^{-iEt}. To reproduce a positive energy state we must go over to the complex conjugate.

a function $a(s)$, where s stands for one or more variables (possibly also an infinite number of variables). Then in the Hilbert space we can denote the various axes of coordinates by the different values of s so that the component along the sth axis is given by

$$(\mathbf{A})_s = A_s = a(s). \tag{1.13}$$

The scalar product of two vectors then leads to

$$\mathbf{A}^*\mathbf{B} = \int a^*(s)b(s)\, dm(s), \tag{1.14}$$

where $dm(s)$ is an invariant volume element in the s-space. If s can take only a discrete set of values we have

$$\mathbf{A}^*\mathbf{B} = \sum_s a^*(s)b(s). \tag{1.15}$$

Apart from the unitary representations, the Lorentz group possesses representations of finite dimensions. We have already dealt with the faithful representations of this kind to some extent. They characterize the various four-tensors. Two-valued representations of finite dimensions also exist. They lead to entities called spinors, which play a fundamental role in quantum mechanics. Whereas the unitary transformations are connected primarily with the state vectors, the finite (and not unitary) representations are connected with the dynamical observables which are represented in quantum theory by Hermitian operators in the Hilbert space of state vectors. These variables are therefore by nature either tensors or spinors or, as we shall see later, of mixed character. In the case of certain spinors they transform according to a two-valued representation of the Lorentz group.

To begin with let us investigate the effect of a unitary transformation in the Hilbert space on an operator. Let O be an operator which changes $\mathbf{A}$ into $\mathbf{B}$:

$$O\mathbf{A} = \mathbf{B}. \tag{1.16}$$

Then, if a unitary transformation changes $\mathbf{A}$ into $U_L\mathbf{A}$, $\mathbf{B}$ into $U_L\mathbf{B}$, O necessarily changes into $U_L O U_L^{-1}$. For let the relation which corresponds with (1.16) after transformation be

$$O'\mathbf{A}' = \mathbf{B}'. \tag{1.17}$$

This yields

$$O'U_L\mathbf{A} = U_L\mathbf{B}; \tag{1.18}$$

therefore

$$U_L^{-1}O'U_L\mathbf{A} = U_L^{-1}U_L\mathbf{B} = \mathbf{B}, \tag{1.19}$$

where we have made use of $U_L U_L^{-1} = 1$. A comparison with (1.16) yields

$$U_L^{-1}O'U_L = O \quad \text{or} \quad O' = U_L O U_L^{-1}. \tag{1.20}$$

Suppose now that instead of a single operator O we have n operators O_p $(p = 1,..., n)$ undergoing a twofold transformation:

(1) as operators according to (1.20). This is necessitated because the variables in the functions upon which the operators are acting are transformed, leading to the transformation U_L in the Hilbert space;

(2) as components of a tensor, i.e. according to T_{pq} $(p, q = 1,..., n)$. Then we shall have

$$O'_p = U_L T_{pq} O_q U_L^{-1}. \tag{1.21}$$

In (1.21) T_{pq} is equal to $L_{\mu\nu}$ if the transformation is like that of a four-vector. T_{pq} is equal to $L_{\mu\nu} \times L_{\kappa\lambda}$ if the operator is of the nature of a tensor of second rank, etc.

Scalars, vectors, and more generally tensors have a significance which is independent of the system of coordinates. In quantum theory they are in addition operators, and we can admit only those operators to represent dynamical variables which have the same form independently of the system of coordinates. It will not do to have a given observable, for instance the energy, represented by different operators each depending on the particular Lorentz frame used. Hence in quantum theory we can only admit invariant tensor operators, and we might as well include this condition in the definition of a tensor operator. Thus a scalar operator is defined through the following transformation law:

$$O = U_L O U_L^{-1} \quad \text{(therefore } OU_L = U_L O\text{)}. \tag{1.22}$$

A four-vector operator is defined by

$$O_\mu = U_L L_{\mu\nu} O_\nu U_L^{-1}, \tag{1.23}$$

and more generally, for any tensor,

$$O_p = U_L T_{pq} O_q U_L^{-1}. \tag{1.24}†$$

From (1.22) we conclude that a scalar operator can only be a multiple of a unit matrix, since only such a matrix can commute with all the matrices U_L of a group (U_L is assumed to be an irreducible representation).

From the foregoing it follows that T_{pq} in (1.24) can only belong to a faithful representation of the Lorentz group, because any change in sign through U_L will be cancelled by the same change in sign in U_L^{-1} and no room is left for an ambiguity in sign in the transformation T_{pq}. However, it would be hasty to conclude that finite representations which

† If O_p is a field operator $O_p(x)$, (1.24) is to be used as:

$$O_p(Lx) = U_L T_{pq} O_q(x) U_L^{-1}. \tag{1.24 a}$$

are two-valued have no application in quantum theory. Since observables are represented by operators, they can be factorized. For instance, it may happen that a Hermitian operator O is equal to the product of two operators, say P_1 and P_2:

$$O = P_1 P_2, \tag{1.25}$$

and that the theory refers primarily to P_1 and P_2 and not directly to O. It may even turn out that P_1 and P_2 are not Hermitian and are thus not themselves observables. In such a case P_1 and P_2 may well transform according to a two-valued representation because their product will be free from ambiguity. As mentioned already, entities transforming according to finite two-valued representations of the Lorentz group are called spinors. In the following we shall deal with the spinor representations of the Lorentz group.

2. The Unimodular Group

Let us consider entities with two components $\phi_A = (\phi_1, \phi_2)$ and subject them to a transformation $\phi'_A = T_{AB}\phi_B$, or

$$\begin{aligned} \phi'_1 &= T_{11}\phi_1 + T_{12}\phi_2, \\ \phi'_2 &= T_{21}\phi_1 + T_{22}\phi_2, \end{aligned} \tag{2.1}$$

where the determinant of the matrix T is 1:

$$T_{11}T_{22} - T_{12}T_{21} = 1. \tag{2.2}$$

Transformations of this type are called unimodular. It is assumed that the components T_{AB} of the matrix may have complex values. As a consequence, this also applies to (ϕ_1, ϕ_2).

The product of two unimodular matrices is again a unimodular matrix. Moreover, each such matrix possesses a reciprocal. Consequently, all the unimodular matrices form a group. It is denoted by (C_2), where the suffix 2 indicates that we are dealing with entities with two components.

Entities ϕ_A ($A = 1, 2$) transforming according to the group (C_2) are called spinors of first rank. Let ϕ_A and ψ_A be two such spinors and let each be transformed according to (2.1). Then, because of (2.2), the following invariance holds:

$$\phi_1\psi_2 - \phi_2\psi_1 = \phi'_1\psi'_2 - \phi'_2\psi'_1. \tag{2.3}$$

This may be expressed in a more convenient form by introducing the numerical matrix

$$\epsilon^{AB} = \begin{pmatrix} 0 & 1 \\ -1 & 0 \end{pmatrix} \equiv \epsilon'^{AB}, \tag{2.4}$$

which, by definition, is to have the same elements independently of any transformations. It will later turn out that ϵ^{AB} is an invariant spinor of second rank. Instead of (2.3) we then get

$$\epsilon^{AB}\phi_A\psi_B = \epsilon^{AB}\phi'_A\psi'_B. \tag{2.5}$$

Again we follow the usual convention of dropping the $\sum$ whenever we sum up over equal indices.

Each matrix
$$A = \begin{bmatrix} A_{11} & A_{12} \\ A_{21} & A_{22} \end{bmatrix} \tag{2.6}$$
possesses its complex conjugate
$$A^* = \begin{bmatrix} A^*_{11} & A^*_{12} \\ A^*_{21} & A^*_{22} \end{bmatrix}. \tag{2.7}$$

If A, B, C are three matrices, such that $AB = C$, then also $A^*B^* = C^*$. If A is unimodular, so is A^*. Instead of considering the original group (C_2) of matrices A, B,... we can consider (C_2^*) with matrices A^*, B^*,.... This group contains the same elements as (C_2) and yields a representation of (C_2). (C_2) and (C_2^*) are not equivalent, i.e. there exists no matrix S which transforms all the elements of (C_2) into the corresponding elements of (C_2^*):

$$SAS^{-1} \neq A^* \quad \text{(for all } A\text{'s).} \tag{2.8}$$

To prove this we need only consider any one of the following three unimodular matrices:

$$\begin{bmatrix} i & 0 \\ 0 & -i \end{bmatrix}, \quad \begin{bmatrix} 0 & i \\ i & 0 \end{bmatrix}, \quad \begin{bmatrix} 0 & -1 \\ 1 & 0 \end{bmatrix}.$$

Every one of them transforms into the conjugate matrix by

$$S = S^{-1} = \begin{bmatrix} 0 & i \\ -i & 0 \end{bmatrix},$$

and only by this matrix (apart from a factor). However, when applied to a general unimodular matrix we get

$$\begin{bmatrix} 0 & i \\ -i & 0 \end{bmatrix}\begin{bmatrix} \alpha & \beta \\ \gamma & \delta \end{bmatrix}\begin{bmatrix} 0 & i \\ -i & 0 \end{bmatrix} = \begin{bmatrix} \delta & -\gamma \\ -\beta & \alpha \end{bmatrix}, \tag{2.9}$$

and this is generally different from $\begin{bmatrix} \alpha^* & \beta^* \\ \gamma^* & \delta^* \end{bmatrix}$.

If we wish to emphasize that a spinor transforms according to (C_2^*) we dot the index. Thus, $\phi_{\dot{A}}$ transforms according to

$$\phi'_{\dot{1}} = T^*_{11}\phi_{\dot{1}} + T^*_{12}\phi_{\dot{2}},$$
$$\phi'_{\dot{2}} = T^*_{21}\phi_{\dot{1}} + T_{22}\phi_{\dot{2}}. \tag{2.10}$$

If ϕ_A transforms according to (C_2), $\phi_{\dot{A}}$ transforms according to (C_2^*). We shall always adopt the notational convention according to which ϕ_A and $\phi_{\dot{A}}$ are the complex conjugates of each other:

$$\phi_A = \phi_{\dot{A}}^*, \qquad \phi_A^* = \phi_{\dot{A}}, \qquad \phi_{\dot{A}}^{\cdot} = \phi_A. \tag{2.11}$$

Another matrix which can be associated with A is the contragredient:

$$A^c = \tilde{A}^{-1} = \widetilde{A^{-1}}, \tag{2.12}$$

where $\tilde{A}$ is the transpose of A. If A is given by (2.6) we have

$$\tilde{A} = \begin{bmatrix} A_{11} & A_{21} \\ A_{12} & A_{22} \end{bmatrix}, \quad A^{-1} = \begin{bmatrix} A_{22} & -A_{12} \\ -A_{21} & A_{11} \end{bmatrix}, \quad A^c = \begin{bmatrix} A_{22} & -A_{21} \\ -A_{12} & A_{11} \end{bmatrix}. \tag{2.13}$$

Like A, A^c is unimodular and it is easily verified by carrying out the multiplication that if $AB = C$, it follows that $A^cB^c = C^c$. Again, instead of considering the original group (C_2) with matrices A, B,..., we can consider (C_2^c) with matrices A^c, B^c which is identical with (C_2) except for a rearrangement of the elements. (C_2^c) is a representation of (C_2) but these two sets of matrices are equivalent because we can determine a matrix which transforms all the matrices of (C_2) into those of (C_2^c):

$$SAS^{-1} = \begin{bmatrix} 0 & 1 \\ -1 & 0 \end{bmatrix}\begin{bmatrix} \alpha & \beta \\ \gamma & \delta \end{bmatrix}\begin{bmatrix} 0 & -1 \\ 1 & 0 \end{bmatrix} = \begin{bmatrix} \delta & -\gamma \\ -\beta & \alpha \end{bmatrix}. \tag{2.14}$$

We call a quantity transforming according to (C_2^c) a contravariant spinor and denote it by an upper index, such as ϕ^A. Thus, if ϕ_A transforms according to (2.1), ϕ^A transforms as follows:

$$\begin{aligned} \phi^{1'} &= T_{22}\phi^1 - T_{21}\phi^2, \\ \phi^{2'} &= -T_{12}\phi^1 + T_{11}\phi^2. \end{aligned} \tag{2.15}$$

Apart from A^* and A^c we can also associate with each matrix the transpose $\tilde{A}$, mentioned earlier, and the reciprocal. But in these cases, if $AB = C$, we get $\tilde{B}\tilde{A} = \tilde{C}$ and $B^{-1}A^{-1} = C^{-1}$. Therefore, although the transposed and reciprocal matrices are also unimodular, they do not yield representations of the original group (C_2).

If the matrices (C_2) are a two-dimensional representation of some group, then (C_2^*) and (C_2^c) are also representations of this group. We shall see later that (C_2) represents the Lorentz group proper (which does not involve inversions of space axes from right-handed to left-handed systems or time reversal).

Let us form the direct product $\phi_A\psi_B$. Then, exactly as in the case of tensors, it can be directly verified that the column with four components $\phi_1\psi_1$, $\phi_1\psi_2$, $\phi_2\psi_1$, $\phi_2\psi_2$ transforms according to the direct product of the group $(C_2 \times C_2)$ which consists of matrices $(A \times A)$, $(B \times B)$, $(C \times C)$, etc., and that if

$$AB = C, \tag{2.16}$$

$$(A \times A)(B \times B) = (C \times C). \tag{2.17}$$

In other words, if we form the direct product of each matrix of (C_2) with itself, we get matrices of four rows and columns which represent the original group. As indicated above, this new group is denoted by $(C_2 \times C_2)$. In a similar way we can form the direct products $\phi_A \psi_{\dot{B}}, \phi_A \psi^B$ which transform according to $(C_2 \times C_2^*)$ and $(C_2 \times C_2^c)$ respectively. If (2.16) holds, we also have

$$(A \times A^*)(B \times B^*) = (C \times C^*), \tag{2.18}$$

$$(A \times A^c)(B \times B^c) = (C \times C^c). \tag{2.19}$$

Other direct products are $\phi^A \psi^B$ transforming according to $(C_2^c \times C_2^c)$ and $\phi^{\dot{A}} \psi^B$ transforming according to $(C_2^{*c} \times C_2)$.

It may be remarked in passing that the matrix products (AA), (BB), (CC) $(= A^2, B^2, C^2)$ do not form a representation. If $AB = C$,

$$(AA)(BB) \neq (AB)(AB). \tag{2.20}$$

The associative law holds, but both sides are equal only if $AB = BA$. Hence, only in the case of commutative groups can we get new representations by taking the matrix product of each matrix with itself.

From the foregoing it follows that if (C_2) is a representation of some group, this also applies to C_2^*, C_2^c, $(C_2 \times C_2)$, $(C_2^* \times C_2)$, $(C_2^c \times C_2^c)$, $(C_2 \times C_2^*)$, etc.

Entities transforming like $\phi_A \psi_B$, i.e. according to matrices from $(C_2 \times C_2)$, are denoted by symbols with two indices, as ϕ_{AB}. Similarly, entities transforming according to $(C_2 \times C_2^*)$ are denoted by $\phi_{A\dot{B}}$, those transforming according to $(C_2 \times C_2^c)$ by $\phi_A{}^B$, and those transforming according to $(C_2^c \times C_2)$ by $\phi^A{}_B$, etc. All these are called spinors of second rank.

The spinors which we have introduced are all to be treated as columns, and are expressed here as rows only for convenience:

$$\begin{aligned}
&\phi_A \equiv (\phi_1, \phi_2), \quad \phi^A \equiv (\phi^1, \phi^2), \quad \phi_{\dot{A}} = (\phi_{\dot{1}}, \phi_{\dot{2}}),\\
&\phi_{AB} \equiv (\phi_{11}, \phi_{12}, \phi_{21}, \phi_{22}), \quad \phi^{AB} \equiv (\phi^{11}, \phi^{12}, \phi^{21}, \phi^{22}),\\
&\phi_A{}^B \equiv (\phi_1{}^1, \phi_1{}^2, \phi_2{}^1, \phi_2{}^2),\\
&\phi^A{}_B \equiv (\phi^1{}_1, \phi^1{}_2, \phi^2{}_1, \phi^2{}_2).
\end{aligned} \tag{2.21}$$

Denoting the matrices T of (C_2) by $T_A{}^B$ we get for spinors of first rank

$$\phi'_A = T_A{}^B\phi_B, \tag{2.22}$$

$$\phi'_{\dot{A}} = T_A^{*B}\phi_{\dot{B}}, \tag{2.23}$$

$$\phi^{A\prime} = T_A^{cB}\phi^B, \tag{2.24}$$

$$\phi^{\dot{A}\prime} = T_A^{*c\,B}\phi^{\dot{B}}. \tag{2.25}$$

In all these formulae we sum over the index appearing twice, irrespective of one being dotted and the other not and irrespective of their being both upper or lower indices or mixed. If we wish to adhere to a rule of summing only over indices appearing as an upper or a lower index, we must re-write (2.24) and (2.25) as follows:

$$\phi^{A\prime} = T^{-1}_{\;B}{}^A\phi^B, \tag{2.26}$$

$$\phi^{\dot{A}\prime} = T^{-1*}_{\;B}{}^A\phi^{\dot{B}}. \tag{2.27}$$

However, we now get a formalism in which the summation indices do not follow each other (as in (2.22) and (2.23)). Formally we can remedy this by using the convention

$$T^{-1}_{\;B}{}^A \equiv T^A{}_B = T_A^{cB}. \tag{2.28}$$

This convention has already been used when dealing with the Lorentz matrices $L^\mu{}_\nu$ and $L_\mu{}^\nu$. We shall adopt it only when dealing with transformation matrices, but not for spinors written as matrices. In fact, all we can say about $\phi_A{}^B$ is that if we write it as a (2×2) matrix and transpose it we shall get an entity which transforms like $\phi^A{}_B$. This will be proved later.

Eqns. (2.22) to (2.25) may be expressed in matrix formulae where the ϕ's are columns, as follows:

$$\phi' = T\phi, \tag{2.29}$$

$$\phi^{*\prime} = T^*\phi^*, \tag{2.30}$$

$$\phi^{c\prime} = T^c\phi^c, \tag{2.31}$$

$$\phi^{c*\prime} = T^{c*}\phi^{c*}. \tag{2.32}$$

As regards spinors of second rank we have

$$\phi'_{AB} = T_A{}^C T_B{}^D\phi_{CD}, \tag{2.33}$$

$$\phi'_{\dot{A}B} = T_A^{*C} T_B{}^D\phi_{\dot{C}D}, \tag{2.34}$$

$$\phi'^{AB} = T_A^{cC} T_B^{cD}\phi_{CD} \equiv T^A{}_C\, T^B{}_D\phi^{CD}, \tag{2.35}$$

$$\phi'_A{}^B = T_A{}^C T_B^{cD}\phi_C{}^D \equiv T_A{}^C T^B{}_D\,\phi_C{}^D, \tag{2.36}$$

$$\phi'^A{}_B = T_A^{cC} T_B{}^D\phi^C{}_D \equiv T^A{}_C\, T_B{}^D\phi^C{}_D. \tag{2.37}$$

In this case we can express the ϕ's also as two-dimensional matrices. It is then essential to arrange the factors so that summation indices follow each other:

$$(\phi_{AB})' = T_A{}^C \phi_{CD} \tilde{T}_D{}^B = T(\phi_{AB})\tilde{T}, \tag{2.38}$$

$$(\phi_{\dot{A}B})' = \qquad = T^*(\phi_{\dot{A}B})\tilde{T}, \tag{2.39}$$

$$(\phi^{AB})' = \qquad = T^c(\phi^{AB})\tilde{T}^c, \tag{2.40}$$

$$(\phi_A{}^B)' = \qquad = T(\phi_A{}^B)\tilde{T}^c, \tag{2.41}$$

$$(\phi^A{}_B)' = \qquad = T^c(\phi^A{}_B)\tilde{T}. \tag{2.42}$$

The brackets are to indicate that ϕ_{AB}, etc., are to be expressed as two-dimensional matrices and not as columns.

We now see that if we transpose (2.41) we get a relationship which is the same as in (2.42):

$$(\widetilde{\phi_A{}^B})' = \tilde{\tilde{T}}^c(\tilde{\phi}_A{}^B)\tilde{T} = T^c(\tilde{\phi}_A{}^B)\tilde{T}. \tag{2.43}$$

Hence, if $(\phi_A{}^B)$ transforms according to $(T \times T^c)$, $(\phi_B{}^A)$ transforms according to $(T^c \times T)$, i.e. like a $(\phi^A{}_B)$. Similar results are obtained when dealing with spinors of higher rank. A spinor transforming with $(T \times T^c \times T^*)$ differs from a spinor transforming according to $(T^c \times T^* \times T)$ only by a corresponding permutation of indices. We can therefore characterize a spinor essentially only by stating how many indices are covariant (some with dots and some without dots), and how many are contravariant. (2.36), for instance, will then become

$$\phi'^B_A = T_A{}^C T^c_B{}^D \phi^D_C = T_A{}^C T^B{}_D \phi^D_C, \tag{2.44}$$

and this should be identical with the relation obtained from (2.37):

$$\phi'^A_B = T^{cC}_A T_B{}^D \phi^C_D;$$

therefore

$$\phi'^A_B = T^{cD}_A T_B{}^C \phi^D_C, \tag{2.45}$$

and

$$\phi'^B_A = T^c_B{}^D T_A{}^C \phi^D_C = T_A{}^C\, T^c_B{}^D \phi^D_C, \tag{2.46}$$

which is identical with (2.44).

In the general case we shall have

$$\phi^{C\ldots\dot{D}}_{A\ldots\dot{B}} = T_A{}^P \ldots T^*_{\dot{B}}{}^Q T^c_C{}^R \ldots T^{c*}_{\dot{D}}{}^S \phi^{R\ldots\dot{S}}_{P\ldots\dot{Q}} = T_A{}^P \ldots T^*_{\dot{B}}{}^Q T^C{}_R \ldots T^{*\dot{D}}{}_{\dot{S}} \phi^{R\ldots\dot{S}}_{P\ldots\dot{Q}}. \tag{2.47}$$

The spacing of indices can be performed in many ways, as $\phi^A{}_B{}^C{}_D$ or $\phi_A{}^{BC}{}_D$, etc. A distinction becomes necessary only if we revert to matrix formalism (columns, square matrices, etc.).

The reason for the greater complexity of types of spinors encountered now, as compared with our previous tensors, is due to the difference which exists between T, T^*, T^c. In the case of our previous three-dimensional tensors, if R is a rotation of axes, it is real ($R = R^*$) and self-contragredient ($R = R^c$), i.e. orthogonal. Consequently, in the case of our former three-tensors there was no need to distinguish between covariant and contravariant indices, nor did the need for something like dotted indices arise.

Also, the more general rule of direct multiplication applies in the case of spinors as in the case of tensors. Two spinors of rank m and n yield on direct multiplication a spinor of rank $m+n$ which transforms according to $(C_2\times C_2\times\ldots\times C_2)(C_2\times C_2\times\ldots\times C_2) = (C_2\times C_2\times\ldots\times C_2)$ (the first factor has m members, the second n, and the third $m+n$).

However, the rule for contraction requires a certain caution. By turning a dotted and an undotted index into a dummy index and summing up, the spinor character is not maintained. The same applies also if we make two covariant or two contravariant indices equal and sum up. The spinor properties are preserved only as we contract with respect to a covariant and contravariant index, both either undotted or dotted. Thus,

$$\phi_B{}^A{}_A \left(= \sum_{A=1,2} \phi_B{}^A{}_A\right)$$

is a covariant spinor of first rank. Similarly, $\phi_B{}^{\dot A}{}_{\dot A}$ or $\phi_B{}^A{}_A$ are spinors of first rank, but $\phi_B{}^{\dot A}{}_A$ has no invariant meaning.

Although the matrix ϵ^{AB} has been introduced purely algebraically and by definition it remains unaffected when a transformation is performed, we can now prove that we may also look upon it as a spinor of second rank which happens to have invariant components. For convenience let us denote the unimodular matrix T by

$$T = \begin{bmatrix} \alpha & \beta \\ \gamma & \delta \end{bmatrix} \quad (\alpha\delta-\beta\gamma = 1). \tag{2.48}$$

We can then express the transformation of ϵ^{AB} when treated as a spinor as follows:

$$(T\times T)\epsilon^{AB} \equiv \begin{bmatrix} \alpha\begin{bmatrix} \alpha & \beta \\ \gamma & \delta \end{bmatrix} & \beta\begin{bmatrix} \alpha & \beta \\ \gamma & \delta \end{bmatrix} \\ \gamma\begin{bmatrix} \alpha & \beta \\ \gamma & \delta \end{bmatrix} & \delta\begin{bmatrix} \alpha & \beta \\ \gamma & \delta \end{bmatrix} \end{bmatrix} \begin{bmatrix} 0 \\ -1 \\ +1 \\ 0 \end{bmatrix}$$

$$= \begin{bmatrix} \alpha^2 & \alpha\beta & \beta\alpha & \beta^2 \\ \alpha\gamma & \alpha\delta & \beta\gamma & \beta\delta \\ \alpha\gamma & \beta\gamma & \alpha\delta & \beta\delta \\ \gamma^2 & \gamma\delta & \gamma\delta & \delta^2 \end{bmatrix} \begin{bmatrix} 0 \\ -1 \\ +1 \\ 0 \end{bmatrix} = \begin{bmatrix} 0 \\ -1 \\ +1 \\ 0 \end{bmatrix}. \tag{2.49}$$

Alternatively, using (2.38),

$$(\epsilon^{AB})' = T(\epsilon^{AB})\tilde{T} = \begin{bmatrix}\alpha & \beta\\ \gamma & \delta\end{bmatrix}\begin{bmatrix}0 & 1\\ -1 & 0\end{bmatrix}\begin{bmatrix}\alpha & \gamma\\ \beta & \delta\end{bmatrix} = \begin{bmatrix}0 & 1\\ -1 & 0\end{bmatrix} = (\epsilon^{AB}). \tag{2.50}$$

The same applies also if we introduce $\epsilon_{AB} = \epsilon_{\dot{A}\dot{B}}$, which is defined by

$$\epsilon_{AB} = \epsilon_{\dot{A}\dot{B}} = \begin{bmatrix}0 & -1\\ 1 & 0\end{bmatrix}, \tag{2.51}$$

so that

$$\epsilon_{AB} = \epsilon_{\dot{A}\dot{B}} = -\epsilon^{AB} = -\epsilon^{\dot{A}\dot{B}}. \tag{2.52}$$

With the help of ϵ^{AB} and ϵ_{AB} we can lower and raise indices as follows:

$$\phi^A = \epsilon^{AB}\phi_B = \begin{bmatrix}\phi_2\\ -\phi_1\end{bmatrix}, \tag{2.53}$$

$$\phi_A = \epsilon_{AB}\phi^B = \begin{bmatrix}-\phi^2\\ \phi^1\end{bmatrix}. \tag{2.54}$$

By using $\epsilon_{AB} = -\epsilon^{AB}$ we have the advantage that in both cases we sum over consecutive indices. This is slightly offset by the disadvantage of ϵ_{AB} yielding $-\epsilon^{AB}$ on raising the indices.

Raising the index of ϕ_A in (2.54) should again yield the original ϕ^A:

$$\phi^A = \begin{bmatrix}\phi^1\\ -(-\phi^2)\end{bmatrix} = \begin{bmatrix}\phi^1\\ \phi^2\end{bmatrix}. \tag{2.55}$$

It is easily proved that if (ϕ_1, ϕ_2) transform with a matrix T, $(\phi_2, -\phi_1)$ transform with a matrix T^c:

$$\begin{aligned}\phi_2' &= \gamma\phi_1+\delta\phi_2 = \delta\phi_2+\gamma\phi_1 = \delta\phi_2-\gamma(-\phi_1),\\ -\phi_1' &= -(\alpha\phi_1+\beta\phi_2) = -\beta\phi_2-\alpha\phi_1 = -\beta\phi_2+\alpha(-\phi_1).\end{aligned} \tag{2.56}$$

3. The Relation between the Unimodular Group and the Restricted Lorentz Group

Let us consider a spinor of second rank $\phi_{\dot{A}B}$. A transformation matrix of such a spinor is given by

$$(T^*\times T) = \begin{bmatrix}\alpha^*\alpha & \alpha^*\beta & \beta^*\alpha & \beta^*\beta\\ \alpha^*\gamma & \alpha^*\delta & \beta^*\gamma & \beta^*\delta\\ \gamma^*\alpha & \gamma^*\beta & \delta^*\alpha & \delta^*\beta\\ \gamma^*\gamma & \gamma^*\delta & \delta^*\gamma & \delta^*\delta\end{bmatrix} \tag{3.1}$$

By applying this matrix on $(\phi_{\dot{1}1}, \phi_{\dot{1}2}, \phi_{\dot{2}1}, \phi_{\dot{2}2})$ (written as a column) we get the new components and the following relation can be verified:

$$\phi_{\dot{1}1}\phi_{\dot{2}2}-\phi_{\dot{1}2}\phi_{\dot{2}1} = \phi'_{\dot{1}1}\phi'_{\dot{2}2}-\phi'_{\dot{1}2}\phi'_{\dot{2}1}. \tag{3.2}$$

This can also be seen by considering the spinor of zero rank (an invariant):

$$\tfrac{1}{2}\epsilon^{\dot{A}\dot{B}}\epsilon^{CD}\phi_{\dot{A}C}\phi_{\dot{B}D}. \tag{3.3}$$

This fourfold sum contains non-vanishing terms only when $\dot{A} \neq \dot{B}$, $C \neq D$. Hence only the terms $\phi_{\dot{1}1}\phi_{\dot{2}2}$ and $\phi_{\dot{1}2}\phi_{\dot{2}1}$ do not vanish. Each of them occurs twice, the first with positive, the second with negative sign.

Let us denote the invariant in (3.2) or (3.3) by s^2. We are particularly interested in spinors $\phi_{\dot{A}B}$ which yield a real value for s^2. This will certainly be the case if

$$\phi_{\dot{1}1} = \phi^*_{\dot{1}1}, \quad \phi_{\dot{2}2} = \phi^*_{\dot{2}2}, \quad \phi_{\dot{1}2} = \phi^*_{\dot{2}1}, \tag{3.4}$$

which may be summed up as follows:

$$\phi_{\dot{A}B} = \phi^*_{\dot{B}A} = \phi_{B\dot{A}}. \tag{3.5}$$

A second-rank spinor which satisfies the condition (3.5) may be called a Hermitian spinor, for if we order its components as a matrix and use $\phi_{\dot{1}1} = \phi^*_{\dot{1}1}$, etc., we get the following condition satisfied:

$$\begin{bmatrix} \phi_{1\dot{1}} & \phi_{1\dot{2}} \\ \phi_{2\dot{1}} & \phi_{2\dot{2}} \end{bmatrix} = \begin{bmatrix} \phi^*_{1\dot{1}} & \phi^*_{2\dot{1}} \\ \phi^*_{1\dot{2}} & \phi^*_{2\dot{2}} \end{bmatrix}. \tag{3.6}$$

This property of a spinor of second rank with one simple and one dotted index remains unchanged when a transformation (3.1) is performed and this makes it possible to produce a relationship between the unimodular group and the Lorentz group. Lorentz transformations are characterized by the invariance of

$$s^2 = -x^2-y^2-z^2+c^2t^2 = (z+ct)(-z+ct)-(x+iy)(x-iy). \tag{3.7}$$

As will be seen, this invariant contains two real factors and two complex conjugate factors. We can order them as a Hermitian matrix:

$$\begin{bmatrix} z+ct & x+iy \\ x-iy & -z+ct \end{bmatrix}. \tag{3.8}$$

Let

$$\begin{aligned} \phi_{\dot{1}1} &= z+ct, & \phi'_{\dot{1}1} &= z'+ct', \\ \phi_{\dot{2}2} &= -z+ct, & \phi'_{\dot{2}2} &= -z'+ct', \\ \phi_{\dot{1}2} &= x+iy, & \phi'_{\dot{1}2} &= x'+iy', \\ \phi_{\dot{2}1} &= x-iy, & \phi'_{\dot{2}1} &= x'-iy'. \end{aligned} \tag{3.9}$$

Then, if $\phi_{\dot{A}B}$ transforms according to $(C_2^* \times C_2)$ the variables x, y, z, t transform linearly, and from (3.2) it follows that $-(x^2+y^2+z^2)+c^2t^2$ will be an invariant. Moreover, since $\phi_{\dot{1}2} = x+iy$ and $\phi_{\dot{2}1} = x-iy$

remain complex conjugates, the linear relation between x, y, z, t and x', y', z', t' has real coefficients so that this transformation is a Lorentz transformation.

Our next step will be to show that the four-dimensional matrix group $(C_2^* \times C_2)$ contains as many transformations as the Lorentz group proper and that the two groups are isomorphic with each other, so that every Lorentz transformation proper is represented by a corresponding matrix in $(C_2^* \times C_2)$ and vice versa. It will be recalled that the restricted Lorentz group does not contain an inversion of single space axes $(x' = -x)$, $(y' = -y)$, $(z' = -z)$, the simultaneous reflection of all the space axes, or a time reversal $t' = -t$.

We begin by showing that all the space rotations, involving x, y, z only, are represented in $(C_2^* \times C_2)$. A rotation of space as an element of the Lorentz group is obtained if t or ct is not transformed.

In the present case this implies that

$$\phi_{11}+\phi_{\dot{2}2} = (z+ct)+(-z+ct) = 2ct \tag{3.10}$$

must remain unchanged. The corresponding sub-group in $(C_2^* \times C_2)$ is obtained if we consider among the unimodular matrices those which are in addition unitary:

$$U = \begin{bmatrix} \alpha & \beta \\ -\beta^* & \alpha^* \end{bmatrix} \tag{3.11}$$

where

$$\alpha\alpha^*+\beta\beta^* = 1. \tag{3.12}$$

The unitary matrices form a sub-group (U_2) in (C_2) because the product of two unitary matrices is again a unitary matrix. Also, the reciprocal of a unitary matrix is unitary. It may be obtained from the defining relation

$$U^{-1} = \tilde{U}^*, \tag{3.13}$$

or

$$U^{-1} = \begin{bmatrix} \alpha^* & -\beta \\ \beta^* & \alpha \end{bmatrix}. \tag{3.14}$$

Under a unitary transformation not only $\psi_1\phi_2-\psi_2\phi_1$ but also $\phi_1\phi_1^*+\phi_2\phi_2^*$ is an invariant as is easily verified (the same applies to $\psi_1\psi_1^*+\psi_2\psi_2^*$).

Since ϕ_1^* and ϕ_2^* transform like ψ_1^* and ψ_2^* or $\psi_{\dot{1}}$ and $\psi_{\dot{2}}$, it follows that $\psi_{\dot{1}}\phi_1+\psi_{\dot{2}}\phi_2$ is an invariant under the transformations $(U_2^* \times U_2)$. It follows further that $\psi_{\dot{1}1}+\psi_{\dot{2}2}$ is an invariant of the group $(U_2^* \times U_2)$. From (3.10) it thus follows that the unitary sub-group represents space rotations. We have now only to convince ourselves that all the rotations are represented. From (3.9) we have

$$\begin{aligned} x &= \tfrac{1}{2}(\phi_{1\dot{2}}+\phi_{2\dot{1}}), \\ y &= \tfrac{1}{2}i^{-1}(\phi_{1\dot{2}}-\phi_{2\dot{1}}), \\ z &= \tfrac{1}{2}(\phi_{1\dot{1}}-\phi_{2\dot{2}}), \\ ct &= \tfrac{1}{2}(\phi_{1\dot{1}}+\phi_{2\dot{2}}). \end{aligned} \tag{3.15}$$

The invariance of t under unitary transformations implies that we have relations of the following type:

$$\begin{aligned} 2x' &= \phi'_{1\dot{2}}+\phi'_{2\dot{1}} &&= r_{11}(\phi_{1\dot{2}}+\phi_{2\dot{1}})+r_{12}\,i^{-1}(\phi_{1\dot{2}}-\phi_{2\dot{1}})+r_{13}(\phi_{1\dot{1}}-\phi_{2\dot{2}}), \\ 2y' &= i^{-1}(\phi'_{1\dot{2}}-\phi'_{2\dot{1}}) &&= r_{21}(\phi_{1\dot{2}}+\phi_{2\dot{1}})+r_{22}\,i^{-1}(\phi_{1\dot{2}}-\phi_{2\dot{1}})+r_{23}(\phi_{1\dot{1}}-\phi_{2\dot{2}}), \\ 2z' &= \phi'_{1\dot{1}}-\phi'_{2\dot{2}} &&= r_{31}(\phi_{1\dot{2}}+\phi_{2\dot{1}})+r_{32}\,i^{-1}(\phi_{1\dot{2}}-\phi_{2\dot{1}})+r_{33}(\phi_{1\dot{1}}-\phi_{2\dot{2}}), \\ 2ct' &= \phi'_{1\dot{1}}+\phi'_{2\dot{2}} &&= (\phi_{1\dot{1}}+\phi_{2\dot{2}}). \end{aligned} \tag{3.16}$$

We have to prove that the r_{ik} yield all the possible rotations, as we run through all the unitary transformations. Because of (3.12) the unitary matrices depend on three real parameters, say ϕ, θ, ψ, and we may express every unitary transformation as

$$\begin{aligned} U &= \pm\begin{bmatrix} e^{-\frac{1}{2}i\phi} & 0 \\ 0 & e^{\frac{1}{2}i\phi} \end{bmatrix}\begin{bmatrix} \cos\frac{1}{2}\theta & i\sin\frac{1}{2}\theta \\ i\sin\frac{1}{2}\theta & \cos\frac{1}{2}\theta \end{bmatrix}\begin{bmatrix} e^{-\frac{1}{2}i\psi} & 0 \\ 0 & e^{\frac{1}{2}i\psi} \end{bmatrix} \\ &= \pm\begin{bmatrix} e^{-\frac{1}{2}i\phi}\cos\frac{1}{2}\theta\, e^{-\frac{1}{2}i\psi} & ie^{-\frac{1}{2}i\phi}\sin\frac{1}{2}\theta\, e^{\frac{1}{2}i\psi} \\ ie^{+\frac{1}{2}i\phi}\sin\frac{1}{2}\theta\, e^{-\frac{1}{2}i\psi} & e^{\frac{1}{2}i\phi}\cos\frac{1}{2}\theta\, e^{\frac{1}{2}i\psi} \end{bmatrix}. \end{aligned} \tag{3.17}$$

For the direct product $(U^*\times U)$ we get

$$(U^*\times U) = \begin{bmatrix} \cos^2\frac{1}{2}\theta & i\sin\frac{1}{2}\theta\cos\frac{1}{2}\theta e^{i\psi} & -i\sin\frac{1}{2}\theta\cos\frac{1}{2}\theta e^{-i\psi} & \sin^2\frac{1}{2}\theta \\ i\sin\frac{1}{2}\theta\cos\frac{1}{2}\theta e^{i\phi} & \cos^2\frac{1}{2}\theta e^{i(\phi+\psi)} & \sin^2\frac{1}{2}\theta e^{+i(\phi-\psi)} & -i\sin\frac{1}{2}\theta\cos\frac{1}{2}\theta e^{i\phi} \\ -i\sin\frac{1}{2}\theta\cos\frac{1}{2}\theta e^{-i\phi} & \sin^2\frac{1}{2}\theta e^{-i(\phi-\psi)} & \cos^2\frac{1}{2}\theta e^{-i(\phi+\psi)} & i\sin\frac{1}{2}\theta\cos\frac{1}{2}\theta e^{-i\phi} \\ \sin^2\frac{1}{2}\theta & -i\sin\frac{1}{2}\theta\cos\frac{1}{2}\theta\, e^{i\psi} & i\sin\frac{1}{2}\theta\cos\frac{1}{2}\theta e^{-i\psi} & \cos^2\frac{1}{2}\theta \end{bmatrix}. \tag{3.18}$$

With this transformation $\phi'_{1\dot{1}}$ is given by

$$\phi'_{1\dot{1}} = \cos^2\tfrac{1}{2}\theta\phi_{1\dot{1}}+i\sin\tfrac{1}{2}\theta\cos\tfrac{1}{2}\theta\, e^{i\psi}\phi_{1\dot{2}}-i\sin\tfrac{1}{2}\theta\cos\tfrac{1}{2}\theta\, e^{-i\psi}\phi_{2\dot{1}}+\sin^2\tfrac{1}{2}\theta\,\phi_{2\dot{2}}, \tag{3.19}$$

and corresponding relations follow for the other components of $\phi_{A\dot{B}}$. If we then evaluate the transformations for the linear combinations in (3.16) we get the following result:

$$\begin{aligned} x' &= (\cos\phi\cos\psi-\sin\phi\sin\psi\cos\theta)x+(\cos\phi\sin\psi+\sin\phi\cos\psi\cos\theta)y-\sin\phi\sin\theta\, z, \\ y' &= -(\sin\phi\cos\psi+\cos\phi\sin\psi\cos\theta)x+(\sin\phi\sin\psi+\cos\phi\cos\psi\cos\theta)y-\cos\phi\sin\theta\, z, \\ z' &= -\sin\psi\sin\theta\, x+\cos\psi\sin\theta\, y+\cos\theta\, z. \end{aligned} \tag{3.20}$$

This transformation is, however, a rotation from x, y, z to x', y', z' in terms of the Euler angles ϕ, θ, ψ. The rotation is obtained by rotating the frame first around the z-axis by an angle ϕ, then around the new x-axis by an angle θ, and finally again around the new z-axis by an angle ψ. We thus find that the group $(U_2^* \times U_2)$ is isomorphic with the groups of rotations and vice versa. The inverse rotation to (3.20) is obtained by first rotating around the z-axis by an angle $-\psi$, then around the new x-axis by an angle $-\theta$, and finally around the new z-axis by an angle $-\phi$.

Hence, to obtain the reciprocal matrix we have to replace ϕ by $-\psi$, θ by $-\theta$, and ψ by $-\phi$ in (3.20). It is then easily verified that the orthogonality condition $r_{ik} = r_{ki}^{-1}$ holds.

The matrices $(U_2^* \times U_2)$ are four-dimensional and contain two irreducible representations of the group of rotations, according to the scheme

$$(U^* \times U) \sim \begin{bmatrix} R & 0 \\ 0 & 1 \end{bmatrix}, \tag{3.21}$$

where R is the matrix (r_{ik}). The change from ϕ_{11}, ϕ_{12}, $\phi_{\dot{2}1}$, $\phi_{\dot{2}2}$ to the linear combinations of (3.15) amounts to a change of frame in the space of spinors $\phi_{\dot{A}B}$ according to

$$\begin{aligned} \psi_{11} &= \tfrac{1}{2}\phi_{12} + \tfrac{1}{2}\phi_{\dot{2}1}, \\ \psi_{12} &= -\frac{1}{2i}\phi_{12} + \frac{1}{2i}\phi_{\dot{2}1}, \\ \psi_{\dot{2}1} &= \tfrac{1}{2}\phi_{11} - \tfrac{1}{2}\phi_{\dot{2}2}, \\ \psi_{\dot{2}2} &= \tfrac{1}{2}\phi_{11} + \tfrac{1}{2}\phi_{\dot{2}2}. \end{aligned} \tag{3.22}$$

We may express this transformation in vector form as

$$\Psi = S\Phi, \tag{3.23}$$

where S is the matrix

$$S = \tfrac{1}{2}\begin{bmatrix} 0 & 1 & 1 & 0 \\ 0 & i & -i & 0 \\ 1 & 0 & 0 & -1 \\ 1 & 0 & 0 & 1 \end{bmatrix}. \tag{3.24}$$

This change in frame also transforms every $\Phi' = (U^* \times U)\Phi$ into an $S\Phi'$ and changes $(U^* \times U)$ into another matrix which may be denoted by $(U^* \times U)^t$. Instead of the old relation

$$\Phi' = (U^* \times U)\Phi, \tag{3.25}$$

we get

$$\begin{aligned} S\Phi' &= S\{(U^* \times U)\Phi\} = S\{(U^* \times U)S^{-1}S\Phi\} \\ &= S(U^* \times U)S^{-1}S\Phi = (U^* \times U)^t S\Phi, \end{aligned} \tag{3.26}$$

where $$(U^* \times U)^t = S(U^* \times U)S^{-1} \tag{3.27}$$

is the new form of the matrix $(U^* \times U)$. The multiplication of $(U^* \times U)$ by S and S^{-1} as indicated in (3.27) reduces all the matrices $(U^* \times U)$ to the form indicated in (3.21).

For S^{-1} we get, by reversing (3.22),

$$S^{-1} = \begin{bmatrix} 0 & 0 & 1 & 1 \\ 1 & -i & 0 & 0 \\ 1 & i & 0 & 0 \\ 0 & 0 & -1 & 1 \end{bmatrix}, \tag{3.28}$$

and for $S(U^* \times U)S^{-1}$ we get (3.21) if we use (3.18), while on using (3.11) and $(U^* \times U)$ which results from it, we get

$$(U^* \times U)^t = \tfrac{1}{2}\begin{bmatrix} \alpha^2+\alpha^{*2}-\beta^2-\beta^{*2} & i\alpha^2-i\alpha^{*2}+i\beta^2-i\beta^{*2} & -2(\alpha\beta+\alpha^*\beta^*) & 0 \\ -i\alpha^2+i\alpha^{*2}+i\beta^2-i\beta^{*2} & \alpha^2+\alpha^{*2}+\beta^2+\beta^{*2} & +2i(\alpha\beta-\alpha^*\beta^*) & 0 \\ 2(\alpha\beta^*+\alpha^*\beta) & 2i(\alpha\beta^*-\alpha^*\beta) & 2(\alpha\alpha^*-\beta\beta^*) & 0 \\ 0 & 0 & 0 & 2 \end{bmatrix} \tag{3.29}$$

The upper, three-dimensional part of $(U^* \times U)^t$ represents a rotation in the same way as (3.20). However, instead of expressing the coefficients r_{ik} in terms of ϕ, θ, ψ they are now expressed in terms of two complex numbers which satisfy the relation $\alpha\alpha^*+\beta\beta^* = 1$. In this context these numbers are known as the Klein–Cayley parameters. Two complex numbers define four real parameters, and one condition between them would usually reduce their numbers to two. However, in our present case $(\alpha\alpha^*+\beta\beta^*)^* = \alpha\alpha^*+\beta\beta^*$, so that the condition $\alpha\alpha^*+\beta\beta^* = 1$ imposed on the coefficients involves only one relation between the four real parameters, and the Klein–Cayley parameters involve three real numbers as is required for specifying rotations in space.

One can easily verify that the upper part of (3.29) satisfies the orthogonality relation as is required from a rotational matrix.

From a more algebraic approach we can sum up the situation as follows. The four-dimensional matrices $(U_2^* \times U_2)$ (they depend on three real parameters) can be reduced by a transformation S (3.24) into a reduced form (3.21). The upper matrices happen to satisfy the orthogonality relation ($r_{ik} = r_{ki}^{-1}$) and are therefore identical with rotations. The determinant of these matrices is 1, hence they can represent only proper rotations, but not improper rotations which involve space inversions with a determinant -1.

The group of matrices $(U_2^* \times U_2)$ is isomorphic with the group of proper rotations and each of these groups is a faithful representation of the other. On the other hand, the group (U_2) forms a two-valued representation of the group $(U_2^* \times U_2)$ since the matrices U and $-U$ yield the same direct product $(U^* \times U)$. It can also be seen from (3.17) that apart from the factor ± 1, if we make $\phi = 2\pi$, $\theta = 0$, $\psi = 0$, we get a matrix which differs from a unit matrix by a factor -1. The two-valuedness cannot be removed. Nor does the group (U_2) form a two-valued representation of the full rotational group which contains also space inversions; (U_2) is a two-valued representation of the group of rotations proper only. A direct way of proving this statement is by trying to determine a matrix U, such that $U^* \times U$ would represent a space inversion. For example, let us consider the inversion $x' = x$, $y' = y$, $z' = -z$. In terms of the ψ_{AB} (see (3.15) where we must also use $t' = t$), the inversion is given by

$$\begin{aligned}
\phi'_{1\dot{1}} &= 0.\phi_{1\dot{1}}+0.\phi_{1\dot{2}}+0.\phi_{2\dot{1}}+1.\phi_{2\dot{2}}, \\
\phi'_{1\dot{2}} &= 0.\phi_{1\dot{1}}+1.\phi_{1\dot{2}}+0.\phi_{2\dot{1}}+0.\phi_{2\dot{2}}, \\
\phi'_{2\dot{1}} &= 0.\phi_{1\dot{1}}+0.\phi_{1\dot{2}}+1.\phi_{2\dot{1}}+0.\phi_{2\dot{2}}, \\
\phi'_{2\dot{2}} &= 1.\phi_{1\dot{1}}+0.\phi_{1\dot{2}}+0.\phi_{2\dot{1}}+0.\phi_{2\dot{2}}.
\end{aligned} \tag{3.30}$$

Hence, the matrix is

$$U^* \times U = \begin{bmatrix} 0 & 0 & 0 & 1 \\ 0 & 1 & 0 & 0 \\ 0 & 0 & 1 & 0 \\ 1 & 0 & 0 & 0 \end{bmatrix}. \tag{3.31}$$

On the other hand the general form of a matrix $U^* \times U$ is given by

$$U^* \times U = \begin{bmatrix} \alpha^*\alpha & \alpha^*\beta & \beta^*\alpha & \beta^*\beta \\ -\alpha^*\beta^* & \alpha^{*2} & -\beta^{*2} & \beta^*\alpha^* \\ -\beta\alpha & -\beta^2 & \alpha^2 & \alpha\beta \\ \beta\beta^* & -\beta\alpha^* & -\alpha\beta^* & \alpha\alpha^* \end{bmatrix}. \tag{3.32}$$

Hence to conform with (3.31) we must have $\alpha^2 = \alpha^{*2} = 1$ and $\beta\beta^* = 1$. It follows that $\alpha = \pm 1$ and we get $\alpha\alpha^* + \beta\beta^* = 2$ which contradicts (3.12).

We have now to turn our attention again to the Lorentz group. Every Lorentz transformation proper can be obtained by combining a simple Lorentz transformation (parallel axes and motion along one axis, say the z-axis) with suitable rotations. We must therefore prove that all the simple Lorentz transformations are represented in the group $(C_2^* \times C_2)$.

Let us consider the group of transformations of the type

$$A = \begin{bmatrix} \alpha & 0 \\ 0 & \alpha^{-1} \end{bmatrix}, \tag{3.33}$$

$$(A^* \times A) = \begin{bmatrix} \alpha^2 & 0 & 0 & 0 \\ 0 & 1 & 0 & 0 \\ 0 & 0 & 1 & 0 \\ 0 & 0 & 0 & \alpha^{-2} \end{bmatrix}, \tag{3.34}$$

where α is a real parameter. Then

$$\phi'_{\dot{1}1} = \alpha^2 \phi_{\dot{1}1}, \quad \phi'_{\dot{1}2} = \phi_{\dot{1}2}, \quad \phi'_{\dot{2}1} = \phi_{\dot{2}1}, \quad \phi'_{\dot{2}2} = \alpha^{-2}\phi_{\dot{2}2}. \tag{3.35}$$

The second and the third of these relations yield according to (3.8): $x'+iy' = x+iy$ and $x'-iy' = x-iy$ or $x' = x$, $y' = y$. Hence, the transformation (3.34) can involve only z and t. The first and last of (3.35) yield

$$z'+ct' = \alpha^2(z+ct),$$

$$-z'+ct' = \alpha^{-2}(-z+ct). \tag{3.36}$$

Consequently,

$$z' = \tfrac{1}{2}(\alpha^2+\alpha^{-2})z+\tfrac{1}{2}(\alpha^2-\alpha^{-2})ct,$$

$$ct' = \tfrac{1}{2}(\alpha^2-\alpha^{-2})z+\tfrac{1}{2}(\alpha^2+\alpha^{-2})ct. \tag{3.37}$$

This is a simple Lorentz transformation along the z-axis, provided we identify α as related to the relative velocity according to

$$v = c\frac{\alpha^4-1}{\alpha^4+1}. \tag{3.38}$$

On the other hand, a time reversal cannot be represented by a matrix of the form $A^* \times A$, where A is unimodular. According to (3.15), $x' = x$, $y' = y$, $z' = z$, $t' = -t$ is represented by $\phi'_{\dot{1}1} = -\phi_{\dot{2}2}$, $\phi'_{\dot{1}2} = \phi_{\dot{1}2}$, $\phi'_{\dot{2}1} = \phi_{\dot{2}1}$, $\phi'_{\dot{2}2} = -\phi_{\dot{1}1}$. On comparing with the general form as given by (3.1), we get $\beta\beta^* = -1$ and $\gamma\gamma^* = -1$. Neither of these relations can be satisfied.

The restricted Lorentz transformations (without inversion of axes or time reversal) are thus a true representation of the group $(C_2^* \times C_2)$ and vice versa. The restricted group also represents the unimodular group C_2. The latter is, however, only a two-valued representation of the restricted group, since this already applies to the group of rotations $((U_2)$ which is a sub-group of $(C_2))$.

From an algebraic point of view we can say that the group $(C_2^* \times C_2)$ which is obtained from the unimodular matrices is four-dimensional and involves six real parameters just as the Lorentz group does. After transformation with the matrix S (3.24) $\{(C_2^* \times C_2) \to S(C_2^* \times C_2)S^{-1}\}$, all the matrices become identical with the matrices of the Lorentz group proper. This transformation has already been required to produce identity with space rotations and is also required now. The group of transformed matrices $S(C_2^* \times C_2)S^{-1}$ is different from the group of original matrices $(C_2^* \times C_2)$ although equivalent to it.

4. The Invariant Spin-vectors

The relationship between the $\phi_{\dot{A}B}$ and x, y, z, ct as introduced in (3.9) can be expressed in the form

$$\phi_{\dot{A}B} = g_{\kappa \dot{A}B} x^{\kappa}, \tag{4.1}$$

where the $g_{\kappa \dot{A}B}$ may be written as four matrices:

$$g_{1\dot{A}B} = \begin{bmatrix} 0 & 1 \\ 1 & 0 \end{bmatrix}, \qquad g_{2\dot{A}B} = \begin{bmatrix} 0 & i \\ -i & 0 \end{bmatrix},$$

$$g_{3\dot{A}B} = \begin{bmatrix} 1 & 0 \\ 0 & -1 \end{bmatrix}, \qquad g_{4\dot{A}B} = \begin{bmatrix} 1 & 0 \\ 0 & 1 \end{bmatrix}. \tag{4.2}$$

As will be seen, these matrices are Hermitian, as required if the $\phi_{\dot{A}B}$ are to be Hermitian.

On a purely algebraic basis, if we look upon $\phi = x^{\kappa} g_{\kappa}$ and $\phi' = x^{\kappa'} g_{\kappa}$ where the g_{κ} are given by (4.2), as 2×2 Hermitian matrices ($\phi^{\dagger} = \phi$, $\phi'^{\dagger} = \phi'$, $\dagger \equiv \sim *$), we have

$$\det \phi = -(x^2+y^2+z^2)+c^2t^2 = -(x'^2+y'^2+z'^2)+c^2t'^2 = \det \phi'.$$

The only transformation of a matrix ϕ into a matrix ϕ' which will retain the hermiticity and leave the determinant invariant is

$$\phi' = T\phi T^{\dagger}, \tag{4.3}$$

where $\det T = \pm 1$. If $\phi^{\dagger} = \phi$, we get

$$\phi'^{\dagger} = T^{\dagger\dagger}\phi^{\dagger}T^{\dagger} = T\phi T^{\dagger} = \phi'.$$

The transformation also satisfies the group property

$$T_2 T_1 \phi T_1^{\dagger} T_2^{\dagger} = T_2 T_1 \phi (T_2 T_1)^{\dagger}$$

provided the T's form a group. This is the case for the set with $\det T = 1$, so that the T's must be unimodular. And so, each Lorentz transformation is accompanied by a transformation of a two-dimensional Hermitian matrix by a unimodular matrix $T(L)$. This two-dimensional representation is two-valued, because T and $-T$ represent the same Lorentz transformation. The group C_2 is therefore a two-valued representation of the Lorentz group. That only the restricted group is represented, follows from previous considerations and also on considerations of connectivity. The full Lorentz group is not simply connected; the identity transformation is not continuously transformable within the group into a space inversion, a time reversal, or their combination. On the other hand, C_2 is a simply connected group. For instance the unit matrix 1 is connected with -1 via the continuous set of unimodular matrices

$$\begin{bmatrix} \cos\phi & i\sin\phi \\ i\sin\phi & \cos\phi \end{bmatrix}.$$

Since the identity Lorentz transformation is represented by $\begin{bmatrix} 1 & 0 \\ 0 & 1 \end{bmatrix}$ or $\begin{bmatrix} -1 & 0 \\ 0 & -1 \end{bmatrix}$, it follows that as we move continuously from a matrix T to ± 1, we reach in a continuous way the identity transformation of the Lorentz group. Consequently only the restricted group is represented.

The result obtained previously that $(C_2^* \times C_2)$ is a representation of the restricted Lorentz group can be expressed by saying that if we transform x_κ into x'_κ by means of a Lorentz transformation and $\phi_{\dot{A}B}$ by the corresponding second rank spinor transformation, we should get the same relation again as (4.1):

$$\phi'_{\dot{A}B} = g_{\kappa\dot{A}B}x'^{\kappa}. \tag{4.1'}$$

This may be written as

$$S\Phi = g_\kappa(L^\kappa{}_\lambda x^\lambda), \tag{4.4}$$

where S is a matrix which transforms the column Φ into Φ' and $L^\kappa{}_\lambda$ the one transforming x^κ into $x^{\kappa'}$; g_κ is a column in the same way as $\phi_{\dot{A}B}$, S is given by

$$S = (T^* \times T) \equiv (T^*_{\dot{A}}{}^{\dot{C}} T_B{}^D). \tag{4.5}$$

On the other hand, from (4.1) we get

$$S\Phi = Sg_\sigma(L^{-1\sigma}{}_\kappa L^\kappa{}_\lambda)x^\lambda. \tag{4.6}$$

(We have simply multiplied (4.1) by S and inserted the unit matrix $L^{-1}L = 1$ between $g_{\kappa\dot{A}B}$ and x^κ.)

On comparing (4.4) and (4.6) we get the result which is already indicated in the notation: the $g_{\kappa\dot{A}B}$ are entities transforming both as covariant four-vectors as well as second rank spinors. As indicated in (4.1′) they are numerically invariant:

$$\begin{aligned} Sg_{\sigma\dot{A}B}L^{-1\sigma}{}_\kappa = Sg_{\sigma\dot{A}B}\tilde{L}^{-1\kappa}{}_\sigma = Sg_{\sigma\dot{A}B}L_\kappa{}^\sigma = SL_\kappa{}^\sigma g_{\sigma\dot{A}B} \\ \equiv g'_{\kappa\dot{A}B} = g_{\kappa\dot{A}B}. \end{aligned} \tag{4.7}$$

Entities with mixed indices are called spin-tensors. We can say that a necessary consequence from the isomorphism between the Lorentz group and $(C_2^* \times C_2)$ is the existence of the invariant spin-tensors (4.2) and vice versa.

Using a metrical tensor

$$(g^{\mu\nu}) = (g_{\mu\nu}) = \begin{bmatrix} -1 & 0 & 0 & 0 \\ 0 & -1 & 0 & 0 \\ 0 & 0 & -1 & 0 \\ 0 & 0 & 0 & 1 \end{bmatrix} \tag{4.8}$$

and $\epsilon^{\dot{A}\dot{B}}$, we can raise all the indices in (4.2) and obtain

$$g^{1\dot{A}B} = \begin{bmatrix} 0 & 1 \\ 1 & 0 \end{bmatrix}, \qquad g^{2\dot{A}B} = \begin{bmatrix} 0 & -i \\ i & 0 \end{bmatrix},$$

$$g^{3\dot{A}B} = \begin{bmatrix} 1 & 0 \\ 0 & -1 \end{bmatrix}, \qquad g^{4\dot{A}B} = \begin{bmatrix} 1 & 0 \\ 0 & 1 \end{bmatrix}. \tag{4.9}$$

The $g^{\kappa\dot{A}B}$ are thus seen to be the transposes of $g_{\kappa\dot{A}B}$:

$$g_{\kappa\dot{A}B} = g^{\kappa\dot{B}A}. \tag{4.10}$$

We also have
$$\phi^{\dot{A}B} = g^{\kappa\dot{A}B}x_{\kappa}. \tag{4.11}$$

As already mentioned, the $g_{\kappa\dot{A}B}$ have to be Hermitian if the $\phi_{\dot{A}B}$ are to be Hermitian. This also applies to the $g^{\kappa\dot{A}B}$ and, in fact,

$$g^*_{\kappa\dot{A}B} = g_{\kappa\dot{B}A}, \qquad g^{*\kappa\dot{A}B} = g^{\kappa\dot{B}A}. \tag{4.12}$$

The invariant spin-vectors obey the following relations, which may be verified by carrying out the summation:

$$g^{\kappa\dot{A}B}g_{\lambda\dot{A}B} = 2\delta^{\kappa}{}_{\lambda}, \tag{4.13}$$

$$g^{\kappa\dot{A}B}g_{\kappa\dot{C}D} = 2\delta^{\dot{A}}{}_{\dot{C}}\,\delta^{B}{}_{D}. \tag{4.14}$$

From (4.14) we get after multiplication by $\epsilon^{\dot{C}\dot{P}}\epsilon^{DQ}$ (and subsequent change of P, Q to C, A):

$$g^{\kappa\dot{A}B}g_{\kappa}{}^{\dot{C}D} = 2\epsilon^{\dot{A}\dot{C}}\epsilon^{BD}. \tag{4.15}$$

By lowering one of the spinor indices of the $g^{\kappa\dot{A}B}$ we get

$$g^{1\dot{A}}{}_{B} = \begin{bmatrix} 1 & 0 \\ 0 & -1 \end{bmatrix}, \qquad g^{2\dot{A}}{}_{B} = \begin{bmatrix} -i & 0 \\ 0 & -i \end{bmatrix},$$

$$g^{3\dot{A}}{}_{B} = \begin{bmatrix} 0 & -1 \\ -1 & 0 \end{bmatrix}, \qquad g^{4\dot{A}}{}_{B} = \begin{bmatrix} 0 & -1 \\ 1 & 0 \end{bmatrix}. \tag{4.16}$$

These matrices obey the relations

$$g^{*\kappa\dot{A}}{}_{C}\,g^{\lambda\dot{C}}{}_{B} + g^{*\lambda\dot{A}}{}_{C}\,g^{\kappa\dot{C}}{}_{B} = -2g^{\kappa\lambda}\delta^{A}{}_{B}; \tag{4.17}$$

or, using a more convenient notation,

$$g^{*\kappa}g^{\lambda} + g^{*\lambda}g^{\kappa} = -2g^{\kappa\lambda}.1 = -2g^{\kappa\lambda}, \tag{4.18}$$

where each of the two-dimensional matrices g^{κ} stands for the corresponding matrix in (4.16).

On lowering the first spinor index of $g^{\kappa \dot{A}B}$, we get the following result:

$$g^1{}_{\dot{A}}{}^B = \begin{bmatrix} 1 & 0 \\ 0 & -1 \end{bmatrix}, \qquad g^2{}_{\dot{A}}{}^B = \begin{bmatrix} i & 0 \\ 0 & i \end{bmatrix},$$

$$g^3{}_{\dot{A}}{}^B = \begin{bmatrix} 0 & -1 \\ -1 & 0 \end{bmatrix}, \qquad g^4{}_{\dot{A}}{}^B = \begin{bmatrix} 0 & 1 \\ -1 & 0 \end{bmatrix}. \tag{4.16 a}$$

Hence
$$[g^{\kappa \dot{A}}{}_B] = [g^{\kappa}{}_{\dot{A}}{}^B] \sim *. \tag{4.16b}$$

The $g^{\kappa}{}_{\dot{A}}{}^B$ obey the same relations (4.18).

We are now in position to derive the inverse of (4.1) by using (4.13). We multiply (4.1) by $g^{\kappa \dot{A}B}$ and get, after changing one index κ to λ to avoid summation,
$$g^{\kappa \dot{A}B}\phi_{\dot{A}B} = g^{\kappa \dot{A}B} g_{\lambda \dot{A}B} x^{\lambda}. \tag{4.19}$$

Hence,
$$x^{\kappa} = \tfrac{1}{2} g^{\kappa \dot{A}B}\phi_{\dot{A}B}. \tag{4.20}$$

For x_{κ} we obtain
$$x_{\kappa} = g_{\kappa\lambda} x^{\lambda} = \tfrac{1}{2} g_{\kappa\lambda}(g^{\lambda \dot{A}B}\phi_{\dot{A}B}) = \tfrac{1}{2} g_{\kappa}{}^{\dot{A}B}\phi_{\dot{A}B}. \tag{4.21}$$

This yields
$$x_{\kappa} x^{\kappa} = \tfrac{1}{4} g_{\kappa}{}^{\dot{A}B}\phi_{\dot{A}B}\, g^{\kappa \dot{C}D}\phi_{\dot{C}D} = \tfrac{1}{2}\epsilon^{\dot{A}\dot{C}}\epsilon^{BD}\phi_{\dot{A}B}\phi_{\dot{C}D}, \tag{4.22}$$

$$x_{\kappa} x^{\kappa} = \tfrac{1}{2}\phi^{\dot{C}D}\phi_{\dot{C}D}. \tag{4.23}$$

With the help of the invariant spin-tensors we can derive a formula which enables us to express every Lorentz transformation proper in terms of the representing matrices T of (C_2). Let $L^{\kappa}{}_{\lambda}$ and $\pm T_A{}^B$ be the corresponding matrices. We know already that

$$T^* \times T \equiv (T^*_{\dot{A}}{}^{\dot{C}} T_B{}^D)$$

represents $L^{\kappa}{}_{\lambda}$ faithfully and not two-valuedly. However, $T^* \times T$ is not identical with $L^{\kappa}{}_{\lambda}$. An identity between these two matrices can be achieved only after transformation with the matrix S (3.24). We shall now obtain this result in terms of the invariant spin-tensors.

From (4.20) we get

$$x'_{\kappa} \equiv L^{\kappa}{}_{\lambda} x^{\lambda} = \tfrac{1}{2}(g^{\kappa \dot{A}B})'\phi_{\dot{A}B} = \tfrac{1}{2} g^{\kappa \dot{A}B} T^*_{\dot{A}}{}^{\dot{C}} T_B{}^D \phi_{\dot{C}D}$$
$$= \tfrac{1}{2} g^{\kappa \dot{A}B} T^*_{\dot{A}}{}^{\dot{C}} T_B{}^D g_{\lambda \dot{C}D} x^{\lambda}, \tag{4.24}$$

where we have made use of (4.1) in the last step. Consequently,

$$L^{\kappa}{}_{\lambda} = \tfrac{1}{2} g^{\kappa \dot{A}B} T^*_{\dot{A}}{}^{\dot{C}} T_B{}^D g_{\lambda \dot{C}D} = \tfrac{1}{2}\operatorname{tr}(g^{\kappa} T \tilde{g}_{\lambda} \tilde{T}^*). \tag{4.25}$$

This formula enables us to obtain a Lorentz transformation uniquely from every unimodular matrix $\pm T$ (we know it involves six real parameters). Since the only condition which T in (4.25) has to satisfy is unimodularity we can also express the relationship between T and a Lorentz transformation as

$$L^{\kappa}{}_{\lambda} = \tfrac{1}{2} g^{\kappa \dot{A} B} T^{*C}{}_{A} T^{D}{}_{B} g_{\lambda \dot{C} D} = \tfrac{1}{2} \operatorname{tr}(g^{\kappa} \tilde{T} \tilde{g}_{\lambda} T^{*}). \tag{4.26}$$

In formula (4.26) (or (4.25)) $L^{\kappa}{}_{\lambda}$ may be also looked upon as the product of three matrices:

$$L^{\kappa}{}_{\lambda} = (S(T^{*} \times T) S^{-1})^{\kappa}{}_{\lambda}, \tag{4.27}$$

where

$$(T^{*} \times T)^{CD}{}_{AB} = T^{*C}{}_{A} T^{D}{}_{B} \quad \begin{pmatrix} AB = 11, 21, 12, 22 \\ CD = 11, 21, 12, 22 \end{pmatrix}, \tag{4.28}$$

$$S^{\kappa, CD} = \tfrac{1}{2} g^{\kappa, \dot{C} D} \quad (\kappa = 1, 2, 3, 4), \tag{4.29}$$

$$(S^{-1})_{AB, \lambda} = g_{\dot{A} B, \lambda} \quad (\lambda = 1, 2, 3, 4). \tag{4.30}$$

In all these formulae we arrange $g^{\kappa, \dot{C} D}$ and $g_{\dot{A} B, \lambda}$ as four-dimensional matrices. In (4.30) $g_{\dot{A} B, \lambda}$ is the transpose of the matrix $g_{\lambda, \dot{A} B}$. From (4.9) we get

$$\tfrac{1}{2}[g^{\kappa, \dot{C} D}] = \tfrac{1}{2} \begin{bmatrix} 0 & 1 & 1 & 0 \\ 0 & i & -i & 0 \\ 1 & 0 & 0 & -1 \\ 1 & 0 & 0 & 1 \end{bmatrix} = S, \tag{4.31}$$

$$[g_{\lambda, \dot{A} B}] = \begin{bmatrix} 0 & 1 & 1 & 0 \\ 0 & -i & i & 0 \\ 1 & 0 & 0 & -1 \\ 1 & 0 & 0 & 1 \end{bmatrix} = \tilde{S}^{-1}, \tag{4.32}$$

$$[g_{\dot{A} B, \lambda}] = \begin{bmatrix} 0 & 0 & 1 & 1 \\ 1 & -i & 0 & 0 \\ 1 & i & 0 & 0 \\ 0 & 0 & -1 & 1 \end{bmatrix} = S^{-1}. \tag{4.33}$$

The unitary matrix S is identical with the one introduced earlier in (3.24), and we have thus regained our previous result.

Eqn. (4.27) may also be expressed as follows:

$$\begin{aligned} [L^{\kappa}{}_{\lambda}] &= S(T^{*} \times 1)(1 \times T) S^{-1} = S(T^{*} \times 1) S^{-1} . S(1 \times T) S^{-1} \\ &\equiv S(1 \times T)(T^{*} \times 1) S^{-1} = S(1 \times T) S^{-1} . S(T^{*} \times 1) S^{-1}. \end{aligned} \tag{4.34}$$

Hence every Lorentz transformation can be expressed as a product of two factors which commute, $S(T^{*} \times 1) S^{-1}$ and $S(1 \times T) S^{-1}$. All the matrices of the form $S(T^{*} \times 1) S^{-1}$ (or $(T^{*} \times 1)$) form a group which is isomorphic to C_2^{*} and homomorphic to the Lorentz group. The same applies to the matrices $S(1 \times T) S^{-1}$ or $(1 \times T)$. However, none of these matrices is a Lorentz matrix. The situation here is different from that encountered when dealing with the four-dimensional (complex) orthogonal group. In this case each matrix can be split into two commuting factors in such a way that all first factors again represent orthogonal complex transformations, forming an invariant sub-group of the orthogonal group and the same applying to the second factors.

The matrices $S(T^*\times 1)S^{-1}$ and $S(1\times T)S^{-1}$ are the complex conjugates of each other as can be verified by direct evaluation as follows. Let

$$\begin{aligned} \tfrac{1}{2}(T_{11}+T_{22}) &= a, & T_{11} &= a+ib, \\ \tfrac{1}{2}(T_{11}-T_{22}) &= ib, & T_{22} &= a-ib, \\ \tfrac{1}{2}(T_{12}-T_{21}) &= d, & T_{12} &= d-ic, \\ \tfrac{1}{2}(T_{12}+T_{21}) &= -ic, & T_{21} &= -d-ic, \end{aligned} \tag{4.35}$$

and

$$\begin{vmatrix} T_{11} & T_{12} \\ T_{21} & T_{22} \end{vmatrix} = a^2+b^2+c^2+d^2 = 1. \tag{4.36}$$

Then we get

$$S(1\times T)S^{-1} = \begin{bmatrix} a & -b & c & id \\ b & a & -d & ic \\ -c & d & a & ib \\ id & ic & ib & a \end{bmatrix}. \tag{4.37}$$

Similarly,

$$S(T^*\times 1)S^{-1} = \begin{bmatrix} a^* & -b^* & c^* & -id^* \\ b^* & a^* & -d^* & -ic^* \\ -c^* & d^* & a^* & -ib^* \\ -id^* & -ic^* & -ib^* & a^* \end{bmatrix}. \tag{4.38}$$

Moreover, the product of the matrices (4.37) and (4.38) yields a matrix with real elements, as should be, because the Lorentz matrices we are now using are based on the metrical form (4.8) which uses the real ct and not the imaginary ict. Since the determinants of the two matrices (4.37) and (4.38) are the complex conjugates of each other, the matrix of their product has real and positive determinant. In fact in view of (4.36) it is equal to $+1$. Moreover, the element $L^4{}_4$ of the Lorentz matrix is given by

$$\begin{aligned} L^4{}_4 &= (id)^*id+(ic)^*ic+(ib)^*ib+a^*a \\ &= aa^*+bb^*+cc^*+dd^* \geqslant 1. \end{aligned} \tag{4.39}$$

This confirms that we are dealing here with Lorentz matrices which do not reverse the time.

On multiplying (4.37) by (4.38) we get the Lorentz matrix expressed as follows:

$$[L^{\kappa}{}_{\lambda}] = \begin{bmatrix} aa^*-bb^*-cc^*+dd^* & -(ab^*+a^*b)+(cd^*+c^*d) & (ac^*+a^*c)+(bd^*+b^*d) & i(a^*d-ad^*)+i(bc^*-b^*c) \\ ab^*+a^*b+cd^*+c^*d & aa^*-bb^*+cc^*-dd^* & (bc^*+b^*c)-(ad^*+a^*d) & i(a^*c-ac^*)+i(b^*d-bd^*) \\ -(ac^*+a^*c)+(bd^*+b^*d) & (ad^*+a^*d)+(bc^*+b^*c) & aa^*+bb^*-(cc^*+dd^*) & i(a^*b-ab^*)+i(cd^*-c^*d) \\ i(a^*d-ad^*)+i(b^*c-bc^*) & i(bd^*-b^*d)+i(a^*c-ac^*) & i(a^*b-ab^*)+i(c^*d-cd^*) & aa^*+bb^*+cc^*+dd^* \end{bmatrix}. \tag{4.40}$$

On comparing (4.40) with (3.29) we see that the sub-group of space rotation is obtained by taking the four parameters a, b, c, and d to be real. The condition (4.36) will then involve only one relation between four real numbers (and not two as for four complex numbers) and we are left essentially with three parameters as required for the space rotations. The Klein–Cayley parameters α and β are obtained through

$$\alpha = a+ib,$$

$$\beta = -ic+d \quad (a, b, c, d \text{ all real}; \; a^2+b^2+c^2+d^2 = 1). \tag{4.41}$$

With this identification (4.40) is identical with (3.29). We can also confirm that the matrix T is now unitary. From (4.35) we get

$$\begin{bmatrix} T^1{}_1 & T^1{}_2 \\ T^2{}_1 & T^2{}_2 \end{bmatrix} = \begin{bmatrix} a+ib & d-ic \\ -(d+ic) & a-ib \end{bmatrix} = \begin{bmatrix} \alpha & \beta \\ -\beta^* & \alpha^* \end{bmatrix} \quad (\alpha\alpha^*+\beta\beta^* = 1). \tag{4.42}$$

Eqn. (4.26) may also be expressed as the product of two matrices,

$$L^{\kappa}{}_{\lambda} = \tfrac{1}{2} M^{\kappa,AD} N_{AD,\lambda}, \tag{4.43}$$

where

$$M^{\kappa,AD} = g^{\kappa\dot{A}B} T^{D}{}_{B}, \tag{4.44}$$

$$N_{AD,\lambda} = T^{*C}{}_{A}\, g_{\lambda\dot{C}D}. \tag{4.45}$$

Using (4.2) and (4.9), we get

$$[M^{\kappa,AD}] = \begin{bmatrix} T^1{}_2 & T^2{}_2 & T^1{}_1 & T^2{}_1 \\ -iT^1{}_2 & -iT^2{}_2 & iT^1{}_1 & iT^2{}_1 \\ T^1{}_1 & T^2{}_1 & -T^1{}_2 & -T^2{}_2 \\ T^1{}_1 & T^2{}_1 & T^1{}_2 & T^2{}_2 \end{bmatrix}, \tag{4.46}$$

$$[N_{AD,\lambda}] = \begin{bmatrix} T^{*2}{}_1 & -iT^{*2}{}_1 & T^{*1}{}_1 & T^{*1}{}_1 \\ T^{*1}{}_1 & iT^{*1}{}_1 & -T^{*2}{}_1 & T^{*2}{}_1 \\ T^{*2}{}_2 & -iT^{*2}{}_2 & T^{*1}{}_2 & T^{*1}{}_2 \\ T^{*1}{}_2 & iT^{*1}{}_2 & -T^{*2}{}_2 & T^{*2}{}_2 \end{bmatrix}. \tag{4.47}$$

Since we know already (see (4.39)) that $L^4{}_4 \geqslant 1$, we now get the result

$$L^4{}_4 = M^{4,AD}N_{AD,4} = T^1{}_1\, T^{*1}{}_1 + T^2{}_1\, T^{*2}{}_1 + T^1{}_2\, T^{*1}{}_2 + T^2{}_2\, T^{*2}{}_2 \geqslant 1. \tag{4.48}$$

So far we have used the invariant spin-tensors $g_{\kappa\dot{A}B}$ or $g^{\lambda\dot{C}D}$ to express the world vectors x^κ or x_κ in terms of second-rank spinors (4.20), (4.21), (4.1). Similar relations can be introduced between any four-vectors F_κ or F^λ and second-rank Hermitian spinors $\phi_{\dot{A}B}$:

$$F_\kappa = \tfrac{1}{2}g_{\kappa\dot{A}B}\,\phi^{\dot{A}B}, \qquad \phi^{\dot{A}B} = g^{\kappa\dot{A}B}F_\kappa, \tag{4.49}$$

$$F^\lambda = \tfrac{1}{2}g^{\lambda\dot{C}D}\phi_{\dot{C}D}, \qquad \phi_{\dot{C}D} = g_{\lambda\dot{C}D}F^\lambda. \tag{4.50}$$

In particular, if x, y, z, ct are a four-vector x^κ, $\partial/\partial x^\kappa$ will be the component of a covariant vector which we shall denote by ∂_κ,

$$\frac{\partial}{\partial x^\kappa} \equiv \partial_\kappa. \tag{4.51}$$

Similarly,

$$\frac{\partial}{\partial x_\kappa} \equiv \partial^\kappa. \tag{4.52}$$

If we deal with functions of spinors we can introduce the differential operators

$$\frac{\partial}{\partial \phi_A} = \left(\frac{\partial}{\partial \phi_1}, \frac{\partial}{\partial \phi_2}\right), \tag{4.53}$$

$$\frac{\partial}{\partial \phi^A} = \left(\frac{\partial}{\partial \phi^1}, \frac{\partial}{\partial \phi^2}\right). \tag{4.54}$$

More important are the differential operators

$$\frac{\partial}{\partial \phi_{\dot{A}B}} = \left(\frac{\partial}{\partial \phi_{\dot{1}1}}, \frac{\partial}{\partial \phi_{\dot{1}2}}, \frac{\partial}{\partial \phi_{\dot{2}1}}, \frac{\partial}{\partial \phi_{\dot{2}2}}\right), \tag{4.55}$$

$$\frac{\partial}{\partial \phi^{\dot{A}B}} = \left(\frac{\partial}{\partial \phi^{\dot{1}1}}, \frac{\partial}{\partial \phi^{\dot{1}2}}, \frac{\partial}{\partial \phi^{\dot{2}1}}, \frac{\partial}{\partial \phi^{\dot{2}2}}\right). \tag{4.56}$$

When the spinor happens to be connected with the four-vector x^κ in the way described in (4.49), (4.50), it will be convenient to introduce the notation

$$\frac{\partial}{\partial \phi_A} = \partial^A, \qquad \frac{\partial}{\partial \phi^A} = \partial_A, \tag{4.57}$$

$$\frac{\partial}{\partial \phi_{\dot{A}B}} = \tfrac{1}{2}\partial^{\dot{A}B}, \qquad \frac{\partial}{\partial \phi^{\dot{A}B}} = \tfrac{1}{2}\partial_{\dot{A}B}. \tag{4.58}$$

Then the following relations hold:

$$\partial_\kappa = \tfrac{1}{2}g_{\kappa\dot{A}B}\,\partial^{\dot{A}B}, \qquad \partial^{\dot{A}B} = g^{\kappa\dot{A}B}\partial_\kappa, \tag{4.59}$$

$$\partial^\kappa = \tfrac{1}{2}g^{\kappa\dot{A}B}\partial_{\dot{A}B}, \qquad \partial_{\dot{A}B} = g_{\kappa\dot{A}B}\,\partial^\kappa. \tag{4.60}$$

They may be directly verified from (4.49), (4.50) if F_κ and F_λ are replaced by x_κ and x_λ. For instance,

$$\frac{\partial}{\partial x_\kappa} = \frac{\partial}{\partial \phi^{\dot{A}B}}\frac{\partial \phi^{\dot{A}B}}{\partial x_\kappa} = \frac{\partial}{\partial \phi^{\dot{A}B}}\frac{\partial}{\partial x^\kappa}(g^{\lambda\dot{A}B}x_\lambda) = g^{\lambda\dot{A}B}\frac{\partial x_\lambda}{\partial x_\kappa}\frac{\partial}{\partial \phi^{\dot{A}B}}$$

$$= g^{\lambda\dot{A}B}\delta^\lambda{}_\kappa\frac{\partial}{\partial \phi^{\dot{A}B}} = g^{\kappa\dot{A}B}\,\frac{\partial}{\partial \phi^{\dot{A}B}} = \tfrac{1}{2}g^{\kappa\dot{A}B}\partial_{\dot{A}B}. \tag{4.61}$$

The replacement of x^κ by the corresponding spinor leads further to the relation

$$\partial_\kappa F^\kappa = \tfrac{1}{2}\partial^{\dot{A}B}\Phi_{\dot{A}B}, \tag{4.62}$$

where $\Phi_{\dot{A}B}$ is the spinor corresponding to the four-vector F^κ according to (4.49), (4.50). Similarly, if F and Φ are corresponding scalars,

$$F(x^\kappa) = \Phi(\phi_{\dot{A}B}), \tag{4.63}$$

we have

$$\partial_\kappa\partial^\kappa F \equiv \Box F = \tfrac{1}{2}\partial_{\dot{A}B}\,\partial^{\dot{A}B}\Phi. \tag{4.64}$$

This relation follows from (4.59), (4.60), and (4.14):

$$\partial_\kappa\partial^\kappa = \tfrac{1}{4}g_{\kappa\dot{A}B}\,g^{\lambda\dot{C}D}\partial^{\dot{A}B}\partial_{\dot{C}D} = \tfrac{1}{2}\delta^{\dot{C}}{}_{\dot{A}}\,\delta^D{}_B\,\partial^{\dot{A}B}\partial_{\dot{C}D} = \tfrac{1}{2}\partial^{\dot{A}B}\partial_{\dot{A}B}. \tag{4.65}$$

All these formulae help to replace relations between tensors by corresponding relations between spinors: for instance, to express Maxwell's equations in spinor form. (For this the reader is referred to an article by Laporte and Uhlenbeck.)

To give a simple example, let the energy-momentum vector of a particle be given by

$$P_\kappa = \left(P_x, P_y, P_z, \frac{E}{c}\right) \tag{4.66}$$

and let Q^λ be the four-vector

$$Q^\lambda = (x, y, z, -ct). \tag{4.67}$$

Then, in quantum mechanics these entities are replaced by the operators

$$P_\kappa = \frac{\hbar}{i}\left(\frac{\partial}{\partial x}, \frac{\partial}{\partial y}, \frac{\partial}{\partial z}, \; -\frac{1}{c}\frac{\partial}{\partial t}\right) = \frac{\hbar}{i}\frac{\partial}{\partial x^\kappa}, \tag{4.68}$$

$$Q^\lambda = (x, y, z, -ct) = x^\lambda, \tag{4.69}$$

which obey the relation

$$P_\kappa Q^\lambda - Q^\lambda P_\kappa = \frac{\hbar}{i}\delta_\kappa{}^\lambda . 1. \tag{4.70}$$

In terms of $$P_{\dot{A}B} = \frac{\hbar}{i}\partial_{\dot{A}B}, \qquad Q^{\dot{C}D} = \phi^{\dot{C}D}, \tag{4.71}$$

we get $$P_{\dot{A}B}Q^{\dot{C}D} - Q^{\dot{C}D}P_{\dot{A}B} = \frac{2\hbar}{i}\delta^{\dot{C}}{}_{\dot{A}}\delta^{D}{}_{B}.1. \tag{4.72}$$

5. Tensor and Two-Dimensional Spinor Representations of the Four-Group

So far we have been dealing with the restricted group only and found that it can be represented two-valuedly by the group C_2. The full Lorentz group which is composed of four parts (see p. 47) can be obtained from the restricted group by combining it with the space reflection **P**, the time reversal **T**, and the combined operation **PT** = **TP**. The four parts of the full Lorentz group thus obtained form separate sets which cannot be reached from one another by continuous paths. It is customary to denote them by $L_+^\uparrow$, $L_-^\uparrow$, $L_-^\downarrow$, $L_+^\downarrow$. When no time reversal is involved the arrow points upwards and when time reversal is involved, downwards; + indicates a positive determinant and − one which is negative. $L_+^\uparrow$ is itself a group, the restricted group. $L_+^\uparrow$ combined with $L_-^\uparrow$ is also a group, the orthochronous group, because none of its elements reverses the time. $L_+^\uparrow$ and $L_+^\downarrow$ also form a group. The determinant of all its elements is +1 and this sub-group is called the proper Lorentz group. Finally, $L_+^\uparrow$ together with $L_-^\downarrow$ combine to form a group.

P, **T**, and **PT** = **TP** correspond to the following transformations:

$$\begin{aligned} &\mathbf{P} \quad (x^k \to -x^k, \quad x^4 \to x^4), \\ &\mathbf{T} \quad (x^k \to x^k, \quad x^4 \to -x^4), \\ &\mathbf{PT} \quad (x^\mu \to -x^\mu). \end{aligned} \tag{5.1}$$

Hence the corresponding matrices are

$$P = \begin{bmatrix} -1 & & & \\ & -1 & 0 & \\ & 0 & -1 & \\ & & & 1 \end{bmatrix} \qquad T = \begin{bmatrix} 1 & & & \\ & 1 & 0 & \\ & 0 & 1 & \\ & & & -1 \end{bmatrix},$$

$$PT = TP = \begin{bmatrix} -1 & & & \\ & -1 & 0 & \\ & 0 & -1 & \\ & & & -1 \end{bmatrix}. \tag{5.2}$$

These three matrices, together with the unit matrix, form a group called the four-group. Since all four elements commute with each other, this is an Abelian group and as a consequence it possesses only one-dimensional irreducible faithful representations. They are given by

E	**P**	**T**	**PT = TP**
1	1	1	1
1	1	−1	−1
1	−1	1	−1
1	1	−1	1

(5.3)

All these representations are inequivalent. Entities transforming according to the first representation are the true scalars. They do not change value either on space reflection or on time reversal. In the third representation an entity would change sign on space reflection. Such an entity is called a pseudo-scalar. The other two representations yield other modes of transformation. Let ϕ be an entity transforming according to the following representation of the restricted group

$$L_+^\uparrow \to \phi' = 1.\phi. \tag{5.4}$$

Then the following four possibilities arise under the four-group:

$$\begin{array}{lll} \quad\mathbf{P} & \quad\mathbf{T} & \\ \phi' = \phi & \phi' = \phi & (\text{case } 1, 1), \\ \phi' = \phi & \phi' = -\phi & (\text{case } 1, -1), \\ \phi' = -\phi & \phi' = \phi & (\text{case } -1, 1), \\ \phi' = -\phi & \phi' = -\phi & (\text{case } -1, -1). \end{array} \tag{5.5}$$

Next to one-component entities we must consider four-vectors and corresponding four-dimensional representations of the four-group. From (5.3) we can form higher-dimensional representations by considering direct sums in an arbitrary order. These sums will be diagonal matrices because the irreducible representations are one-dimensional. However, in view of the condition that the four-group is to combine with the restricted group to form the full Lorentz group, certain limitations arise on the four-dimensional representation of the four-group and only the following can be admitted:

$$\overset{\mathbf{P}}{\begin{bmatrix} -1 & & & 0 \\ & -1 & & \\ & & -1 & \\ 0 & & & 1 \end{bmatrix}} \qquad \overset{\mathbf{T}}{\begin{bmatrix} 1 & & & 0 \\ & 1 & & \\ & & 1 & \\ 0 & & & -1 \end{bmatrix}} \quad (\text{case } 1, 1),$$

$$\begin{matrix} \mathbf{P} & & \mathbf{T} \\ \begin{bmatrix} -1 & & & 0 \\ & -1 & & \\ & & -1 & \\ 0 & & & 1 \end{bmatrix} & & \begin{bmatrix} -1 & & & 0 \\ & -1 & & \\ & & -1 & \\ 0 & & & 1 \end{bmatrix} & \text{(case } 1, -1), \\ \begin{bmatrix} 1 & & & 0 \\ & 1 & & \\ & & 1 & \\ 0 & & & -1 \end{bmatrix} & & \begin{bmatrix} 1 & & & 0 \\ & 1 & & \\ & & 1 & \\ 0 & & & -1 \end{bmatrix} & \text{(case } -1, 1), \\ \begin{bmatrix} 1 & & & 0 \\ & 1 & & \\ & & 1 & \\ 0 & & & -1 \end{bmatrix} & & \begin{bmatrix} -1 & & & 0 \\ & -1 & & \\ & & -1 & \\ 0 & & & 1 \end{bmatrix} & \text{(case } -1, -1). \end{matrix} \tag{5.6}$$

In all representations **E** is represented by the unit matrix and **PT** by the product of the matrices as given in (5.6).

The case (1, 1) is identical with the matrices P and T themselves as given in (5.2) and is identical with the transformation mode of the coordinates x^κ. The second case is obtained by multiplying the matrices P and T by 1 and -1 respectively, the third by multiplying P and T by -1 and 1, and the last case by -1, -1. Case $(1, -1)$ is typical for the way the four-velocity u_μ transforms.

All these representations have the property of yielding for **PT** or **TP** matrices which are multiples of a unit matrix. This is necessary because if we take the actual space and time inversions P, T as given in case (1, 1), we have

$$PT = TP = \begin{bmatrix} -1 & 0 & 0 & 0 \\ 0 & -1 & 0 & 0 \\ 0 & 0 & -1 & 0 \\ 0 & 0 & 0 & -1 \end{bmatrix}. \tag{5.7}$$

Since this is a multiple of a unit matrix it follows that **PT** = **TP** commutes with all the Lorentz transformations proper. This property must reflect itself in any representation of the Lorentz group. Now the Lorentz group proper forms an irreducible set and consequently, by a theorem of Schur, any matrix which commutes with the whole set can only be a multiple of a unit matrix. As pointed out, this is the case with all four representations in (5.6). A pair of matrices as follows cannot form a representation because their product is not a multiple of a unit matrix:

$$\overset{\mathbf{P}}{\begin{bmatrix} 1 & 0 & 0 & 0 \\ 0 & 1 & 0 & 0 \\ 0 & 0 & 1 & 0 \\ 0 & 0 & 0 & -1 \end{bmatrix}} \qquad \overset{\mathbf{T}}{\begin{bmatrix} 1 & 0 & 0 & 0 \\ 0 & 1 & 0 & 0 \\ 0 & 0 & -1 & 0 \\ 0 & 0 & 0 & 1 \end{bmatrix}} \quad (PT \neq m1). \tag{5.8}$$

On the other hand,

$$\overset{\mathbf{P}}{\begin{bmatrix} 1 & & & \\ & -1 & & O \\ & O & 1 & \\ & & & -1 \end{bmatrix}} \qquad \overset{\mathbf{T}}{\begin{bmatrix} 1 & & & \\ & -1 & & O \\ & O & 1 & \\ & & & -1 \end{bmatrix}} \tag{5.9}$$

produce a PT which is a multiple of a unit matrix. However, these matrices are obtainable from those of (5.6) when combined with suitable proper Lorentz transformations. Consequently they do not yield anything new.

The situation illustrated here in the case of scalars and vectors can be proved to hold quite generally for any irreducible tensors.

Vectors transforming under the full Lorentz group in accordance with case (1, 1) are called regular four-vectors. All other cases are referred to as various pseudo-vectors (more generally pseudo-tensors).

In terms of spinors of second rank we obtain the representations of the four-group, by using (3.15) as has already been done when obtaining (3.31). We get for vectors transforming like x, y, z, t,

$$\overset{\mathbf{P}}{A_P = \begin{bmatrix} 0 & 0 & 0 & 1 \\ 0 & -1 & 0 & 0 \\ 0 & 0 & -1 & 0 \\ 1 & 0 & 0 & 0 \end{bmatrix}} \qquad \overset{\mathbf{T}}{A_T = \begin{bmatrix} 0 & 0 & 0 & -1 \\ 0 & 1 & 0 & 0 \\ 0 & 0 & 1 & 0 \\ -1 & 0 & 0 & 0 \end{bmatrix}}. \tag{5.10}$$

The representations which correspond to the three types of pseudo-vectors are obtained by multiplying these matrices by 1, -1; -1, 1; and -1, -1 respectively. As can be seen, $A_P A_T = A_T A_P = -1$ and is thus a multiple of a unit matrix. It can also be verified that the matrix S of (3.24) which reduces all the matrices $(U^* \times U)$ to the reduced form (3.21) also reduces A_P and A_T to P and T, respectively, although neither A_P nor A_T is of the form $(C^* \times C)$, where C is a unimodular matrix.

Next we turn to two-valued representations by means of two-component spinors. As already emphasized on several occasions, the unimodular group C_2 provides a representation only for the restricted group $L_+^\uparrow$. Each of these groups is simply connected and as the transformations belonging to $L_+^\downarrow$, $L_-^\uparrow$, $L_-^\downarrow$ cannot be reached along a continuous path from $L_+^\uparrow$, C_2 cannot cover them as well. Nevertheless, it will be shown in this paragraph that with a suitable modification of the concept of a representation by means of matrices, it is possible to use the matrices of C_2 to cover also the orthochronous group $L^\uparrow$ $(= L_+^\uparrow, L_-^\uparrow)$, which includes the space reflections. Moreover, with the modified concept of a representation, the sets $L_+^\downarrow$ and $L_-^\downarrow$, which involve time reversals, can also be represented if we use the set of matrices C_2^-, consisting of all the matrices iC_2 and having a determinant which is equal to -1.

Although the connectivity consideration provides us with the simplest prerequisite for deciding whether two groups can be mapped into one another, it is useful to possess also a specific algebraic criterion. We have already seen that the possibility of representing the transformations of x^κ by means of transformations of $\phi_{\dot{A}B}$ must rest on the invariance of the relations

$$\phi_{\dot{A}B} = g_{\kappa\dot{A}B}x^\kappa, \tag{5.11}$$

$$x^\kappa = \tfrac{1}{2}g^{\kappa\dot{A}B}\phi_{\dot{A}B} \tag{5.12}$$

within the groups of transformations envisaged. In other words a necessary and sufficient condition for the representation to be possible is the invariance of the spin tensors $g^{\kappa\dot{A}B}$ against combined vector transformation with regard to the index κ and the representing (spinor) transformation with regard to $\dot{A}$, B. On the other hand, if no invariance is established then a representation does not exist. On using this criterion, we shall now confirm that in fact **P** and **T** cannot be represented by any two-dimensional matrices.

The invariance of the $g^{\kappa\dot{A}}{}_B$ against a combined Lorentz transformation L and the spinor transformation which is to represent it, is given by the relation

$$g^{\kappa\dot{A}}{}_B = L^\kappa{}_\sigma T^{*\dot{A}}{}_{\dot{C}} T_B{}^D g^{\sigma\dot{C}}{}_D. \tag{5.13}$$

In matrix form this is

$$g^\kappa = L^\kappa{}_\sigma T^* g^\sigma T^{-1}. \tag{5.14}$$

Now, in the case of the space inversion,

$$\mathbf{P} \to L^1{}_1 = L^2{}_2 = L^3{}_3 = -1, \quad L^4{}_4 = 1, \quad L^\kappa{}_\lambda = 0 \quad (\kappa \neq \lambda); \tag{5.15}$$

while, for time reversal,

$$\mathbf{T} \to L^1{}_1 = L^2{}_2 = L^3{}_3 = 1, \quad L^4{}_4 = -1, \quad L^\kappa{}_\lambda = 0 \quad (\kappa \neq \lambda). \tag{5.16}$$

Let the matrices T which are to represent **P** and **T**, be A and B respectively. We then get from (5.14)

$$\mathbf{P} \to A^*g^aA^{-1} = -g^a \quad (a = 1, 2, 3), \quad A^*g^4A^{-1} = g^4, \tag{5.17}$$

$$\mathbf{T} \to B^*g^aB^{-1} = g^a \quad (a = 1, 2, 3), \quad B^*g^4B^{-1} = -g^4. \tag{5.18}$$

Since the behaviour of **T** with regard to sign is just opposed to that of **P**, it necessarily follows that

$$B = \pm iA, \tag{5.19}$$

so that if A is unimodular this cannot be the case with B. Consequently, it is not possible for the invariance (5.14) to hold both under **P** and **T**. Actually neither of them is possible. Among the matrices g^a, g^2 is a multiple of 1 (see (4.16)). Consequently from (5.17) it would follow that

$$A^*A^{-1} = -1. \tag{5.20}$$

Hence

$$A^* = -A \tag{5.21}$$

and A is purely imaginary,

$$A = iR, \tag{5.22}$$

where R is a real matrix. Consequently, B is a real matrix ($B = B^*$) and from (5.18) we get

$$Bg^a = g^aB. \tag{5.23}$$

The only matrix which can commute with all three matrices g^a is the unit matrix or a multiple of it. It would then follow from (5.18) that $g^4 = 0$. Since none of the g's can vanish, the impossibility of the representation is demonstrated. It follows further that no two-dimensional matrices can exist for the combination of **P** and **T** with $L_-^\uparrow$.

We shall now introduce the modified mode of transformation to be defined by

$$\phi'_A = T_A{}^B\phi^*_B. \tag{5.24}$$

This transformation is called antilinear. As can be seen, to perform the transformation, the entity ϕ_B is first changed into its complex conjugate and then transformed linearly as a spinor. On subjecting ϕ_A to a further antilinear transformation with a matrix $S_C{}^A$, the result is equivalent to an ordinary linear transformation

$$\phi''_C = S_C{}^A(T_A{}^B\phi^*_B)^* = S_C{}^AT^{*B}_A\phi_B = R_C{}^B\phi_B, \tag{5.25}$$

where R is the unimodular matrix ST^*.

For spinors of second rank the anti-linear transformation, as follows from (5.24), is given by

$$\phi'_{AB} = T^{*C}_A T_B{}^D\phi^*_{CD} = T^{*C}_A T^{-1D}{}_B\phi^*_{CD}. \tag{5.26}$$

Hence on using matrix formalism, this gives

$$\phi' = T^*\phi^*T^{-1}. \tag{5.27}$$

The same mode of transformation applies also to spinors $\phi^A{}_B$.

When it is applied to the $g^{\kappa\dot{A}}{}_{B}$, we get instead of (5.17) and (5.18),

$$\mathbf{P} \to A^*g^{a*}A^{-1} = -g^a, \qquad A^*g^{4*}A^{-1} = g^4, \tag{5.28}$$

$$\mathbf{T} \to B^*g^{a*}B^{-1} = g^a, \qquad B^*g^{4*}B^{-1} = -g^4, \tag{5.29}$$

where again (5.19) holds. The second of the relations (5.28), will be satisfied identically if we let

$$A = \pm g^{4*}, \qquad A^* = \pm g^4. \tag{5.30}$$

Since $B = iA$, the second of the relations (5.29) will also be identically satisfied. As to the relations in (5.28) and (5.29) for $a = 1, 2, 3$, they are satisfied in view of the relation (4.18). (5.30) is the only solution for the conditions implied in (5.28) and (5.29). In the representation used for the g's, we get

$$\mathbf{P} \to A = \pm\begin{bmatrix} 0 & 1 \\ -1 & 0 \end{bmatrix}, \qquad \mathbf{T} \to B = \pm\begin{bmatrix} 0 & i \\ -i & 0 \end{bmatrix}, \qquad \mathbf{PT} \to \pm\begin{bmatrix} i & 0 \\ 0 & i \end{bmatrix}. \tag{5.31}$$

Because **PT** commutes with the Lorentz group it is represented by a matrix which is a multiple of 1. Whenever time reversal is involved the representing matrix belongs to C_2^-. However, although **PT** is represented by a matrix belonging to C_2^-, the transformation is linear, because the product of two anti-linear transformations is a linear transformation.

With the invariance of the g's now established for the full Lorentz group, we have the following relations for the various types of Lorentz transformations:

$$L_+^\uparrow \quad L^\kappa{}_\lambda x^\lambda = x^{\kappa'} = \tfrac{1}{2} g^{\kappa\dot{A}B} T^{*\dot{C}}_{\dot{A}} T_B{}^D \phi_{\dot{C}D}, \tag{5.32}$$

$$L_-^\uparrow \quad L^\kappa{}_\lambda x^\lambda = x^{\kappa'} = \tfrac{1}{2} g^{\kappa\dot{A}B} T^{*\dot{C}}_{\dot{A}} T_B{}^D \phi^*_{\dot{C}D}, \tag{5.33}$$

$$L_+^\downarrow \quad L^\kappa{}_\lambda x^\lambda = x^{\kappa'} = \tfrac{1}{2} g^{\kappa\dot{A}B} T^{*-\dot{C}}_{\dot{A}} T^-_B{}^D \phi^*_{\dot{C}D}, \tag{5.34}$$

$$L_-^\downarrow \quad L^\kappa{}_\lambda x^\lambda = x^{\kappa'} = \tfrac{1}{2} g^{\dot{A}B} T^{*-\dot{C}}_{\dot{A}} T^-_B{}^D \phi_{\dot{C}D}, \tag{5.35}$$

where T^- indicates a matrix belonging to C_2^-.

In all these relations, we can replace $\phi_{\dot{C}D}$ in accordance with (4.1) by

$$\phi_{\dot{C}D} = g_{\lambda\dot{C}D} x^\lambda, \tag{5.36}$$

which holds for every Lorentz transformation and merely serves to change from the x^κ to the $\phi_{\dot{C}D}$. From (5.36) we have, since the x^κ are real,

$$\overset{*}{\phi}_{\dot{C}D} = g^*_{\lambda\dot{C}D} x^\lambda. \tag{5.37}$$

On inserting (5.36) or (5.37) in (5.32)–(5.35) we get, since x^λ is arbitrary, the following expressions for Lorentz matrices:

$$L_+^\uparrow:\quad L^\kappa{}_\lambda = \tfrac{1}{2}g^{\kappa\dot{A}B}T^{*C}_{A}T_B{}^D g_{\lambda\dot{C}D} = \tfrac{1}{2}\operatorname{tr}(g^\kappa T\tilde{g}_\lambda \tilde{T}^*), \tag{5.38}$$

$$L_-^\uparrow:\quad L^\kappa{}_\lambda = \tfrac{1}{2}g^{\kappa\dot{A}B}T^{*C}_{A}T_B{}^D g_{\lambda\dot{D}C} = \tfrac{1}{2}\operatorname{tr}(g^\kappa T g_\lambda \tilde{T}^*),\dagger \tag{5.39}$$

$$L_+^\downarrow:\quad L^\kappa{}_\lambda = \tfrac{1}{2}g^{\kappa\dot{A}B}T^{*-C}_{A}T^{-D}_{B} g_{\lambda\dot{D}C} = \tfrac{1}{2}\operatorname{tr}(g^\kappa T^- g_\lambda \tilde{T}^{-*}), \tag{5.40}$$

$$L_-^\downarrow:\quad L^\kappa{}_\lambda = \tfrac{1}{2}g^{\kappa\dot{A}B}T^{*-C}_{A}T^{-D}_{B} g_{\lambda\dot{C}D} = \tfrac{1}{2}\operatorname{tr}(g^\kappa T^- \tilde{g}_\lambda \tilde{T}^{-*}). \tag{5.41}$$

In particular, if we use these expressions to express a Lorentz matrix as a product of two matrices $M^{\kappa,AD}$ and $N_{AD,\lambda}$ as was done in (4.43)–(4.47), we get the following results for the matrices 1, P, T, and PT which belong to the different segments of the Lorentz group:

$$1 = \tfrac{1}{2}\begin{bmatrix} 0 & 1 & 1 & 0 \\ 0 & -i & i & 0 \\ 1 & 0 & 0 & -1 \\ 1 & 0 & 0 & 1 \end{bmatrix}\begin{bmatrix} 0 & 0 & 1 & 1 \\ 1 & i & 0 & 0 \\ 1 & -i & 0 & 0 \\ 0 & 0 & -1 & 1 \end{bmatrix} \left(\text{using the matrix } T = \pm\begin{bmatrix} 1 & 0 \\ 0 & 1 \end{bmatrix}\right), \tag{5.42}$$

$$P = \tfrac{1}{2}\begin{bmatrix} -1 & 0 & 0 & 1 \\ i & 0 & 0 & i \\ 0 & 1 & 1 & 0 \\ 0 & 1 & -1 & 0 \end{bmatrix}\begin{bmatrix} 1 & i & 0 & 0 \\ 0 & 0 & -1 & 1 \\ 0 & 0 & -1 & -1 \\ -1 & i & 0 & 0 \end{bmatrix} \left(\text{using } T = \pm\begin{bmatrix} 0 & -1 \\ 1 & 0 \end{bmatrix}\right), \tag{5.43}$$

$$T = \tfrac{1}{2}\begin{bmatrix} i & 0 & 0 & -i \\ 1 & 0 & 0 & 1 \\ 0 & -i & -i & 0 \\ 0 & -i & i & 0 \end{bmatrix}\begin{bmatrix} -i & 1 & 0 & 0 \\ 0 & 0 & i & -i \\ 0 & 0 & i & i \\ i & 1 & 0 & 0 \end{bmatrix} \left(\text{using } T^- = \pm\begin{bmatrix} 0 & -i \\ i & 0 \end{bmatrix}\right), \tag{5.44}$$

$$PT = \tfrac{1}{2}\begin{bmatrix} i & 0 & 0 & -i \\ 1 & 0 & 0 & 1 \\ 0 & -i & -i & 0 \\ 0 & -i & i & 0 \end{bmatrix}\begin{bmatrix} i & -1 & 0 & 0 \\ 0 & 0 & -i & i \\ 0 & 0 & -i & -i \\ -i & -1 & 0 & 0 \end{bmatrix} \left(\text{using } T^- = \pm\begin{bmatrix} i & 0 \\ 0 & i \end{bmatrix}\right). \tag{5.45}$$

On using (5.24) and (5.31) we get:

$$\mathbf{P} \to \begin{matrix} \phi_1' = \pm\phi_2^* \\ \phi_2' = \mp\phi_1^*, \end{matrix} \qquad \mathbf{T} \to \begin{matrix} \phi_1' = \pm i\phi_2^* \\ \phi_2' = \mp i\phi_1^*. \end{matrix} \tag{5.46}$$

If we first perform a Lorentz transformation, which is represented by Λ, and then follow it by **P** or **T**, we get

$$\mathbf{P} \to \phi'' = A\phi'^* = A(\Lambda\phi)^* = A\Lambda^*\phi^*, \tag{5.47}$$

$$\mathbf{T} \to \phi'' = B\phi'^* = B(\Lambda\phi)^* = B\Lambda^*\phi^*. \tag{5.48}$$

The full inversion of the coordinates **PT** is represented by

$$\mathbf{PT} \to \phi'' = A(B\phi^*)^* = AB^*\phi, \tag{5.49}$$

$$\mathbf{TP} \to \phi'' = B(A\phi^*)^* = BA^*\phi. \tag{5.50}$$

† Using $g^*_{\lambda\dot{C}D} = g_{\lambda\dot{D}C}$, see (4.2).

Although **PT** = **TP**, the representing matrices differ in sign:

$$AB^* = -BA^*. \tag{5.51}$$

The representation of the four-group is thus two-valued. In this connexion it is essential that $A^* = A$, $B^* = -B$. The law of multiplication is completely determined from the following relations:

$$A = A^*, \quad B = -B^*, \quad AA^* = -1, \quad BB^* = -1, \quad AB^* = -BA^*, \tag{5.52}$$

and is given by Table II which should be compared with the law of multiplication of the four-group itself, where every minus sign in the table is replaced by a plus sign.

TABLE II

	1	A (P)	B (T)	AB^* (PT = TP)
1	1	A	B	AB^*
A	A	-1	$-AB^*$	B
B	B	$-AB^*$	1	$-A$
AB^*	AB^*	B	$-A$	-1

Every product in the table is obtainable from (5.52), for instance, $AB = A(-B^*) = -AB^*$. Alternatively,

$$AB = A^*B = -B^*A = BA = BA^* = -AB^*. \tag{5.53}$$

The assumption $AA^* = 1$ ($BB^* = 1$) does not lead to an invariance of the g's.

The matrices in Table II constitute a two-valued representation of the four-group. However, adjoining them to the unimodular group which represents the restricted Lorentz group does not lead to a two-valued representation of the full group. We must revert to anti-linear transformations. (In any case, some of the matrices in the table are already contained in the unimodular group.)

As an example of a relation in two-component spinors which remains invariant under an anti-linear scheme with respect to space and time inversion, we may cite

$$g^{\kappa\dot{A}}{}_{B}\,\partial_\kappa \phi^B + m\phi^{\dot{A}} = 0, \tag{5.54}$$

where $\phi^{\dot{A}}$ is the complex conjugate of ϕ^A and m is a constant.

It has been shown by J. A. McLennan, Jr., that the field ϕ^A which satisfies eqn. (5.54) is equivalent to a Majorana field for a neutrino of mass m and spin $\frac{1}{2}$. Majorana's theory is a special case of Dirac's theory which is based on a four-spinor formalism. This formalism will be developed in the next section. Four-spinors allow covariance with respect to the full Lorentz group through linear transformations and do not necessitate anti-linear transformations. However, not every equation in four-spinors can be expressed in terms of two-spinors and their anti-linear mode of transformations. This only applies in special cases such as in Majorana's theory. (A Majorana field is a self-charge conjugate Dirac field. Such a condition reduces the number of independent components effectively from four to two and it is therefore not surprising that Majorana's theory can also be expressed in terms of two-component spinors.)

In (5.54) if $m = 0$, we get the Weyl equation. Like (5.54) it is not invariant with regard to extended Lorentz transformations if these are to be represented by linear transformations, but invariance holds under anti-linear transformations. Interest in this equation has been revived recently by L. Landau, T. D. Lee and C. N. Yang, and A. Salam in connexion with the breakdown of the law of conservation of parity in β-decay and other processes involving weak interactions. We have already mentioned that in the case of neutrinos the spin always points in one way only along the line of motion. This statement could not be Lorentz invariant if the neutrinos did not always travel with the velocity of light. If they could travel with a velocity which was smaller than c we could always find a frame of reference in which the neutrino would appear to travel in an opposite direction and thus to have the wrong spin. However, a particle with finite mass cannot travel with the velocity of light, hence we must attribute to the neutrino a zero mass.

Multiplying (5.54) by $g^{\lambda C}{}_{\dot{A}}$ we get

$$g^{\lambda C}{}_{\dot{A}}\, g^{\kappa \dot{A}}{}_{B}\, \partial_\kappa \phi^B + m g^{\lambda C}{}_{\dot{A}}\, \phi^{\dot{A}} = 0. \tag{5.55}$$

In this expression we can exchange the summation indices κ and λ. Also, from (5.54),

$$g^{\lambda C}{}_{\dot{A}}\, \partial_\lambda \phi^{\dot{A}} = -m\phi^C. \tag{5.56}$$

Consequently, if we rewrite (5.55) with κ and λ exchanged and use (5.56) we get

$$\tfrac{1}{2}(g^{\kappa C}{}_{\dot{A}}\, g^{\lambda \dot{A}}{}_{B} + g^{\lambda C}{}_{\dot{A}}\, g^{\kappa \dot{A}}{}_{B})\partial_\kappa \partial_\lambda \phi^B - m^2\phi^C = 0. \tag{5.57}$$

On using (4.17) we get

$$g^{\kappa\lambda}\delta^C{}_B\, \partial_\kappa \partial_\lambda \phi^B + m^2\phi^C = 0; \tag{5.58}$$

therefore,†
$$\Box\phi^C - m^2\phi^C = 0. \tag{5.59}$$
Thus, ϕ^C satisfies the wave equation.

6. Four-spinors

From the preceding section it follows that the full Lorentz group involving also space inversion and time reversal does not possess a two-valued representation by means of linear transformations of two dimensions. We must revert to anti-linear transformations when dealing with improper Lorentz transformations. Matters are, however, different if we proceed to consider four-dimensional representations. Let us consider the matrices

$$T_a^{\ b} = \begin{bmatrix} A_A^{\ B} & 0 \\ 0 & A_A^{\ B} \end{bmatrix} \equiv (A_A^{\ B}) \oplus (A_A^{\ B}) \quad (a,b = 1,2,3,4). \tag{6.1}$$

They form a four-dimensional two-valued representation of the proper group. It is not, of course, an 'irreducible' representation since it is already expressed in a fully reduced form. We have simply stacked in the manner shown, known as the direct sum, two two-dimensional representations of the Lorentz group.

To extend this representation to the full Lorentz group we could use the result of the previous section and represent $\mathbf{P}$ and $\mathbf{T}$ (anti-linearly) as follows:

$$\mathbf{P} \rightarrow \pm A_P, \qquad \mathbf{T} \rightarrow \pm A_T, \qquad \mathbf{PT} \rightarrow \pm A_P A_T^*,$$

$$A_P = \begin{bmatrix} 0 & -1 & & \\ 1 & 0 & & \\ & & 0 & -1 \\ & & 1 & 0 \end{bmatrix}, \qquad A_T = \begin{bmatrix} 0 & -i & & \\ i & 0 & & \\ & & 0 & -i \\ & & i & 0 \end{bmatrix},$$

$$A_P A_T^* = \begin{bmatrix} i & 0 & & \\ 0 & i & & \\ & & i & 0 \\ & & 0 & i \end{bmatrix}. \tag{6.2}$$

Here again we do not really get a new type of a representation, but merely a pair of two-dimensional representations considered together, i.e. the direct sum of two two-dimensional representations.

The use of four-dimensional matrices offers, however, a greater number of inequivalent two-valued representations of the four-group, and the question arises whether among these some new possibilities exist which are suitable for a representation of the full Lorentz group

† $\Box$ is defined as $-\partial_\kappa \partial^\kappa = \dfrac{\partial^2}{\partial x^2}+\dfrac{\partial^2}{\partial y^2}+\dfrac{\partial^2}{\partial z^2}-\dfrac{1}{c^2}\dfrac{\partial^2}{\partial t^2}$.

by linear transformations. As an example of a new representation of the four-group let us consider

$$\mathbf{P} \to \pm A_P, \qquad \mathbf{T} \to \pm A_T, \qquad \mathbf{PT} \to \pm A_{PT},$$

$$A_P = \begin{bmatrix} O & \begin{matrix} 1 & 0 \\ 0 & 1 \end{matrix} \\ \begin{matrix} 1 & 0 \\ 0 & 1 \end{matrix} & O \end{bmatrix}, \qquad A_T = \begin{bmatrix} O & \begin{matrix} -1 & 0 \\ 0 & -1 \end{matrix} \\ \begin{matrix} 1 & 0 \\ 0 & 1 \end{matrix} & O \end{bmatrix},$$

$$A_{PT} = \begin{bmatrix} \begin{matrix} 1 & 0 \\ 0 & 1 \end{matrix} & O \\ O & \begin{matrix} -1 & 0 \\ 0 & -1 \end{matrix} \end{bmatrix},$$

$$A_P = A_P^{-1} = \tilde{A}_P, \qquad A_T = -A_T^{-1} = -\tilde{A}_T. \tag{6.3}$$

Since A_{PT} is the direct sum of two matrices which are multiples of a unit matrix, A_{PT} commutes with all the matrices $T_a^{\ b}$ (6.1) or any others which are direct sums of two matrices. This is an essential condition in subsequent developments. The matrices $\pm A_P$, $\pm A_T$, $\pm A_{PT}$ of (6.3) together with ± 1 form a group of eight elements and its multiplication table is easily established. They constitute a two-valued representation of the four-group because

$$A_P A_T = -A_T A_P, \qquad A_P^2 = +1, \qquad A_T^2 = -1. \tag{6.4}$$

If (6.3) is combined with (6.1) we get a set of four-dimensional matrices which are not reducible. However, in the previous section we have already seen that in general we do not get representations of the full Lorentz group by merely combining a representation of the proper group with a representation of the four-group. An additional criterion is required, which, as we have seen, in the case of two-spinors consisted in the invariance of the g's. In the present case we shall have to introduce a similar criterion. This, as will be shown in the next section, will again take the form of an invariance of certain spin-vectors to be denoted by γ^κ. The γ's will depend in a certain way on the g's so that the invariance of the former will entail the invariance of the latter. For the moment, let us disregard this aspect and assume that (6.3) will in fact lead to a two-valued linear representation of the full Lorentz group, when combined with a suitable representation of the proper group. For the latter we cannot use (6.1), for the following reason. $\mathbf{P}^{-1}L_{\text{rest}}\mathbf{P}$ and $\mathbf{T}^{-1}L_{\text{rest}}\mathbf{T}$ are in general different from L_{rest};† on the

† L_{rest} stands here for an individual restricted Lorentz transformation. The restricted

other hand, if we form $A_P^{-1}TA_P$ and $A_T^{-1}TA_T$, where T is given by (6.1), we get in each case the original matrix T again:†

$$\mathbf{P}\rightarrow\begin{bmatrix}0&1\\1&0\end{bmatrix}\begin{bmatrix}A&0\\0&A\end{bmatrix}\begin{bmatrix}0&1\\1&0\end{bmatrix}=\begin{bmatrix}A&0\\0&A\end{bmatrix},$$

$$\mathbf{T}\rightarrow\begin{bmatrix}0&-1\\1&0\end{bmatrix}\begin{bmatrix}A&0\\0&A\end{bmatrix}\begin{bmatrix}0&1\\-1&0\end{bmatrix}=\begin{bmatrix}A&0\\0&A\end{bmatrix}. \tag{6.5}$$

Hence we shall get the result that each unimodular matrix represents two Lorentz transformations and not one (two-valuedly). We can modify (6.1) so as to deal with this state of affairs, in three ways. In the first way we let $T_a{}^b$ be given by

$$T_a{}^b=\begin{bmatrix}A_A{}^B&0\\0&A^c_A{}^B\end{bmatrix}\equiv(A_A{}^B)\oplus(A^c_A{}^B), \tag{6.6}$$

where A^c is the contragredient of A. We now get

$$\mathbf{P}\rightarrow\begin{bmatrix}0&1\\1&0\end{bmatrix}\begin{bmatrix}A&0\\0&A^c\end{bmatrix}\begin{bmatrix}0&1\\1&0\end{bmatrix}=\begin{bmatrix}A^c&0\\0&A\end{bmatrix},$$

$$\mathbf{T}\rightarrow\begin{bmatrix}0&-1\\1&0\end{bmatrix}\begin{bmatrix}A&0\\0&A^c\end{bmatrix}\begin{bmatrix}0&1\\-1&0\end{bmatrix}=\begin{bmatrix}A^c&0\\0&A\end{bmatrix}, \tag{6.7}$$

so that in each case the result differs from the original matrix T.

From (6.7) it follows that we get the same result whether T is transformed with A_P or A_T:

$$A_P\,TA_P^{-1}=A_T\,TA_T^{-1}. \tag{6.8}$$

This conforms with a similar property of the Lorentz transformations. Since $\mathbf{PT}$ is given by the diagonal matrix -1 $(x^{\kappa'}=-x^{\kappa})$ it commutes with every Lorentz transformation. Also $\mathbf{P}=\mathbf{P}^{-1}$, $\mathbf{T}=\mathbf{T}^{-1}$. Hence,

$$L\mathbf{PT}=\mathbf{PT}L,$$

therefore

$$L\mathbf{P}=\mathbf{PT}L\mathbf{T}^{-1},$$

$$\mathbf{P}^{-1}L\mathbf{P}=\mathbf{P}L\mathbf{P}^{-1}=\mathbf{T}L\mathbf{T}^{-1}. \tag{6.9}$$

In (6.7) the resulting matrices are of the same form as (6.6) because $(A^c)^c=A$. This is a necessary feature. In this connexion it is interesting to note that if we represent $\mathbf{P}$ by itself, (6.7) will not transform into

group is an invariant sub-group of the full Lorentz group so that $\mathbf{P}^{-1}L_{\text{rest}}\,\mathbf{P}$ and $\mathbf{T}^{-1}L_{\text{rest}}\,\mathbf{T}$ are again restricted Lorentz transformations, but in general they are different from the original one.

† Thick numbers indicate two-dimensional matrices.

a matrix of the same form:

$$\begin{bmatrix} -1 & & & \\ & -1 & & \\ & & -1 & \\ & & & 1 \end{bmatrix} \begin{bmatrix} A & 0 \\ 0 & A^c \end{bmatrix} \begin{bmatrix} -1 & & & \\ & -1 & & \\ & & -1 & \\ & & & 1 \end{bmatrix} = \begin{bmatrix} A & & 0 & \\ & & A^{c1}_1 & -A^{c2}_1 \\ 0 & & -A^{c1}_2 & A^{c2}_2 \end{bmatrix}. \tag{6.10}$$

The second way which is open to us in modifying (6.1) is given by

$$T_a{}^b = \begin{bmatrix} A_A{}^B & 0 \\ 0 & A^{*B}_A \end{bmatrix} \equiv (A_A{}^B) \oplus (A^{*B}_A). \tag{6.11}$$

Here again A_P and A_T will transform this matrix into a matrix of the same form, but different from the original matrix.

Finally, we can also use

$$T_a{}^b = \begin{bmatrix} A_A{}^B & 0 \\ 0 & A^{c*B}_A \end{bmatrix} \equiv (A_A{}^B) \oplus (A^{c*B}_A). \tag{6.12}$$

Entities transforming linearly under a four-dimensional two-valued representation of the full Lorentz group are called four-spinors. (The names 'undors' and 'bi-spinors' are also used to indicate four-spinors.) Since we are dealing with four-dimensional matrices, four-spinors must have four components which we may write as a column matrix:

$$\Psi_a = \begin{bmatrix} \psi_1 \\ \psi_2 \\ \psi_3 \\ \psi_4 \end{bmatrix}, \qquad \Psi_{\dot a} = \begin{bmatrix} \psi_{\dot 1} \\ \psi_{\dot 2} \\ \psi_{\dot 3} \\ \psi_{\dot 4} \end{bmatrix}. \tag{6.13}$$

If Ψ_a transforms with a matrix T, $\Psi_{\dot a}$ transforms with a matrix T^*.

Because of the special form of the matrices (6.6), (6.11), and (6.12), whenever we deal with proper Lorentz transformations, the components ψ_1, ψ_2 will transform among themselves, in the same way as the components of a two-spinor, say a covariant spinor. The pair ψ_3, ψ_4 will then transform as components of a contravariant spinor in the case (6.6), as a complex conjugate in the case (6.11), and as both contravariant and complex conjugate in the case (6.12).

$$\text{(Case 1)} \quad \Psi_a = \begin{bmatrix} \psi_1 \\ \psi_2 \\ \phi^1 \\ \phi^2 \end{bmatrix}, \qquad \Psi_{\dot a} = \begin{bmatrix} \psi_{\dot 1} \\ \psi_{\dot 2} \\ \phi^{\dot 1} \\ \phi^{\dot 2} \end{bmatrix}. \tag{6.14}$$

$$(\text{Case 2})\quad \Psi_a = \begin{bmatrix} \psi_1 \\ \psi_2 \\ \phi_{\dot{1}} \\ \phi_{\dot{2}} \end{bmatrix}, \qquad \Psi_{\dot{a}} = \begin{bmatrix} \psi_{\dot{1}} \\ \psi_{\dot{2}} \\ \phi_1 \\ \phi_2 \end{bmatrix}. \tag{6.15}$$

$$(\text{Case 3})\quad \Psi_a = \begin{bmatrix} \psi_1 \\ \psi_2 \\ \phi^{\dot{1}} \\ \phi^{\dot{2}} \end{bmatrix}, \qquad \Psi_{\dot{a}} = \begin{bmatrix} \psi_{\dot{1}} \\ \psi_{\dot{2}} \\ \phi^1 \\ \phi^2 \end{bmatrix}. \tag{6.16}$$

In all three cases (6.14), (6.15), (6.16) the effect of the space inversion is simply to exchange the pair of components. For instance, using (6.14),

$$\Psi'_a = \begin{bmatrix} \mathbf{0} & \mathbf{1} \\ \mathbf{1} & \mathbf{0} \end{bmatrix} \begin{bmatrix} \psi_1 \\ \psi_2 \\ \phi^1 \\ \phi^2 \end{bmatrix} = \begin{bmatrix} \phi^1 \\ \phi^2 \\ \psi_1 \\ \psi_2 \end{bmatrix}. \tag{6.17}$$

Ψ'_a has the same nature as Ψ_a because the bottom pair is again contravariant with respect to the top pair. This applies also to the other two cases. With respect to time-reversal the pairs are again exchanged, but in addition the bottom pair changes sign. Here again the nature of the spinor is not changed, because multiplying a two-spinor by any factor does not change its character.

We have yet to examine whether (6.3) combined with any of these three types of columns in fact yields a representation of the full Lorentz group. This question is deferred to the next section.

As in the case of two-spinors, we introduce contravariant entities defined by:

$$\Psi^a = \begin{bmatrix} \psi^1 \\ \psi^2 \\ \phi_1 \\ \phi_2 \end{bmatrix}, \qquad \Psi^{\dot{a}} = \begin{bmatrix} \psi^{\dot{1}} \\ \psi^{\dot{2}} \\ \phi_{\dot{1}} \\ \phi_{\dot{2}} \end{bmatrix}, \tag{6.18}$$

$$\Psi^a = \begin{bmatrix} \psi^1 \\ \psi^2 \\ \phi^{\dot{1}} \\ \phi^{\dot{2}} \end{bmatrix}, \qquad \Psi^{\dot{a}} = \begin{bmatrix} \psi^{\dot{1}} \\ \psi^{\dot{2}} \\ \phi^1 \\ \phi^2 \end{bmatrix}, \tag{6.19}$$

$$\Psi^a = \begin{bmatrix} \psi^1 \\ \psi^2 \\ \phi_{\dot{1}} \\ \phi_{\dot{2}} \end{bmatrix}, \qquad \Psi^{\dot{a}} = \begin{bmatrix} \psi^{\dot{1}} \\ \psi^{\dot{2}} \\ \phi_1 \\ \phi_2 \end{bmatrix}. \tag{6.20}$$

We also introduce numerical matrices, ϵ_{ab}, ϵ^{ab} $(a, b = 1, 2, 3, 4)$, which enable us to lower and raise indices. In view of the fact that the four-spinor is composed of a pair of two-spinors, ϵ^{ab} and ϵ_{ab} have the following forms for the three cases (6.18), (6.19), and (6.20) respectively:

$$\epsilon^{ab} = \begin{bmatrix} \epsilon^{AB} & 0 \\ 0 & \epsilon_{AB} \end{bmatrix}, \qquad \epsilon_{ab} = \begin{bmatrix} \epsilon_{AB} & 0 \\ 0 & \epsilon^{AB} \end{bmatrix}, \tag{6.21}$$

$$\epsilon^{ab} = \begin{bmatrix} \epsilon^{AB} & 0 \\ 0 & \epsilon^{\dot{A}\dot{B}} \end{bmatrix}, \qquad \epsilon_{ab} = \begin{bmatrix} \epsilon_{AB} & 0 \\ 0 & \epsilon_{\dot{A}\dot{B}} \end{bmatrix}, \tag{6.22}$$

$$\epsilon^{ab} = \begin{bmatrix} \epsilon^{AB} & 0 \\ 0 & \epsilon_{\dot{A}\dot{B}} \end{bmatrix}, \qquad \epsilon_{ab} = \begin{bmatrix} \epsilon_{AB} & 0 \\ 0 & \epsilon^{\dot{A}\dot{B}} \end{bmatrix}. \tag{6.23}$$

We note that

$$\epsilon_{ab} = -\epsilon^{ab}, \quad \epsilon_{ab} = -\epsilon_{ba}, \quad \epsilon^{ab} = -\epsilon^{ba}, \tag{6.24}$$

$$\epsilon_{ab}^{-1} \equiv (\epsilon^{-1})_{ab} = -\epsilon^{ba} = \epsilon_{ba}^{c}. \tag{6.25}$$

Consequently also
$$\epsilon_{ab}^{-1} = \epsilon^{ab}. \tag{6.26}$$

We shall later identify ϵ^{ab} and ϵ_{ab} as invariant four-spinors of second rank as far as Lorentz transformations proper are concerned. The behaviour under **P** and **T** will also be examined.

Using ϵ_{ab}, we have
$$\Psi_a = \epsilon_{ab}\Psi^b. \tag{6.27}$$

The inversion of this relation is

$$\Psi^a = (\epsilon^{-1})_{ab}\Psi_b \quad \text{or} \quad \Psi^a = \epsilon^{ab}\Psi_b. \tag{6.28}$$

7. Invariant Spin-tensors

In the previous section we have introduced (tentatively) four-spinors of first rank. A four-spinor of second rank is defined as a quantity Ψ_{ab} transforming according to

$$\Psi'^{ab} = T^a{}_c T^b{}_d \Psi^{cd}, \qquad \Psi'_{ab} = T_a{}^c T_b{}^d \Psi_{cd}. \tag{7.1}$$

The spinor Ψ_{ab} has sixteen components and for convenience may be ordered as a matrix with four rows and four columns:

$$\Psi_{ab} = \begin{bmatrix} \Psi_{11} & \Psi_{12} & \Psi_{13} & \Psi_{14} \\ \cdot & \cdot & \cdot & \cdot \\ \Psi_{41} & \Psi_{42} & \Psi_{43} & \Psi_{44} \end{bmatrix}. \tag{7.2}$$

Just as a four-spinor of first rank is related to two-spinors of first rank, so is also a four-spinor of second rank related to two-spinors of second rank. The relationship is obtained by forming first of all the products $\psi_A\psi_B$, $\psi_A\phi^B$, $\phi^A\psi_B$, $\phi^A\phi^B$, in the special spinor $\Psi_{ab} = \Psi_a \times \Psi_b$.

$$(\Psi_a \times \Psi_b) = \begin{bmatrix} \psi_1\psi_1 & \psi_1\psi_2 & \psi_1\phi^1 & \psi_1\phi^2 \\ \psi_2\psi_1 & \psi_2\psi_2 & \psi_2\phi^1 & \psi_2\phi^2 \\ \phi^1\psi_1 & \phi^1\psi_2 & \phi^1\phi^1 & \phi^1\phi^2 \\ \phi^2\psi_1 & \phi^2\psi_2 & \phi^2\phi^1 & \phi^2\phi^2 \end{bmatrix}, \tag{7.3}$$

where instead of ordering the components as a column we have ordered them as a matrix, which is always more convenient when dealing with second-rank tensors and spinors.

In the general case we thus get

$$
\begin{aligned}
(\Psi_{ab}) &= \begin{bmatrix} \psi_{AB} & \chi_{A}{}^{B} \\ \rho^{A}{}_{B} & \phi^{AB} \end{bmatrix}, &\quad (\Psi^{ab}) &= \begin{bmatrix} \psi^{AB} & \chi^{A}{}_{B} \\ \rho_{A}{}^{B} & \phi_{AB} \end{bmatrix}, \\
(\Psi^{a}{}_{b}) &= \begin{bmatrix} \psi^{A}{}_{B} & \chi^{AB} \\ \rho_{AB} & \phi_{A}{}^{B} \end{bmatrix}, &\quad (\Psi_{a}{}^{b}) &= \begin{bmatrix} \psi_{A}{}^{B} & \chi_{AB} \\ \rho^{AB} & \phi^{A}{}_{B} \end{bmatrix}, \\
(\Psi_{\dot{a}b}) &= \begin{bmatrix} \psi_{\dot{A}B} & \chi_{\dot{A}}{}^{B} \\ \rho^{\dot{A}}{}_{B} & \phi^{\dot{A}B} \end{bmatrix}, &\quad (\Psi_{a\dot{b}}) &= \begin{bmatrix} \psi_{A\dot{B}} & \chi_{A}{}^{\dot{B}} \\ \rho^{A}{}_{\dot{B}} & \phi^{A\dot{B}} \end{bmatrix}, \text{ etc.}
\end{aligned} \tag{7.4}
$$

All this applies to the type (6.14). In the other cases, (6.15) and (6.16), we have respectively

$$
\begin{aligned}
(\Psi_{ab}) &= \begin{bmatrix} \psi_{AB} & \chi_{A\dot{B}} \\ \rho_{\dot{A}B} & \phi_{\dot{A}\dot{B}} \end{bmatrix}, &\quad (\Psi^{ab}) &= \begin{bmatrix} \psi^{AB} & \chi^{A\dot{B}} \\ \rho^{\dot{A}B} & \phi^{\dot{A}\dot{B}} \end{bmatrix}, \\
(\Psi^{a}{}_{b}) &= \begin{bmatrix} \psi^{A}{}_{B} & \chi^{A}{}_{\dot{B}} \\ \rho^{\dot{A}}{}_{B} & \phi^{\dot{A}}{}_{\dot{B}} \end{bmatrix}, &\quad (\Psi_{a}{}^{b}) &= \begin{bmatrix} \psi_{A}{}^{B} & \chi_{A}{}^{\dot{B}} \\ \rho_{\dot{A}}{}^{B} & \phi_{\dot{A}}{}^{\dot{B}} \end{bmatrix}, \\
(\Psi_{\dot{a}b}) &= \begin{bmatrix} \psi_{\dot{A}B} & \chi_{\dot{A}\dot{B}} \\ \rho_{AB} & \phi_{A\dot{B}} \end{bmatrix}, &\quad (\Psi_{a\dot{b}}) &= \begin{bmatrix} \psi_{A\dot{B}} & \chi_{AB} \\ \rho_{\dot{A}\dot{B}} & \phi_{\dot{A}B} \end{bmatrix}, \text{ etc.}
\end{aligned} \tag{7.5}
$$

$$
\begin{aligned}
(\Psi_{ab}) &= \begin{bmatrix} \psi_{AB} & \chi_{A}{}^{\dot{B}} \\ \rho^{\dot{A}}{}_{B} & \phi^{\dot{A}\dot{B}} \end{bmatrix}, &\quad (\Psi^{ab}) &= \begin{bmatrix} \psi^{AB} & \chi^{A}{}_{\dot{B}} \\ \rho_{\dot{A}}{}^{B} & \phi_{\dot{A}\dot{B}} \end{bmatrix}, \\
(\Psi^{a}{}_{b}) &= \begin{bmatrix} \psi^{A}{}_{B} & \chi^{A\dot{B}} \\ \rho_{\dot{A}B} & \phi_{\dot{A}}{}^{\dot{B}} \end{bmatrix}, &\quad (\Psi_{a}{}^{b}) &= \begin{bmatrix} \psi_{A}{}^{B} & \chi_{\dot{A}\dot{B}} \\ \rho^{AB} & \phi^{A}{}_{\dot{B}} \end{bmatrix}, \\
(\Psi_{\dot{a}b}) &= \begin{bmatrix} \psi_{\dot{A}B} & \chi_{\dot{A}}{}^{\dot{B}} \\ \rho^{A}{}_{B} & \phi^{A\dot{B}} \end{bmatrix}, &\quad (\Psi_{a\dot{b}}) &= \begin{bmatrix} \psi_{A\dot{B}} & \chi_{A}{}^{B} \\ \rho^{A}{}_{\dot{B}} & \phi^{\dot{A}B} \end{bmatrix}, \text{ etc.}
\end{aligned} \tag{7.6}
$$

If we look upon Ψ^{ab} as was done above as a matrix (Ψ^{ab}), we can express (7.1) as the following matrix product:

$$
\Psi'^{ab} = T^{a}{}_{c} T^{b}{}_{d} \Psi^{cd} = T^{a}{}_{c} \Psi^{cd} \tilde{T}^{d}{}_{b}. \tag{7.7}
$$

Hence,

$$
(\Psi^{ab})' = T(\Psi^{cd})\tilde{T}, \qquad (\Psi^{\dot{a}b})' = T^{*}(\Psi^{\dot{c}d})\tilde{T}. \tag{7.8}
$$

Similarly,

$$
(\Psi_{ab})' = \tilde{T}^{-1}(\Psi_{cd})T^{-1}, \qquad (\Psi_{\dot{a}b})' = \tilde{T}^{*-1}(\Psi_{\dot{c}d})T^{-1}, \tag{7.9}
$$

$$
(\Psi_{a}{}^{b})' = \tilde{T}^{-1}(\Psi_{c}{}^{d})\tilde{T}, \qquad (\Psi_{\dot{a}}{}^{b})' = \tilde{T}^{*-1}(\Psi_{\dot{c}}{}^{d})\tilde{T}, \tag{7.10}
$$

$$
(\Psi^{a}{}_{b})' = T(\Psi^{a}{}_{b})T^{-1}, \qquad (\Psi^{\dot{a}}{}_{b})' = T^{*}(\Psi^{\dot{c}}{}_{d})T^{-1}. \tag{7.11}
$$

In particular, if we take for T the matrices A_P and A_T as given by (6.3) and transform the Ψ_{ab} of (7.4), (7.5), and (7.6), we again get in every case matrices of the same form.

We now return to the question of whether in fact the four-spinors yield two-valued representations of the full Lorentz group. In analogy with the case of the two-spinors and the restricted Lorentz group (see in particular section 4) we raise the question of whether a Lorentz invariant relationship exists between a four-vector x^κ and a four-spinor of second rank Ψ^{ab} of the form

$$x^\kappa = \gamma^\kappa{}_{ab}\Psi^{ab} \quad (a, b = 1, 2, 3, 4), \tag{7.12}$$

where the γ^κ are numerical (invariant) coefficients yet to be determined. In other words, we ask whether there exists for each Lorentz transformation L a corresponding 16-dimensional faithful transformation $S = T \times T$, so that if (7.12) holds it also follows that

$$L^\kappa{}_\mu x^\mu = \gamma^\kappa{}_{cd} S^{cd}{}_{ab} \Psi^{ab}. \tag{7.13}$$

Inserting (7.12) in (7.13), we get

$$L^\kappa{}_\mu \gamma^\mu{}_{ab} \Psi^{ab} = \gamma^\kappa{}_{cd} S^{cd}{}_{ab} \Psi^{ab}, \tag{7.14}$$

or

$$L^\kappa{}_\mu \gamma^\mu{}_{ab} = \gamma^\kappa{}_{cd} S^{cd}{}_{ab}. \tag{7.15}$$

Expressing S in terms of $T \times T$, we get

$$L^\kappa{}_\mu \gamma^\mu{}_{ab} = \gamma^\kappa{}_{cd} T^c{}_a T^d{}_b, \tag{7.16}$$

or

$$L^\kappa{}_\mu \gamma^\mu{}_{ab} = \tilde{T}^a{}_c \gamma^\kappa{}_{cd} T^d{}_b. \tag{7.17}$$

Consequently,

$$\gamma^\kappa = L^\kappa{}_\mu \tilde{T}^{-1} \gamma^\mu T^{-1} = L^\kappa{}_\mu (\gamma^\mu)', \tag{7.18}$$

where

$$(\gamma^\mu)' = \tilde{T}^{-1} \gamma^\mu T^{-1}. \tag{7.19}$$

(The matrices L and the matrices T operate in different spaces, the Minkowski space and the spinor space respectively, and therefore they can be rearranged as was done in (7.18).)

On comparing (7.19) with (7.9) we see that the $\gamma^\mu{}_{ab}$ behave like spinors of second rank. Moreover, they are also vectors in the Minkowski space and under a combined Lorentz and spinor transformation they remain unchanged. Our problem will be to find out whether such invariant spin vectors exist with respect to the full Lorentz group.

From (7.18) it easily follows that the condition for a group representation is satisfied when considering two successive transformations:

$$\gamma = L_2 \tilde{T}_2^{-1} L_1 \tilde{T}_1^{-1} \gamma T_1^{-1} T_2^{-1} = L_2 L_1 (\widetilde{T_2 T_1})^{-1} \gamma (T_2 T_1)^{-1}. \tag{7.20}$$

Since we now recognize the $\gamma^\kappa{}_{ab}$ as spinors of second rank, we see from (7.12) that a change of frame in the spinor space only ($\Psi' = W\Psi$, $T' = WTW^{-1}$) leaves the right-hand side of this equation invariant. We shall get the same x^κ but with a new set of γ's and Ψ_{ab}. The new representation of the Lorentz group will differ from the first one only by a similarity transformation.

From (7.18) we see that for each Lorentz transformation L, the corresponding transformation T in the spinor space is determined only up to a sign, so that the resulting representation $L \to \pm T$ is two-valued. This does not apply to the 16-dimensional representation $S = T \times T$ by the spinors Ψ^{ab} of second rank.

The task now is to prove that a set of γ's exists (determinable up to a similarity transformation in the space of four-spinors) which is invariant with regard to the full Lorentz group, with no restrictions on L in (7.18). We shall find that such γ's exist only with regard to the entities of the form given in (6.16) and the corresponding (6.20), so that only these can be accepted four-spinors.

It may be guessed that the γ's should have a close relationship to the g's which we encountered when dealing with the restricted Lorentz group. Consequently we resolve the γ's into sums of such terms as

$$[\gamma^\kappa] = \begin{bmatrix} \mathbf{g}^\kappa & \mathbf{0} \\ \mathbf{0} & \mathbf{f}^\kappa \end{bmatrix} \quad \text{and} \quad [\gamma^\kappa] = \begin{bmatrix} \mathbf{0} & \mathbf{g}^\kappa \\ \mathbf{f}^\kappa & \mathbf{0} \end{bmatrix}, \tag{7.21}$$

where $\mathbf{f}^\kappa$ is correlated to $\mathbf{g}^\kappa$ in a way to be determined yet. However, we can easily convince ourselves that the first part in (7.21) cannot be invariant with respect to space reflections and time reversal. Let us express (7.19) for $T = A_P$ and $T = A_T$ respectively. Apart from the transformation in the Lorentz frame with $L = \mathbf{P}$ and $L = \mathbf{T}$, we get for the transformation in the spinor space

$$\tilde{A}_P^{-1}\gamma^\kappa A_P^{-1} = \begin{bmatrix} \mathbf{0} & \mathbf{1} \\ \mathbf{1} & \mathbf{0} \end{bmatrix}\begin{bmatrix} \mathbf{g}^\kappa & \mathbf{0} \\ \mathbf{0} & \mathbf{f}^\kappa \end{bmatrix}\begin{bmatrix} \mathbf{0} & \mathbf{1} \\ \mathbf{1} & \mathbf{0} \end{bmatrix} = \begin{bmatrix} \mathbf{f}^\kappa & \mathbf{0} \\ \mathbf{0} & \mathbf{g}^\kappa \end{bmatrix}, \tag{7.22}$$

$$\tilde{A}_T^{-1}\gamma^\kappa A_T^{-1} = \begin{bmatrix} \mathbf{0} & -\mathbf{1} \\ \mathbf{1} & \mathbf{0} \end{bmatrix}\begin{bmatrix} \mathbf{g}^\kappa & \mathbf{0} \\ \mathbf{0} & \mathbf{f}^\kappa \end{bmatrix}\begin{bmatrix} \mathbf{0} & \mathbf{1} \\ -\mathbf{1} & \mathbf{0} \end{bmatrix} = \begin{bmatrix} \mathbf{f}^\kappa & \mathbf{0} \\ \mathbf{0} & \mathbf{g}^\kappa \end{bmatrix}. \tag{7.23}$$

Now, with the final transformation in the four-vector space we should get the original γ's ,but this is not possible because in (7.22) and (7.23) we have the same result and consequently the transformations $L = \mathbf{P}$ and $L = \mathbf{T}$ could not possibly both lead to the same original γ's. In fact neither of them leads to an invariance. In the case of $\mathbf{P}$ we get

$$\begin{bmatrix} -\mathbf{f}^k & \mathbf{0} \\ \mathbf{0} & -\mathbf{g}^k \end{bmatrix}, \qquad \begin{bmatrix} \mathbf{f}^4 & \mathbf{0} \\ \mathbf{0} & \mathbf{g}^4 \end{bmatrix} \quad (k = 1, 2, 3) \tag{7.24}$$

and in the case of $\mathbf{T}$ we get the opposite result:

$$\begin{bmatrix} \mathbf{f}^k & \mathbf{0} \\ \mathbf{0} & \mathbf{g}^k \end{bmatrix}, \qquad \begin{bmatrix} -\mathbf{f}^4 & \mathbf{0} \\ \mathbf{0} & -\mathbf{g}^4 \end{bmatrix}. \tag{7.25}$$

On the other hand, if we take the second expression in (7.21) we get

$$\tilde{A}_P^{-1}\gamma^\kappa A_P^{-1} = \begin{bmatrix} \mathbf{0} & \mathbf{1} \\ \mathbf{1} & \mathbf{0} \end{bmatrix}\begin{bmatrix} \mathbf{0} & \mathbf{g}^\kappa \\ \mathbf{f}^\kappa & \mathbf{0} \end{bmatrix}\begin{bmatrix} \mathbf{0} & \mathbf{1} \\ \mathbf{1} & \mathbf{0} \end{bmatrix} = \begin{bmatrix} \mathbf{0} & \mathbf{f}^\kappa \\ \mathbf{g}^\kappa & \mathbf{0} \end{bmatrix}, \tag{7.26}$$

$$\tilde{A}_T^{-1}\gamma^\kappa A_T^{-1} = \begin{bmatrix} \mathbf{0} & -\mathbf{1} \\ \mathbf{1} & \mathbf{0} \end{bmatrix}\begin{bmatrix} \mathbf{0} & \mathbf{g}^\kappa \\ \mathbf{f}^\kappa & \mathbf{0} \end{bmatrix}\begin{bmatrix} \mathbf{0} & \mathbf{1} \\ -\mathbf{1} & \mathbf{0} \end{bmatrix} = -\begin{bmatrix} \mathbf{0} & \mathbf{f}^\kappa \\ \mathbf{g}^\kappa & \mathbf{0} \end{bmatrix}. \tag{7.27}$$

The difference in sign between the results in (7.26) and (7.27) makes it possible to achieve invariance both for **P** and **T** $= -$**P**, although as we shall see this is not possible for all the types of γ's arising out of (7.4), (7.5), and (7.6) respectively. From (7.26) and (7.27) it may also be anticipated that $\mathbf{f}^\kappa$ can at most only be some $\mathbf{g}^\kappa$ multiplied by a factor $m = \pm 1$. Consequently, the type of γ's arising out of (7.4)–(7.6) will be

$$[\gamma^\kappa{}_{\dot{a}b}] = \begin{bmatrix} 0 & g^\kappa{}_A{}^B \\ mg^{\kappa\dot{A}}{}_B & 0 \end{bmatrix}, \tag{7.28 a}$$

$$[\gamma^\kappa{}_{ab}] = \begin{bmatrix} 0 & g^\kappa{}_{A\dot{B}} \\ mg^\kappa{}_{\dot{A}B} & 0 \end{bmatrix}, \tag{7.28 b}$$

$$[\gamma^\kappa{}_{ab}] = \begin{bmatrix} 0 & g^\kappa{}_A{}^{\dot{B}} \\ mg^{\kappa\dot{A}}{}_B & 0 \end{bmatrix}. \tag{7.28 c}$$

Let us consider the case $m = 1$. Then none of the γ's will remain invariant when the final transformation is carried out, because both from (7.26) and (7.27) it follows that for these to remain invariant, we must have

$$\mathbf{g}^k = -\mathbf{f}^k, \qquad \mathbf{g}^4 = \mathbf{f}^4. \tag{7.29}$$

This follows from the effect of **P** to change the sign of the three space components and leave the time component unchanged, and from the effect of **T** to leave the space components unchanged and change the sign of the fourth time component. Now, in none of the cases (7.28 a, b, c) can (7.29) be satisfied. In fact, we have

$$g^k{}_A{}^B = g^{k*\dot{A}}{}_B \neq -g^{k\dot{A}}{}_B, \qquad g^4{}_A{}^B = -g^{4\dot{A}}{}_B \neq +g^{4\dot{A}}{}_B, \tag{7.30 a}$$

$$g^k{}_{\dot{A}B} = g^{k*}{}_{A\dot{B}} \neq -g^k{}_{A\dot{B}}, \qquad g^4{}_{\dot{A}B} = g^{4*}{}_{A\dot{B}} = g^4{}_{A\dot{B}}, \tag{7.30 b}$$

$$g^k{}_A{}^{\dot{B}} = g^{k\dot{A}}{}_B, \qquad g^4{}_A{}^{\dot{B}} = -g^{4\dot{A}}{}_B. \tag{7.30 c}$$

In the last case we get exactly the opposite result from that according to (7.29). This shows that if, in (7.28 c) we take $m = -1$, we shall achieve the required invariance with regard to reflections and time reversal.

It now only remains to consider the proper Lorentz transformations. In the only permissible case, (6.12), which led to (6.16) and finally to (7.28 c) (with $m = -1$), the matrices representing the proper Lorentz transformations have the form $T = A \oplus A^{c*}$, where A is a two-dimensional unimodular matrix representing a Lorentz transformation. We have

$$T^{-1} = A^{-1} \oplus A^{c*-1} = A^{-1} \oplus \tilde{A}^*, \tag{7.31}$$

$$\tilde{T}^{-1} = \qquad A^c \oplus A^*. \tag{7.32}$$

Now the invariance of the g's against restricted Lorentz transformations (see (5.1) for $g^{\kappa\dot{A}}{}_B$, for instance) may be expressed as follows

$$L^\kappa{}_\mu A^c g^\mu{}_A{}^{\dot{B}} \tilde{A}^* = g^\kappa{}_A{}^{\dot{B}}, \tag{7.33}$$

$$L^\kappa{}_\mu A^* g^{\mu\dot{A}}{}_B A^{-1} = g^{\kappa\dot{A}}{}_B. \tag{7.34}$$

Hence

$$L^\kappa{}_\mu \tilde{T}^{-1}\gamma^\mu T^{-1} = L^\kappa{}_\mu \begin{bmatrix} A^c & \mathbf{0} \\ \mathbf{0} & A^* \end{bmatrix} \begin{bmatrix} \mathbf{0} & g^\mu{}_A{}^{\dot{B}} \\ -g^{\mu\dot{A}}{}_B & \mathbf{0} \end{bmatrix} \begin{bmatrix} A^{-1} & \mathbf{0} \\ \mathbf{0} & \tilde{A}^* \end{bmatrix}$$

$$= \begin{bmatrix} \mathbf{0} & L^\kappa{}_\mu A^c g^\mu{}_A{}^{\dot{B}} \tilde{A}^* \\ -L^\kappa{}_\mu A^* g^{\mu\dot{A}}{}_B A^{-1} & \mathbf{0} \end{bmatrix} = \begin{bmatrix} \mathbf{0} & g^\kappa{}_A{}^{\dot{B}} \\ -g^{\kappa\dot{A}}{}_B & \mathbf{0} \end{bmatrix}. \tag{7.35}$$

This concludes the proof that the γ's of the type (7.28 c) are invariants with respect to the full Lorentz group.

To sum up the situation, we find that only the following entities are eligible as four-spinors if covariance with the full Lorentz group is to result:

$$\Psi_a = \begin{bmatrix} \psi_1 \\ \psi_2 \\ \phi^{\dot{1}} \\ \phi^{\dot{2}} \end{bmatrix}, \qquad \Psi_{\dot{a}} = \begin{bmatrix} \psi_{\dot{1}} \\ \psi_{\dot{2}} \\ \phi^1 \\ \phi^2 \end{bmatrix}, \qquad \Psi^a = \begin{bmatrix} \psi^1 \\ \psi^2 \\ \phi_{\dot{1}} \\ \phi_{\dot{2}} \end{bmatrix}, \qquad \Psi^{\dot{a}} = \begin{bmatrix} \psi^{\dot{1}} \\ \psi^{\dot{2}} \\ \phi_1 \\ \phi_2 \end{bmatrix}. \tag{7.36}$$

The corresponding transformations in the spinor space, which represent the restricted Lorentz group, are

$$\pm\begin{bmatrix} A & \mathbf{0} \\ \mathbf{0} & A^{c*} \end{bmatrix}, \qquad \pm\begin{bmatrix} A^* & \mathbf{0} \\ \mathbf{0} & A^c \end{bmatrix}, \qquad \pm\begin{bmatrix} A^c & \mathbf{0} \\ \mathbf{0} & A^* \end{bmatrix}, \qquad \pm\begin{bmatrix} A^{c*} & \mathbf{0} \\ \mathbf{0} & A \end{bmatrix}. \tag{7.37}$$

For the representation of the four-group we have used

$$\pm A_P = \pm\begin{bmatrix} \mathbf{0} & \mathbf{1} \\ \mathbf{1} & \mathbf{0} \end{bmatrix}, \qquad \pm A_T = \pm\begin{bmatrix} \mathbf{0} & -\mathbf{1} \\ \mathbf{1} & \mathbf{0} \end{bmatrix}, \qquad \pm A_{PT} = \pm\begin{bmatrix} \mathbf{1} & \mathbf{0} \\ \mathbf{0} & -\mathbf{1} \end{bmatrix}. \tag{7.38}$$

All the columns of (7.36) are essentially of the same type, since in every case the upper two components differ from the lower two components, as far as character is concerned, by two steps: complex conjugation and contragredience. As a result of this, the matrices in (7.37) are all of the same type in the sense that every unimodular matrix at the top corner differs from the one in the bottom corner by the two steps:

c, $*$. Consequently, we may expect all the representations in (7.37) and (7.38) to be equivalent. In fact, the first and fourth of the matrices (7.37), as well as the second and third, involve only an exchange of the corner matrices and this can be achieved, for instance, with the matrix

$$S_1 = \begin{bmatrix} 0 & 1 \\ 1 & 0 \end{bmatrix}, \qquad S_1^{-1} = \begin{bmatrix} 0 & 1 \\ 1 & 0 \end{bmatrix}. \tag{7.39}$$

Thus

$$\begin{bmatrix} 0 & 1 \\ 1 & 0 \end{bmatrix} \begin{bmatrix} A & 0 \\ 0 & A^{c*} \end{bmatrix} \begin{bmatrix} 0 & 1 \\ 1 & 0 \end{bmatrix} = \begin{bmatrix} A^{c*} & 0 \\ 0 & A \end{bmatrix}. \tag{7.40}$$

In a similar way

$$S_2 = \begin{bmatrix} 0 & 0 & 0 & 1 \\ 0 & 0 & -1 & 0 \\ 0 & 1 & 0 & 0 \\ -1 & 0 & 0 & 0 \end{bmatrix}, \qquad S_2^{-1} = \begin{bmatrix} 0 & 0 & 0 & -1 \\ 0 & 0 & 1 & 0 \\ 0 & -1 & 0 & 0 \\ 1 & 0 & 0 & 0 \end{bmatrix} \tag{7.41}$$

transform the first of the matrices in (7.37) into the second. The situation here is different from that encountered when dealing with two-component spinors. In this case ϕ_A and $\phi_{\dot{A}}$ transform according to C_2 and C_2^* and these are not equivalent. By combining S_1 and S_2, all the matrices of (7.37) are transformable into each other. It remains to consider the matrices (7.38). Both under S_1 as well as S_2, $\pm A_P$ transforms into $\pm A_P$ while $\pm A_T$ transforms into $\mp A_T$. However, the change in sign of A_T is irrelevant because we are dealing with two-valued representations in which each matrix has a $\pm$ sign and these are not coupled to each other.

The Ψ^{ab} transform according to a faithful representation of the full Lorentz group. From (7.8) we see that a change of sign in T is cancelled by the corresponding change of sign in $\tilde{T}$. The Ψ^{ab} transform like direct products $\Psi^a \times \Phi^b$ of two four-spinors of first rank which belong to the type (6.16). It follows further, therefore, that these four-spinors of first rank transform according to the two-valued representation of the full Lorentz group which is given in (7.37) and (7.38). The situation here is quite similar to the one we encountered before when dealing with two-component spinors and when introducing the invariant g's. (7.12) is the analogue of (4.20) (or (4.1) which is its inverse). The difference between the two cases is that now it covers the full Lorentz group, whereas before it applied to the restricted group only.

We must next consider the question whether (6.3) is the only possible representation of the four-group which when coupled with (6.12) yields the invariance of the γ's. In the case of tensors, we know there exist four different possibilities which are obtained from $A_P = \mathbf{P}$, $A_T = \mathbf{T}$

by multiplication respectively with 1, 1; 1, −1; −1, +1; −1, −1. If we apply this method to A_P and A_T of (6.3) we get the following set of matrices:

$$
\begin{array}{c|c|c|c|c}
 & (1,1) & (1,-1) & (-1,1) & (-1,-1) \\
A_P = & \begin{bmatrix} 0 & 1 \\ 1 & 0 \end{bmatrix} & \begin{bmatrix} 0 & 1 \\ 1 & 0 \end{bmatrix} & \begin{bmatrix} 0 & -1 \\ -1 & 0 \end{bmatrix} & \begin{bmatrix} 0 & -1 \\ -1 & 0 \end{bmatrix} \\
A_T = & \begin{bmatrix} 0 & -1 \\ 1 & 0 \end{bmatrix} & \begin{bmatrix} 0 & 1 \\ -1 & 0 \end{bmatrix} & \begin{bmatrix} 0 & -1 \\ 1 & 0 \end{bmatrix} & \begin{bmatrix} 0 & 1 \\ -1 & 0 \end{bmatrix}
\end{array}
\tag{7.42}
$$

All these matrices satisfy the condition of invariance of the γ's since $A\gamma\tilde{A}$ is not affected by a change of sign of A. However, for this same reason they do not really yield four different possibilities, because we are dealing with two-valued representations and a factor ± 1 in front of each matrix does not change the representation.

Since we are dealing with two-valued representations we can represent the elements **P** and **T** of the four-group not only by **P**, **T** → 1, 1; 1, −1; −1, 1; and −1, −1, but also as shown in the following table:

$$
\begin{array}{rr|rr|rr}
\mathbf{P} & \mathbf{T} & \mathbf{P} & \mathbf{T} & \mathbf{P} & \mathbf{T} \\
1 & i & i & 1 & i & i \\
-1 & -i & -i & -1 & -i & -i \\
1 & -i & i & -1 & i & -i \\
-1 & i & -i & 1 & -i & i
\end{array}
\tag{7.43}
$$

As regards spinors of second rank, they represent three additional possibilities (and not twelve) because all those belonging to the same column yield matrices which differ from each other only by a factor ± 1. Hence, altogether we must consider the following representation of the four-group

$$
\begin{array}{ccc}
 & \pm A_P & \pm A_T \\
\text{(i)} & \pm\begin{bmatrix} 0 & 1 \\ 1 & 0 \end{bmatrix} & \pm\begin{bmatrix} 0 & -1 \\ 1 & 0 \end{bmatrix} \\
\text{(ii)} & \pm\begin{bmatrix} 0 & 1 \\ 1 & 0 \end{bmatrix} & \pm\begin{bmatrix} 0 & -\mathrm{i} \\ \mathrm{i} & 0 \end{bmatrix} \\
\text{(iii)} & \pm\begin{bmatrix} 0 & \mathrm{i} \\ \mathrm{i} & 0 \end{bmatrix} & \pm\begin{bmatrix} 0 & -1 \\ 1 & 0 \end{bmatrix} \\
\text{(iv)} & \pm\begin{bmatrix} 0 & \mathrm{i} \\ \mathrm{i} & 0 \end{bmatrix} & \pm\begin{bmatrix} 0 & -\mathrm{i} \\ \mathrm{i} & 0 \end{bmatrix}
\end{array}
\tag{7.44}
$$

The first of these sets, (i), yields as we know already a representation through γ's with $m = -1$. Restricting ourselves for the moment again to the case $m = -1$, none of the other three sets shown in (7.44) leads to a representation. However, in the foregoing we have dealt only with a correlation (7.12) in which x_κ is a regular four-vector (the case (1, 1) according to the classification in (5.6)). We must consider also correlations

$$F^\kappa = \gamma^\kappa{}_{ab}\Psi^{ab}, \tag{7.45}$$

where F^κ is a four-vector of any of the following types: $(1, -1)$, $(-1, 1)$, and $(-1, -1)$. In the case $(1, -1)$, the matrix **T** is identical with the matrix **P** ($F^k \to -F^k$, $F^4 \to F^4$). In this case invariance results for γ's with $m = -1$ provided **P** and **T** are represented according to (ii) in (7.44). In the case $(-1, 1)$, **P** and **T** are again identical matrices but the sign is opposite to that of $(1, -1)$. Consequently we now get invariant γ's if we use the set (iii) in (7.44). Finally, the case $(-1, -1)$ differs from the case $(1, 1)$ by a change of sign in both **P** and **T**. For spinors we shall now need the set (iv) ($A_P\gamma\tilde{A}_P$ and $A_T\gamma\tilde{A}_T$ differ from $iA_P\gamma i\tilde{A}_P$ and $iA_T\gamma i\tilde{A}_T$, just by the required $-$ sign).

The representations (7.44) of the four-group has led us to the conclusion that only $m = -1$ in (7.28 c) leads to invariant γ's. But it is easily seen on inspection that (7.28 c) with $m = +1$ may also be used in the basic correlation (7.45), provided we exchange A_P and A_T in all the previous considerations. This leads to a result which is described in the following table, in which A_P and A_T stand for the matrices shown in (i) of (7.44):

	$m = -1$		$m = +1$	
	P	**T**	**P**	**T**
(i)	A_P	A_T	A_T	A_P
(ii)	A_P	iA_T	A_T	iA_P
(iii)	iA_P	A_T	iA_T	A_P
(iv)	iA_P	iA_T	iA_T	iA_P

(7.46)

At first sight it might appear that we are getting eight different representations. But this is not the case, since the difference between the matrices belonging to $m = -1$ and $m = +1$ consists only in an exchange of A_P and A_T. More specifically any pair of matrices with $m = -1$ can be transformed into a pair with $m = +1$ provided we disregard $\pm$ signs, which is permissible since we are dealing with two-valued representations. Thus, consider the following transformation in the four-spinor space

$$\begin{bmatrix} \psi^1 \\ \psi^2 \\ \phi_{\dot 1} \\ \phi_{\dot 2} \end{bmatrix}' = \begin{bmatrix} i & 0 & 0 & 0 \\ 0 & i & 0 & 0 \\ 0 & 0 & 1 & 0 \\ 0 & 0 & 0 & 1 \end{bmatrix} \begin{bmatrix} \psi^1 \\ \psi^2 \\ \phi_{\dot 1} \\ \phi_{\dot 2} \end{bmatrix} \qquad (\Psi^{a'} = S\Psi^a). \tag{7.47}$$

This transformation will change the matrices of the first column in (7.46) into matrices of the second column. Disregarding changes of sign we have

$$SA_P S^{-1} = \begin{bmatrix} \mathbf{i} & \mathbf{0} \\ \mathbf{0} & \mathbf{1} \end{bmatrix} \begin{bmatrix} \mathbf{0} & \mathbf{1} \\ \mathbf{1} & \mathbf{0} \end{bmatrix} \begin{bmatrix} -\mathbf{i} & \mathbf{0} \\ \mathbf{0} & \mathbf{1} \end{bmatrix} = \begin{bmatrix} \mathbf{0} & \mathbf{i} \\ -\mathbf{i} & \mathbf{0} \end{bmatrix} = iA_T, \tag{7.48}$$

$$SA_T S^{-1} = \begin{bmatrix} \mathbf{i} & \mathbf{0} \\ \mathbf{0} & \mathbf{1} \end{bmatrix} \begin{bmatrix} \mathbf{0} & -\mathbf{1} \\ \mathbf{1} & \mathbf{0} \end{bmatrix} \begin{bmatrix} -\mathbf{i} & \mathbf{0} \\ \mathbf{0} & \mathbf{1} \end{bmatrix} = -\begin{bmatrix} \mathbf{0} & \mathbf{i} \\ \mathbf{i} & \mathbf{0} \end{bmatrix} = iA_P. \tag{7.49}$$

It may be added that the transformation (7.47) in the spinor frame transforms the γ's as far as their spinor character is concerned according to (7.9),

$$(\gamma)' = \tilde{S}^{-1}(\gamma)S^{-1}. \tag{7.50}$$

On using (7.28 c) with $m = -1$ we get

$$[\gamma^\kappa_{ab}]' = \begin{bmatrix} 0 & i[g^{\kappa}{}_{A}{}^{\dot{B}}] \\ -i[g^{\kappa \dot{A}}{}_{B}] & 0 \end{bmatrix}. \tag{7.51}$$

Thus the effect of the transformation S is to multiply the γ's of $m = -1$ by i. In the next section we shall see that this is making the γ's obey the same commutation rules as those of $m = +1$. Hence algebraically there is no essential difference between the two types of γ's.

To return to the two-valued representations of the full Lorentz group by means of four-spinors of first rank, we see that there exist essentially four different representations which differ from each other in their behaviour with regard to space inversion and time reversal. These are described, for instance, by the first column of (7.46).

Four-spinors of second rank transform according to a 16-dimensional faithful representation of the full Lorentz group. Since this is equivalent to a tensor transformation, there exist only four different modes of transformation with regard to the four-group. The 16-dimensional representation is reducible into three representations of the full Lorentz group of 1, 5, and 10 dimensions respectively.

8. The ϵ_{ab} and Some Properties of the γ's

As mentioned already spinors of second rank Ψ_{ab} transform according to a 16-dimensional representation of the full Lorentz group. The law of transformation is best expressed when we order the components as a matrix. It has then the form

$$(\Psi_{ab})' = T(\Psi_{ab})\tilde{T}, \tag{8.1}$$

where T is any matrix which belongs to the four-dimensional and two-valued representation of the full Lorentz group.

Let us consider in particular ϵ_{ab}. Under restricted Lorentz transformations it is an invariant. It remains therefore only to consider its behaviour under space inversion and time reversal. Taking the first case in (7.46) we get

$$\mathbf{P} \to \begin{bmatrix} 0 & 1 \\ 1 & 0 \end{bmatrix} \begin{bmatrix} (\epsilon_{AB}) & 0 \\ 0 & (\epsilon^{AB}) \end{bmatrix} \begin{bmatrix} 0 & 1 \\ 1 & 0 \end{bmatrix}$$

$$= \begin{bmatrix} (\epsilon^{AB}) & 0 \\ 0 & (\epsilon_{AB}) \end{bmatrix} = -\begin{bmatrix} (\epsilon_{AB}) & 0 \\ 0 & (\epsilon^{AB}) \end{bmatrix}, \tag{8.2}$$

$$\mathbf{T} \to \begin{bmatrix} 0 & -1 \\ 1 & 0 \end{bmatrix} \begin{bmatrix} (\epsilon_{AB}) & 0 \\ 0 & (\epsilon^{AB}) \end{bmatrix} \begin{bmatrix} 0 & 1 \\ -1 & 0 \end{bmatrix}$$

$$= \begin{bmatrix} (\epsilon^{AB}) & 0 \\ 0 & (\epsilon_{AB}) \end{bmatrix} = -\begin{bmatrix} (\epsilon_{AB}) & 0 \\ 0 & (\epsilon^{AB}) \end{bmatrix}. \tag{8.3}$$

Thus, both **P** and **T** cause ϵ_{ab} to change into $-\epsilon_{ab}$. The remaining cases can be treated in a similar way. We get altogether

$$\begin{array}{lllll} & \mathbf{P} & \mathbf{T} & \\ \text{(i)} & \epsilon_{ab} \to -\epsilon_{ab} & \epsilon_{ab} \to -\epsilon_{ab} & \text{case } (-1, -1), \\ \text{(ii)} & \epsilon_{ab} \to -\epsilon_{ab} & \epsilon_{ab} \to \epsilon_{ab} & \text{case } (-1, 1), \\ \text{(iii)} & \epsilon_{ab} \to \epsilon_{ab} & \epsilon_{ab} \to -\epsilon_{ab} & \text{case } (1, -1), \\ \text{(iv)} & \epsilon_{ab} \to \epsilon_{ab} & \epsilon_{ab} \to \epsilon_{ab} & \text{case } (1, 1). \end{array} \tag{8.4}$$

In every case ϵ_{ab} must satisfy the relation

$$(\epsilon_{ab}\Psi^b)' = \epsilon'_{ab}\Psi'^b, \tag{8.5}$$

which expresses the condition that if we first lower the index of a spinor and then transform, we get the same result as obtained when we first transform and then lower the index. In testing this relation we have to take into account that in the first case we transform a covariant spinor while in the second case we transform a contravariant spinor. For proper Lorentz transformations where $\epsilon'_{ab} = \epsilon_{ab}$ we verify (8.5) by using for the transformations of a covariant and a contravariant spinor the following matrices respectively:

$$[T_a{}^b] = \begin{bmatrix} [A_A{}^B] & 0 \\ 0 & [A^{c*}{}_A{}^B] \end{bmatrix}, \qquad T^c{}_a{}^b = \begin{bmatrix} [A^c{}_A{}^B] & 0 \\ 0 & [A^*{}_A{}^B] \end{bmatrix}. \tag{8.6}$$

If Ψ^b is the column $(\psi^A, \phi_{\dot{A}})$ we get for the left-hand side and right-hand side of (8.5) respectively

$$\begin{bmatrix} A_C{}^A \epsilon_{AB} \psi^B \\ A^{*c}{}_C{}^A \epsilon^{AB} \phi_{\dot{B}} \end{bmatrix} \quad \text{and} \quad \begin{bmatrix} \epsilon_{CA} A^c{}_A{}^B \psi^B \\ \epsilon^{CA} A^*{}_A{}^B \phi_{\dot{B}} \end{bmatrix}. \tag{8.7}$$

These two columns are identical because

$$A_C{}^A \epsilon_{AB} \equiv \begin{bmatrix} \alpha & \beta \\ \gamma & \delta \end{bmatrix} \begin{bmatrix} 0 & -1 \\ 1 & 0 \end{bmatrix} = \epsilon_{CA} A^c{}_A{}^B \equiv \begin{bmatrix} 0 & -1 \\ 1 & 0 \end{bmatrix} \begin{bmatrix} \delta & -\gamma \\ -\beta & \alpha \end{bmatrix}, \tag{8.8}$$

and similarly for the lower half of the columns.

Let us now consider **P** and **T** for case (i) of (7.46): $T_a{}^b = A_P = A_P^c$ and $T_a{}^b = A_T = A_T^c$ respectively. Using $\epsilon'_{ab} = -\epsilon_{ab}$ we get for each side of (8.5):

$$\mathbf{P} \to \begin{bmatrix} \epsilon^{AB} \phi_{\dot{B}} \\ \epsilon_{AB} \psi^B \end{bmatrix} \quad \text{and} \quad -\begin{bmatrix} \epsilon_{AB} \phi_{\dot{B}} \\ \epsilon^{AB} \psi^B \end{bmatrix} = \begin{bmatrix} \epsilon^{AB} \phi_{\dot{B}} \\ \epsilon_{AB} \psi^B \end{bmatrix}, \tag{8.9}$$

$$\mathbf{T} \to \begin{bmatrix} -\epsilon^{AB} \phi_{\dot{B}} \\ \epsilon_{AB} \psi^B \end{bmatrix} \quad \text{and} \quad -\begin{bmatrix} \epsilon_{AB}(-\phi_{\dot{B}}) \\ \epsilon^{AB} \psi^B \end{bmatrix} = \begin{bmatrix} -\epsilon^{AB} \phi_{\dot{B}} \\ \epsilon_{AB} \psi^B \end{bmatrix}. \tag{8.10}$$

Thus, (8.5) is confirmed both for **P** and **T**. In case (iv) of (7.46) we have $\epsilon'_{ab} = \epsilon_{ab}$, $A_P = -A_P^c$, $A_T = -A_T^c$ and this yields for each side of (8.5):

$$\mathbf{P} \to \begin{bmatrix} i\epsilon^{AB} \phi_{\dot{B}} \\ i\epsilon_{AB} \psi^B \end{bmatrix} \quad \text{and} \quad -\begin{bmatrix} \epsilon_{AB} i\phi_{\dot{B}} \\ \epsilon^{AB} i\psi^B \end{bmatrix} = \begin{bmatrix} i\epsilon^{AB} \phi_{\dot{B}} \\ i\epsilon_{AB} \psi^B \end{bmatrix}, \tag{8.11}$$

$$\mathbf{T} \to \begin{bmatrix} -i\epsilon^{AB} \phi_{\dot{B}} \\ i\epsilon_{AB} \psi^B \end{bmatrix} \quad \text{and} \quad -\begin{bmatrix} \epsilon_{AB}(-i\phi_{\dot{B}}) \\ \epsilon^{AB} i\psi^B \end{bmatrix} = \begin{bmatrix} -i\epsilon^{AB} \phi_{\dot{B}} \\ i\epsilon_{AB} \psi^B \end{bmatrix}. \tag{8.12}$$

Thus again (8.5) is confirmed. The same result holds in the remaining cases of (7.46).

Another example in which the transformation is particularly simple is offered by the $\gamma^\kappa{}_{ab}$. Generally $\Psi^\kappa{}_{ab}$ transforms according to the direct product of a four-vector transformation with that of a four-spinor of second rank. In the case of the γ's, if we couple with each type of spinor transformation a suitable mode of vector transformations, the γ's behave as invariants.

In (7.28 c) the $\gamma^\kappa{}_{ab}$ are expressed in terms of the $g^\kappa{}_A{}^{\dot{B}}$ and $g^{\kappa \dot{A}}{}_B$. The relationship between the g's is given by

$$g^\kappa{}_C{}^{\dot{D}} = \epsilon_{CA} \epsilon^{\dot{D}\dot{B}} g^{\kappa A}{}_{\dot{B}} \quad \text{or} \quad [g^\kappa{}_C{}^{\dot{D}}] = [\epsilon_{CA}][g^{\kappa A}{}_{\dot{B}}][\tilde{\epsilon}^{BC}]$$
$$= \begin{bmatrix} 0 & -1 \\ 1 & 0 \end{bmatrix} [g^{\kappa A}{}_{\dot{B}}] \begin{bmatrix} 0 & -1 \\ 1 & 0 \end{bmatrix}. \tag{8.13}$$

Consequently, taking the complex conjugate of (4.16), transforming them according to (8.13), and using (7.28c) with $m = -1$, we get explicitly for the γ's

$$[\gamma^1_{ab}] = \begin{bmatrix} O & \begin{matrix}1 & 0\\ 0 & -1\end{matrix} \\ \begin{matrix}-1 & 0\\ 0 & 1\end{matrix} & O \end{bmatrix}, \quad [\gamma^2_{ab}] = \begin{bmatrix} O & \begin{matrix}-i & 0\\ 0 & -i\end{matrix} \\ \begin{matrix}i & 0\\ 0 & i\end{matrix} & O \end{bmatrix},$$

$$[\gamma^3_{ab}] = \begin{bmatrix} O & \begin{matrix}0 & -1\\ -1 & 0\end{matrix} \\ \begin{matrix}0 & 1\\ 1 & 0\end{matrix} & O \end{bmatrix}, \quad [\gamma^4_{ab}] = \begin{bmatrix} O & \begin{matrix}0 & 1\\ -1 & 0\end{matrix} \\ \begin{matrix}0 & 1\\ -1 & 0\end{matrix} & O \end{bmatrix}. \tag{8.14}$$

To obtain $\gamma^{\kappa ab}$ we use

$$\gamma^{\kappa cd} = \epsilon^{ca}\epsilon^{db}\gamma^{\kappa}{}_{ab} \quad \text{or} \quad [\gamma^{\kappa}{}_{cd}] = [\epsilon^{ca}][\gamma^{\kappa}{}_{ab}][\tilde{\epsilon}^{bd}]. \tag{8.15}$$

This yields

$$[\gamma^{1ab}] = \begin{bmatrix} O & \begin{matrix}1 & 0\\ 0 & -1\end{matrix} \\ \begin{matrix}-1 & 0\\ 0 & 1\end{matrix} & O \end{bmatrix}, \quad [\gamma^{2ab}] = \begin{bmatrix} O & \begin{matrix}i & 0\\ 0 & i\end{matrix} \\ \begin{matrix}-i & 0\\ 0 & -i\end{matrix} & O \end{bmatrix},$$

$$[\gamma^{3ab}] = \begin{bmatrix} O & \begin{matrix}0 & -1\\ -1 & 0\end{matrix} \\ \begin{matrix}0 & 1\\ 1 & 0\end{matrix} & O \end{bmatrix}, \quad [\gamma^{4ab}] = \begin{bmatrix} O & \begin{matrix}0 & -1\\ 1 & 0\end{matrix} \\ \begin{matrix}0 & -1\\ 1 & 0\end{matrix} & O \end{bmatrix}. \tag{8.16}$$

To obtain $\gamma^{\kappa a}{}_b$ and $\gamma^{\kappa}{}_a{}^b$ we have to raise or lower one index only either in $\gamma^{\kappa}{}_{ab}$ or $\gamma^{\kappa ab}$:

$$\gamma^{\kappa a}{}_b = \epsilon^{ac}\gamma^{\kappa}{}_{cb}, \quad \text{therefore} \quad [\gamma^{\kappa a}{}_b] = [\epsilon^{ac}][\gamma^{\kappa}{}_{cb}], \tag{8.17}$$

$$\gamma^{\kappa a}{}_b = \epsilon_{bc}\gamma^{\kappa ac}, \quad \text{therefore} \quad [\gamma^{\kappa a}{}_b] = [\gamma^{\kappa ac}][\tilde{\epsilon}_{cb}], \tag{8.18}$$

$$\gamma^{\kappa}{}_a{}^b = \epsilon^{bc}\gamma^{\kappa}{}_{ac}, \quad \text{therefore} \quad [\gamma^{\kappa}{}_a{}^b] = [\gamma^{\kappa}{}_{ac}][\tilde{\epsilon}^{cb}], \tag{8.19}$$

$$\gamma^{\kappa}{}_a{}^b = \epsilon_{ac}\gamma^{\kappa cb}, \quad \text{therefore} \quad [\gamma^{\kappa}{}_a{}^b] = [\epsilon_{ac}][\gamma^{\kappa cb}]. \tag{8.20}$$

This yields

$$[\gamma^1{}_a{}^b] = \begin{bmatrix} O & \begin{matrix}0 & 1\\ 1 & 0\end{matrix} \\ \begin{matrix}0 & 1\\ 1 & 0\end{matrix} & O \end{bmatrix}, \quad [\gamma^2{}_a{}^b] = \begin{bmatrix} O & \begin{matrix}0 & -i\\ i & 0\end{matrix} \\ \begin{matrix}0 & -i\\ i & 0\end{matrix} & O \end{bmatrix},$$

$$[\gamma^3{}_a{}^b] = \begin{bmatrix} O & \begin{matrix}1 & 0\\ 0 & -1\end{matrix} \\ \begin{matrix}1 & 0\\ 0 & -1\end{matrix} & O \end{bmatrix}, \quad [\gamma^4{}_a{}^b] = \begin{bmatrix} O & \begin{matrix}-1 & 0\\ 0 & -1\end{matrix} \\ \begin{matrix}1 & 0\\ 0 & 1\end{matrix} & O \end{bmatrix}, \tag{8.21}$$

$$[\gamma^{1a}{}_b] = \begin{bmatrix} O & \begin{matrix} 0 & -1 \\ -1 & 0 \end{matrix} \\ \begin{matrix} 0 & -1 \\ -1 & 0 \end{matrix} & O \end{bmatrix}, \quad [\gamma^{2a}{}_b] = \begin{bmatrix} O & \begin{matrix} 0 & -i \\ i & 0 \end{matrix} \\ \begin{matrix} 0 & -i \\ i & 0 \end{matrix} & O \end{bmatrix},$$

$$[\gamma^{3a}{}_b] = \begin{bmatrix} O & \begin{matrix} -1 & 0 \\ 0 & 1 \end{matrix} \\ \begin{matrix} -1 & 0 \\ 0 & 1 \end{matrix} & O \end{bmatrix}, \quad [\gamma^{4a}{}_b] = \begin{bmatrix} O & \begin{matrix} -1 & 0 \\ 0 & -1 \end{matrix} \\ \begin{matrix} 1 & 0 \\ 0 & 1 \end{matrix} & O \end{bmatrix}. \tag{8.22}$$

In the same way, using (7.28 c) with $m = +1$, we get

$$[\gamma^1{}_{ab}] = \begin{bmatrix} O & \begin{matrix} 1 & 0 \\ 0 & -1 \end{matrix} \\ \begin{matrix} 1 & 0 \\ 0 & -1 \end{matrix} & O \end{bmatrix}, \quad [\gamma^2{}_{ab}] = \begin{bmatrix} O & \begin{matrix} -i & 0 \\ 0 & -i \end{matrix} \\ \begin{matrix} -i & 0 \\ 0 & -i \end{matrix} & O \end{bmatrix},$$

$$[\gamma^3{}_{ab}] = \begin{bmatrix} O & \begin{matrix} 0 & -1 \\ -1 & 0 \end{matrix} \\ \begin{matrix} 0 & -1 \\ -1 & 0 \end{matrix} & O \end{bmatrix}, \quad [\gamma^4{}_{ab}] = \begin{bmatrix} O & \begin{matrix} 0 & 1 \\ -1 & 0 \end{matrix} \\ \begin{matrix} 0 & -1 \\ 1 & 0 \end{matrix} & O \end{bmatrix}, \tag{8.23}$$

$$[\gamma^{1ab}] = \begin{bmatrix} O & \begin{matrix} 1 & 0 \\ 0 & -1 \end{matrix} \\ \begin{matrix} 1 & 0 \\ 0 & -1 \end{matrix} & O \end{bmatrix}, \quad [\gamma^{2ab}] = \begin{bmatrix} O & \begin{matrix} i & 0 \\ 0 & i \end{matrix} \\ \begin{matrix} i & 0 \\ 0 & i \end{matrix} & O \end{bmatrix},$$

$$[\gamma^{3ab}] = \begin{bmatrix} O & \begin{matrix} 0 & -1 \\ -1 & 0 \end{matrix} \\ \begin{matrix} 0 & -1 \\ -1 & 0 \end{matrix} & O \end{bmatrix}, \quad [\gamma^{4ab}] = \begin{bmatrix} O & \begin{matrix} 0 & -1 \\ 1 & 0 \end{matrix} \\ \begin{matrix} 0 & 1 \\ -1 & 0 \end{matrix} & O \end{bmatrix}, \tag{8.24}$$

$$[\gamma^1{}_a{}^b] = \begin{bmatrix} O & \begin{matrix} 0 & 1 \\ 1 & 0 \end{matrix} \\ \begin{matrix} 0 & -1 \\ -1 & 0 \end{matrix} & O \end{bmatrix}, \quad [\gamma^2{}_a{}^b] = \begin{bmatrix} O & \begin{matrix} 0 & -i \\ i & 0 \end{matrix} \\ \begin{matrix} 0 & i \\ -i & 0 \end{matrix} & O \end{bmatrix},$$

$$[\gamma^3{}_a{}^b] = \begin{bmatrix} O & \begin{matrix} 1 & 0 \\ 0 & -1 \end{matrix} \\ \begin{matrix} -1 & 0 \\ 0 & 1 \end{matrix} & O \end{bmatrix}, \quad [\gamma^4{}_a{}^b] = \begin{bmatrix} O & \begin{matrix} -1 & 0 \\ 0 & -1 \end{matrix} \\ \begin{matrix} -1 & 0 \\ 0 & -1 \end{matrix} & O \end{bmatrix}, \tag{8.25}$$

$$[\gamma^{1a}{}_b] = \begin{bmatrix} O & \begin{matrix} 0 & -1 \\ -1 & 0 \end{matrix} \\ \begin{matrix} 0 & 1 \\ 1 & 0 \end{matrix} & O \end{bmatrix}, \quad [\gamma^{2a}{}_b] = \begin{bmatrix} O & \begin{matrix} 0 & -i \\ i & 0 \end{matrix} \\ \begin{matrix} 0 & i \\ -i & 0 \end{matrix} & O \end{bmatrix},$$

$$[\gamma^{3a}{}_b] = \begin{bmatrix} O & \begin{matrix} -1 & 0 \\ 0 & 1 \end{matrix} \\ \begin{matrix} 1 & 0 \\ 0 & -1 \end{matrix} & O \end{bmatrix}, \quad [\gamma^{4a}{}_b] = \begin{bmatrix} O & \begin{matrix} -1 & 0 \\ 0 & -1 \end{matrix} \\ \begin{matrix} -1 & 0 \\ 0 & -1 \end{matrix} & O \end{bmatrix}. \tag{8.26}$$

In all cases, i.e. both for $m = -1$ and $m = +1$, we have

$$[\gamma^{ka}{}_b] = -[\gamma^{k*}{}_a{}^b] \quad (k = 1, 2, 3), \qquad [\gamma^{4a}{}_b] = [\gamma^{4*}{}_a{}^b], \tag{8.27}$$

$$[\gamma^{kab}] = -[\gamma^{k*}{}_{ab}] \quad (k = 1, 2, 3), \qquad [\gamma^{4ab}] = [\gamma^{4*}{}_{ab}]. \tag{8.28}$$

The γ's obey the following commutation rules for $m = -1$:

$$[\gamma^{\kappa ab}][\gamma^{\lambda}{}_{bc}] + [\gamma^{\lambda ab}][\gamma^{\kappa}{}_{bc}] = 2g^{\kappa\lambda}.1, \tag{8.29}$$

$$[\gamma^{\kappa a}{}_b][\gamma^{\lambda b}{}_c] + [\gamma^{\lambda a}{}_b][\gamma^{\kappa b}{}_c] = -2g^{\kappa\lambda}.1, \tag{8.30}$$

$$[\gamma^{\kappa}{}_a{}^b][\gamma^{\lambda}{}_b{}^c] + [\gamma^{\lambda}{}_a{}^b][\gamma^{\kappa}{}_b{}^c] = -2g^{\kappa\lambda}.1, \tag{8.31}$$

$$[\gamma^{\kappa}{}_{ab}][\gamma^{\lambda bc}] + [\gamma^{\lambda}{}_{ab}][\gamma^{\kappa bc}] = 2g^{\kappa\lambda}.1, \tag{8.32}$$

where

$$g^{11} = g^{22} = g^{33} = -1, \quad g^{44} = 1, \quad g^{\kappa\lambda} = 0 \quad (\kappa \neq \lambda). \tag{8.33}$$

For the γ's of $m = +1$ we have

$$[\gamma^{\kappa ab}][\gamma^{\lambda}{}_{bc}] + [\gamma^{\lambda ab}][\gamma^{\kappa}{}_{bc}] = -2g^{\kappa\lambda}.1, \tag{8.34}$$

$$[\gamma^{\kappa a}{}_b][\gamma^{\lambda b}{}_c] + [\gamma^{\lambda a}{}_b][\gamma^{\kappa b}{}_c] = 2g^{\kappa\lambda}.1, \tag{8.35}$$

$$[\gamma^{\kappa}{}_a{}^b][\gamma^{\lambda}{}_b{}^c] + [\gamma^{\lambda}{}_a{}^b][\gamma^{\kappa}{}_b{}^c] = 2g^{\kappa\lambda}.1, \tag{8.36}$$

$$[\gamma^{\kappa}{}_{ab}][\gamma^{\lambda bc}] + [\gamma^{\lambda}{}_{ab}][\gamma^{\kappa bc}] = -2g^{\kappa\lambda}.1. \tag{8.37}$$

The γ matrices of $m = -1$ can be made to obey the same commutation rules as those of $m = +1$ by multiplying them by i. We have already seen that this corresponds to a certain transformation in the spinor space (see 7.47).

The matrices $\gamma^{\kappa ab}$ and $\gamma^{\kappa}{}_{ab}$ are anti-symmetrical for $m = -1$ and symmetrical for $m = +1$:

$$\gamma^{\kappa ab} = -\gamma^{\kappa ba} \quad (m = -1), \tag{8.38}$$

$$\gamma^{\kappa}{}_{ab} = -\gamma^{\kappa}{}_{ba} \quad (m = -1), \tag{8.39}$$

$$\gamma^{\kappa ab} = \gamma^{\kappa ba} \quad (m = +1), \tag{8.40}$$

$$\gamma^{\kappa}{}_{ab} = \gamma^{\kappa}{}_{ba} \quad (m = +1). \tag{8.41}$$

The matrices $\gamma^{ka}{}_b$ and $\gamma^k{}_a{}^b$ $(k = 1, 2, 3)$ are Hermitian $(m = -1)$ and skew-Hermitian $(m = 1)$:

$$\gamma^{ka}{}_b = \gamma^{k*b}{}_a \qquad (m = -1), \tag{8.42}$$

$$\gamma^k{}_a{}^b = \gamma^{k*}{}_b{}^a \qquad (m = -1), \tag{8.43}$$

$$\gamma^{ka}{}_b = -\gamma^{k*b}{}_a \quad (m = 1), \tag{8.44}$$

$$\gamma^k{}_a{}^b = -\gamma^{k*}{}_b{}^a \quad (m = 1). \tag{8.45}$$

The matrices $\gamma^{4a}{}_b$ and $\gamma^4{}_a{}^b$ are skew-Hermitian for $m = -1$ and Hermitian for $m = 1$.

From now on we shall assume that all γ's obey the same commutation rules as in (8.34) to (8.37). This implies that the γ's of $m = -1$ have been multiplied by i.

From the invariant spin-vectors γ^κ we can form other invariant spin-tensors. Of particular interest are the anti-symmetrical

$$\gamma^{[\kappa\lambda]} = \tfrac{1}{2}(\gamma^\kappa\gamma^\lambda - \gamma^\lambda\gamma^\kappa), \tag{8.46}$$

$$\gamma^{[\kappa\lambda\mu]} = \tfrac{1}{6}\sum_P \epsilon_P \gamma^\kappa\gamma^\lambda\gamma^\mu, \tag{8.47}$$

where the sum extends over all the permutations of κ, λ, μ and $\epsilon_P = +1$ for even permutations and -1 for odd permutations ($\epsilon_P = 1$ for $\kappa, \lambda, \mu = 1, 2, 3$).

We can also form

$$\gamma^{\kappa\lambda a}{}_b \equiv \gamma^{\kappa a}{}_p \gamma^{\lambda p}{}_b, \qquad \gamma^{\kappa\lambda\mu a}{}_b \equiv \gamma^{\kappa a}{}_p \gamma^{\lambda p}{}_q \gamma^{\mu q}{}_b,$$
$$\gamma^{\kappa\lambda\mu\nu a}{}_b \equiv \gamma^{\kappa a}{}_p \gamma^{\lambda p}{}_q \gamma^{\mu q}{}_r \gamma^{\nu r}{}_b. \tag{8.48}$$

By contraction with respect to all the tensor indices of $\epsilon_{\kappa\lambda\mu\nu}\gamma^{\alpha\beta\gamma\delta}$ and some of the spinor indices we can form the scalar

$$\gamma^{5a}{}_b = \frac{1}{4!}\epsilon_{\kappa\lambda\mu\nu}\gamma^{\kappa\lambda\mu\nu} = \gamma^{1a}{}_u \gamma^{2u}{}_v \gamma^{3v}{}_w \gamma^{4w}{}_b. \tag{8.49}$$

The notation γ^5 must not mislead us to attach to the index 5 any tensor significance. It is used merely for an algebraic reason. It so happens that the matrix γ^5 satisfies a relation similar to the one satisfied by the γ^κ

$$\gamma^5\gamma^\kappa + \gamma^\kappa\gamma^5 = 0. \tag{8.50}$$

This relation follows from the commutation rule. For instance:

$$\begin{aligned} \gamma^5\gamma^4 + \gamma^4\gamma^5 &= (\gamma^1\gamma^2\gamma^3\gamma^4)\gamma^4 + \gamma^4(\gamma^1\gamma^2\gamma^3\gamma^4) \\ &= \gamma^1\gamma^2\gamma^3\gamma^4\gamma^4 - \gamma^1\gamma^4\gamma^2\gamma^3\gamma^4 \\ &= \gamma^1\gamma^2\gamma^3\gamma^4\gamma^4 + \gamma^1\gamma^2\gamma^4\gamma^3\gamma^4 \\ &= \gamma^1\gamma^2\gamma^3\gamma^4\gamma^4 - \gamma^1\gamma^2\gamma^3\gamma^4\gamma^4 = 0. \end{aligned} \tag{8.51}$$

Like ϵ_{ab}, $\gamma^{5a}{}_b$ transforms in four different ways, according to the four representations of the four-group in (7.46). Both when using γ's with $m = +1$ and $m = -1$, we find on using (8.49)

$$\gamma^5 = [\gamma^{5a}{}_b] = [\gamma^5{}_a{}^b] = [\gamma^{5ab}] = [\gamma^5{}_{ab}] = \begin{bmatrix} i & 0 & 0 & 0 \\ 0 & i & 0 & 0 \\ 0 & 0 & -i & 0 \\ 0 & 0 & 0 & -i \end{bmatrix}. \tag{8.52}$$

On transforming, say, $\gamma^5{}_{ab}$ in the same way as was done with ϵ_{ab} in (8.2) and (8.3), we get

$$\begin{aligned} \mathbf{P} &\to \begin{bmatrix} 0 & 1 \\ 1 & 0 \end{bmatrix}\begin{bmatrix} \mathrm{i} & 0 \\ 0 & -\mathrm{i} \end{bmatrix}\begin{bmatrix} 0 & 1 \\ 1 & 0 \end{bmatrix} = -\begin{bmatrix} \mathrm{i} & 0 \\ 0 & -\mathrm{i} \end{bmatrix}, \\ \mathbf{T} &\to \begin{bmatrix} 0 & -1 \\ 1 & 0 \end{bmatrix}\begin{bmatrix} \mathrm{i} & 0 \\ 0 & -\mathrm{i} \end{bmatrix}\begin{bmatrix} 0 & 1 \\ -1 & 0 \end{bmatrix} = -\begin{bmatrix} \mathrm{i} & 0 \\ 0 & -\mathrm{i} \end{bmatrix}. \end{aligned} \tag{8.53}$$

Instead of γ^5 we can also consider

$$\gamma_5 = \gamma_1\gamma_2\gamma_3\gamma_4. \tag{8.54}$$

Since $\qquad \gamma^k = -\gamma_k \quad (k = 1, 2, 3), \qquad \gamma^4 = \gamma_4,$

we get

$$\gamma^5 = -\gamma_5. \tag{8.55}$$

More often, $-i\gamma^5 = i\gamma_5$ is used:

$$i\gamma_5 = \begin{bmatrix} 1 & 0 & 0 & 0 \\ 0 & 1 & 0 & 0 \\ 0 & 0 & -1 & 0 \\ 0 & 0 & 0 & -1 \end{bmatrix}. \tag{8.56}$$

The two-valued representation of the full Lorentz group by means of four-spinors of first rank is irreducible. If we form direct products $\Psi^a \times \Psi^b$ we get entities with sixteen components, which transform according to the direct product of the matrices (6.12) for the proper Lorentz group and of (7.46) for **P** and **T**. The resulting representation of the Lorentz group is no longer two-valued. As mentioned already it is also not irreducible. In fact it is composed of three irreducible representations of dimensions one, five, and ten $(1+5+10 = 16)$. (With respect to the restricted Lorentz group the reduction can be carried out further.) Just as the γ's remain invariant with respect to Lorentz transformations, so there exist invariant spinors of fourth rank $\beta^{\kappa abcd}$, $\beta^{\kappa ab}{}_{cd}$, etc., which play a similar role. The β's are composed of the γ's and are given by

$$(\beta^\kappa) = (\tfrac{1}{2}[1\times\gamma^\kappa + \gamma^\kappa \times 1]). \tag{8.57}$$

For instance, $$\beta^{\kappa ab}{}_{cd} = \tfrac{1}{2}(\delta^a{}_c\gamma^{\kappa b}{}_d+\gamma^{\kappa a}{}_c\delta^b{}_d). \tag{8.58}$$

The β's obey the following Duffin–Kemmer commutation rule:

$$\beta^\kappa\beta^\lambda\beta^\mu+\beta^\mu\beta^\lambda\beta^\kappa = g^{\kappa\lambda}\beta^\mu+g^{\mu\lambda}\beta^\kappa. \tag{8.59}$$

In (8.58) $\delta^a{}_b$ is an invariant spinor of second rank. According to (7.11) it transforms as follows:

$$(T)(\delta^a{}_b)(T^{-1}) = TT^{-1}(\delta^a{}_b) = \delta^a{}_b. \tag{8.60}$$

9. Infinitesimal Transformations

The study of infinitesimal transformations offers great advantages in many cases and plays a particularlr role in quantum mechanics. We have already touched on the subject and mentioned that each continuous group is characterized by a certain number of infinitesimal operators which obey particular commutation laws. We have introduced the operator $P_{(\kappa\lambda)}$ defined by

$$\delta X = \tfrac{1}{2}\sum_{\kappa,\lambda} P_{(\kappa\lambda)}\omega_{\kappa\lambda}X, \tag{9.1}$$

which obeys the conditions

$$P_{(\kappa\lambda)} = -P_{(\lambda\kappa)} \tag{9.2}$$

and $$[P_{(\kappa\lambda)}, P_{(\mu\nu)}] = g_{\lambda\mu}P_{(\kappa\nu)}-g_{\kappa\mu}P_{(\lambda\nu)}+g_{\kappa\nu}P_{(\lambda\mu)}-g_{\lambda\nu}P_{(\kappa\mu)}. \tag{9.2 a}$$

If a tensor has n components $P_{(\kappa\lambda)}$ is a matrix with n columns and n rows. In Chapter 4 we have also derived explicit expressions for the $P_{(\kappa\lambda)}$ assuming an arbitrary tensor.

We shall now follow this line also with respect to spinors. Although spinors yield two-valued representations of the Lorentz group, in the vicinity of the identical transformation the representations can be made to be true representations of the Lorentz group. Consequently the infinitesimal transformations of the Lorentz group and any of its representations, both tensor and spinor, obey the same rules, and the corresponding operators obey in all cases the same commutation rules.

Let us begin with two-spinors. An infinitesimal transformation will have the form $$\phi'_A = \phi_A+\delta\phi_A = \phi_A+\eta_A{}^B\phi_B. \tag{9.3}$$

Since a spinor transformation leaves the ϵ_{AB} invariant we have

$$(1+\eta_A{}^C)(\epsilon_{CD})(\widetilde{1+\eta_D{}^B}) = (\epsilon_{AB}). \tag{9.4}$$

Now, to second order,

$$(\widetilde{1+\eta_D{}^C})(\widetilde{1-\eta_C{}^B}) = 1. \tag{9.5}$$

Consequently, $$(1+\eta_A{}^C)(\epsilon_{CD}) = (\epsilon_{AC})(\widetilde{1-\eta_C{}^B}), \tag{9.6}$$
therefore

$$\begin{bmatrix} 1+\eta_1{}^1 & \eta_1{}^2 \\ \eta_2{}^1 & 1+\eta_2{}^2 \end{bmatrix}\begin{bmatrix} 0 & -1 \\ +1 & 0 \end{bmatrix} = \begin{bmatrix} 0 & -1 \\ +1 & 0 \end{bmatrix}\begin{bmatrix} 1-\eta_1{}^1 & -\eta_2{}^1 \\ -\eta_1{}^2 & 1-\eta_2{}^2 \end{bmatrix}; \tag{9.7}$$

thus $$\begin{bmatrix} \eta_1{}^2 & -1-\eta_1{}^1 \\ 1+\eta_2{}^2 & -\eta_2{}^1 \end{bmatrix} = \begin{bmatrix} \eta_1{}^2 & -1+\eta_2{}^2 \\ 1-\eta_1{}^1 & -\eta_2{}^1 \end{bmatrix}. \tag{9.8}$$

Hence $$\eta_1{}^1 = -\eta_2{}^2. \tag{9.9}$$

Eqn. (9.9) yields the result that the spinor η_{AB} which is obtainable from $\eta_A{}^B$ by lowering the top index, is symmetrical:

$$\begin{bmatrix} \eta_{11} & \eta_{12} \\ \eta_{21} & \eta_{22} \end{bmatrix} = \begin{bmatrix} 0 & 1 \\ -1 & 0 \end{bmatrix}\begin{bmatrix} \eta_1{}^1 & \eta_1{}^2 \\ \eta_2{}^1 & -\eta_1{}^1 \end{bmatrix} = \begin{bmatrix} \eta_2{}^1 & -\eta_1{}^1 \\ -\eta_1{}^1 & -\eta_1{}^2 \end{bmatrix}. \tag{9.10}$$

In the case of four-vectors we have replaced the $\omega_{\kappa\lambda}$ by the following quantities:

$$\tfrac{1}{2}\Big(\sum_{\sigma,\tau} M_{(\sigma\tau)}\,\omega_{\sigma\tau}\Big)_{\kappa\lambda} \equiv \omega_{\kappa\lambda}, \tag{9.11}$$

where the $M_{(\sigma\tau)}$ are four-dimensional matrices.

In the case of a tensor of n components the matrices $M_{(\sigma\tau)}$ are replaced by anti-symmetrical matrices $P_{(\sigma\tau)}$ of n dimensions, and the quantities corresponding to the $\omega_{\kappa\lambda}$ on the right-hand side of (9.11) are replaced by n^2 infinitesimal parameters, say p_{qr} $(= -p_{rq})$,

$$\tfrac{1}{2}\sum_{\sigma\tau} (P_{(\sigma\tau)}\,\omega_{\sigma\tau}) = (p_{qr}). \tag{9.12}$$

The p_{qr} determine the infinitesimal transformation in the n-dimensional tensor space which corresponds with the infinitesimal transformation taking place in the four-dimensional space and determined by the parameters $\omega_{\kappa\lambda}$. (9.12) is just the relation which tells us how to calculate from the parameters $\omega_{\kappa\lambda}$, which determine an infinitesimal transformation in the four-space, the parameters of the corresponding transformation in the tensor space.

The above situation now repeats itself in connexion with spinors. We must seek two-dimensional matrices $\frac{1}{2}S_{(\sigma\tau)A}{}^B$ such that

$$\tfrac{1}{2}\sum_{\sigma,\tau} (S_{(\sigma\tau)}\,\omega_{\sigma\tau})_A{}^B = \eta_A{}^B, \tag{9.13}$$

where $S_{(\sigma\tau)1}{}^1 = -S_{(\sigma\tau)2}{}^2$ (see (9.9)) and obeying the same commutation rules as $M_{(\sigma\tau)}$ or $P_{(\sigma\tau)}$:

$$[S_{(\kappa\lambda)}, S_{(\mu\nu)}] = g_{\lambda\mu}\,S_{(\kappa\nu)} - g_{\kappa\mu}\,S_{(\lambda\nu)} + g_{\kappa\nu}\,S_{(\lambda\mu)} - g_{\lambda\nu}\,S_{(\kappa\mu)}. \tag{9.14}$$

The relationship (9.13) must be independent of any particular choice of a Lorentz frame and it may therefore be expected that $S_{(\sigma\tau)} = -S_{(\tau\sigma)}$ will depend on the invariant spin-vectors $g_\sigma{}^{\dot{A}B}$ and $g_\tau{}^{\dot{A}B}$. We can form a tensor of second rank $g_{\sigma\dot{C}A}\, g_\tau{}^{\dot{C}B}$ which is anti-symmetrical in σ, τ for $\sigma \neq \tau$ but equal to $g_{\sigma\tau}\,\delta_A{}^B$ for $\sigma = \tau$. ($g_{\sigma\tau}$ is the metrical tensor (4.8).) Consequently, we try the solution

$$S_{(\sigma\tau)A}{}^B = \tfrac{1}{2}(g_{\sigma\dot{C}A}\, g_\tau{}^{\dot{C}B} + g_{\sigma\tau}\,\delta_A{}^B), \tag{9.15}$$

$$\{S_{(\sigma\tau)}\} = \tfrac{1}{2}\{(\tilde{g}_\sigma..)(g_\tau{}^{\cdot\cdot}) - g_{\sigma\tau}(1)\}, \tag{9.16}$$

where the dots indicate the position of the spinor indices. Using (4.2) and (4.9) (where we have to lower the indices 1, 2, 3, 4, by changing the sign of the first three matrices and leaving the fourth unchanged) we get explicitly:

$$[S_{(12)A}{}^B] = \tfrac{1}{2}\begin{bmatrix} -i & 0 \\ 0 & i \end{bmatrix} = -S_{(21)}, \qquad [S_{(14)A}{}^B] = \tfrac{1}{2}\begin{bmatrix} 0 & 1 \\ 1 & 0 \end{bmatrix} = -S_{(41)},$$

$$[S_{(13)A}{}^B] = \tfrac{1}{2}\begin{bmatrix} 0 & 1 \\ -1 & 0 \end{bmatrix} = -S_{(31)}, \qquad [S_{(24)A}{}^B] = \tfrac{1}{2}\begin{bmatrix} 0 & -i \\ i & 0 \end{bmatrix} = -S_{(42)},$$

$$[S_{(23)A}{}^B] = \tfrac{1}{2}\begin{bmatrix} 0 & -i \\ -i & 0 \end{bmatrix} = -S_{(32)}, \qquad [S_{(34)A}{}^B] = \tfrac{1}{2}\begin{bmatrix} 1 & 0 \\ 0 & -1 \end{bmatrix} = -S_{(43)},$$

$$S_{(11)} = S_{(22)} = S_{(33)} = S_{(44)} = 0. \tag{9.17}$$

Eqn. (9.14) will be found to hold if the $S_{(\sigma\tau)}$ are assumed to be given by (9.17). We can give (9.15) a more convenient form which exhibits the anti-symmetry in σ, τ by using the commutation rule

$$g_{\sigma\dot{C}A}\, g_\tau{}^{\dot{C}B} + g_{\tau\dot{C}A}\, g_\sigma{}^{\dot{C}B} = 2g_{\sigma\tau}\,\delta^A{}_B = -2g_{\sigma\tau}\,\delta_A{}^B. \tag{9.18}$$

Multiplying (9.18) by $\tfrac{1}{2}$ and substituting for $g_{\sigma\tau}\,\delta_A{}^B$ in (9.15) gives

$$S_{(\sigma\tau)A}{}^B = \tfrac{1}{4}(g_{\sigma\dot{C}A}\, g_\tau{}^{\dot{C}B} - g_{\tau\dot{C}A}\, g_\sigma{}^{\dot{C}B}), \tag{9.19}$$

$$(S_{(\sigma\tau)}) = \tfrac{1}{4}[(\tilde{g}_\sigma..)(g_\tau{}^{\cdot\cdot}) - (\tilde{g}_\tau..)(g_\sigma{}^{\cdot\cdot})]. \tag{9.20}$$

The matrices $S_{(12)}$, $S_{(13)}$, and $S_{(23)}$ in (9.17) refer to infinitesimal rotations of space around the z-, y-, and x-axes respectively. When multiplied by $2i$ we get the Pauli matrices. (In quantum mechanics the spin angular momentum is represented by these matrices when multiplied by half.)

$$\sigma_x = 2iS_{(23)} = \begin{bmatrix} 0 & 1 \\ 1 & 0 \end{bmatrix}, \qquad \sigma_y = 2iS_{(31)} = \begin{bmatrix} 0 & -i \\ i & 0 \end{bmatrix},$$

$$\sigma_z = 2iS_{(12)} = \begin{bmatrix} 1 & 0 \\ 0 & -1 \end{bmatrix}. \tag{9.21}$$

As an example of an infinitesimal transformation let us perform a rotation around the z-axis by an angle $d\phi = \omega_{12}$, all the other ω_{kl} with the exception of $\omega_{21} = -\omega_{12}$ being zero. The corresponding spinor transformation is

$$\phi'_A = [\delta_A{}^B + \eta_A{}^B]\phi_B = [\delta_A{}^B + \tfrac{1}{2}(S_{(21)A}{}^B\omega_{12} + S_{(21)A}{}^B\omega_{21})]\phi_B$$
$$= [\mathbf{1} + [S_{(12)A}{}^B]\, d\phi]\phi_B; \quad (9.22)$$

therefore
$$d\phi_A = S_{(12)A}{}^B\phi_B\, d\phi, \quad (9.23)$$

or
$$\begin{bmatrix} d\phi_1 \\ d\phi_2 \end{bmatrix} = \tfrac{1}{2}\begin{bmatrix} -i & 0 \\ 0 & i \end{bmatrix}\begin{bmatrix} \phi_1 \\ \phi_2 \end{bmatrix} d\phi. \quad (9.24)$$

Expressed for each component this yields

$$d\phi_1 = -\tfrac{1}{2}i\phi_1\, d\phi, \qquad \phi_1 = \phi_{10}\, e^{-\frac{1}{2}i\phi},$$
$$d\phi_2 = \tfrac{1}{2}i\phi_2\, d\phi, \qquad \phi_2 = \phi_{20}\, e^{\frac{1}{2}i\phi} \quad (9.25)$$

The finite transformation is thus given by

$$\begin{bmatrix} \phi'_1 \\ \phi'_2 \end{bmatrix} = \begin{bmatrix} e^{-\frac{1}{2}i\phi} & 0 \\ 0 & e^{\frac{1}{2}i\phi} \end{bmatrix}\begin{bmatrix} \phi_1 \\ \phi_2 \end{bmatrix}. \quad (9.26)$$

This is in agreement with the first of the matrices in (3.17) which corresponds with rotations around the z-axis. In a similar way we get from $S_{(23)}$, $S_{(13)}$ the matrices corresponding with rotations around the x- and y-axes respectively. $S_{(14)}$, $S_{(24)}$, and $S_{(34)}$ yield the matrices corresponding to infinitesimal motion along the x-, y-, and z-axes respectively. All these correspondences refer to two-valued representations of two dimensions.

The infinitesimal transformations of two-spinors of higher rank are directly obtainable from those of first rank, by taking into account their property of transforming as direct products of spinors of first rank. For instance, if we consider the spinor ϕ_{AB}, an infinitesimal transformation will change it as follows:

$$d\phi_{AB} = (-\delta_A{}^C + \eta_A{}^C)(-\delta_B{}^D + \eta_B{}^D)\phi_{CD}. \quad (9.27)$$

Neglecting second-order terms,

$$d\phi_{AB} = (\delta_A{}^C\delta_B{}^D - \delta_A{}^C\eta_B{}^D - \eta_A{}^C\delta_B{}^D)\phi_{CD}. \quad (9.28)$$

Hence
$$[d\phi_{AB}] = [\mathbf{1}_4 + \mathbf{1}_2 \times \boldsymbol{\eta} + \boldsymbol{\eta} \times \mathbf{1}_2][\phi_{CD}], \quad (9.29)$$

where $\mathbf{1}_2$ and $\mathbf{1}_4$ are unit matrices of 2 and 4 dimensions ($\mathbf{1}_4 = \mathbf{1}_2 \times \mathbf{1}_2$). It follows further that the matrices characterizing the infinitesimal transformations in the space of second-rank spinors will be given by

$$S^{(2)}_{(\sigma\tau)} = \mathbf{1} \times S_{(\sigma\tau)} + S_{(\sigma\tau)} \times \mathbf{1}. \quad (9.30)$$

Next we consider four-spinors, i.e. the four-dimensional two-valued representation. An infinitesimal transformation in the space of four-spinors of first rank produces the change

$$\Psi'_a = \Psi_a + \delta\Psi_a = \Psi_a + \rho_a{}^b\Psi_b. \tag{9.31}$$

Since Ψ_a is a direct sum of spinors ψ_A and $\phi^{\dot{A}}$, $\rho_a{}^b$ will also be the direct sum of $\eta_A{}^C$ which yields the change of ψ_A and a corresponding matrix which yields the change of $\phi^{\dot{A}}$. The latter is obtained as follows. As in (9.3), let

$$d\phi_A = \eta_A{}^C\phi_C. \tag{9.32}$$

It follows further that

$$d\phi^A = \eta^{AC}\phi_C = \epsilon^{CD}\eta^A{}_D\epsilon_{CB}\phi^B = \eta^A{}_D\tilde{\epsilon}^{DC}\epsilon_{CB}\phi^B$$
$$= -\eta^A{}_D\delta^D{}_B\phi^B = -\eta^A{}_D\phi^D;$$

therefore

$$d\phi^{\dot{A}} = -\eta^{\dot{A}}{}_{\dot{D}}\phi^{\dot{D}}. \tag{9.33}$$

Hence,

$$[d\Psi_a] = [\rho_a{}^b\Psi_b] = \begin{bmatrix} \eta_A{}^B\psi_B \\ -\eta^{\dot{A}}{}_{\dot{B}}\phi^{\dot{B}} \end{bmatrix}, \tag{9.34}$$

$$[\rho_a{}^b] = \begin{bmatrix} [\eta_A{}^B] & \mathbf{0} \\ \mathbf{0} & -[\eta^{\dot{A}}{}_{\dot{B}}] \end{bmatrix} = \tfrac{1}{2}\begin{bmatrix} S_{(\sigma\tau)A}{}^B\omega_{\sigma\tau} & \mathbf{0} \\ \mathbf{0} & -S_{(\sigma\tau)}{}^{\dot{A}}{}_{\dot{B}}\,\omega_{\sigma\tau} \end{bmatrix}. \tag{9.35}$$

More explicitly, if $\mathscr{S}_{(\sigma\tau)a}{}^b$ are the matrices which characterize the infinitesimal transformations in the space of four-spinors, we have

$$[\mathscr{S}_{(\sigma\tau)a}{}^b] = \begin{bmatrix} \tfrac{1}{4}[g_{\sigma\dot{C}A}\,g_\tau{}^{\dot{C}B} - g_{\tau\dot{C}A}\,g_\sigma{}^{\dot{C}B}] & \mathbf{0} \\ \mathbf{0} & -\tfrac{1}{4}[g_\sigma{}^{C\dot{A}}g_{\tau C\dot{B}} - g_\tau{}^{C\dot{A}}g_{\sigma C\dot{B}}] \end{bmatrix}. \tag{9.36}$$

It should be possible to express the $\mathscr{S}_{(\sigma\tau)a}{}^b$ in terms of the γ's since these play the same role in the four-spinor space as played by the g's in the two-spinor space. From the relation between the γ's and the g's we find

$$\mathscr{S}_{(\sigma\tau)a}{}^b = \tfrac{1}{4}(\gamma_{\sigma a}{}^c\gamma_{\tau c}{}^b - \gamma_{\tau a}{}^c\gamma_{\sigma c}{}^b) \equiv \tfrac{1}{2}\gamma_{(\sigma\tau)}. \tag{9.37}$$

From spinors Ψ_a of first rank we obtain as before with two-spinors, the matrices which characterize the transformations of second rank four-spinors:

$$\mathscr{S}^{(2)}_{(\sigma\tau)} = \mathbf{1}_4\times\mathscr{S}_{(\sigma\tau)} + \mathscr{S}_{(\sigma\tau)}\times\mathbf{1}_4 = \tfrac{1}{2}[\mathbf{1}_4\times\gamma_{(\sigma\tau)} + \gamma_{(\sigma\tau)}\times\mathbf{1}_4]. \tag{9.38}$$

The matrices $\mathscr{S}^{(2)}_{(\sigma\tau)}$ are sixteen-dimensional. Like the matrices $\mathscr{S}_{(\sigma\tau)}$ or $S_{(\sigma\tau)}$ and $S^{(2)}_{(\sigma\tau)}$ they may be expressed in terms of a commutator:

$$\mathscr{S}^{(2)}_{(\sigma\tau)} = \beta_\sigma\beta_\tau - \beta_\tau\beta_\sigma, \tag{9.39}$$

where

$$\beta_\sigma = \tfrac{1}{2}[\mathbf{1}_4\times\gamma_\sigma + \gamma_\sigma\times\mathbf{1}_4]. \tag{9.40}$$

The β's which have been mentioned earlier are sixteen-dimensional matrices which turn out to obey the Duffin–Kemmer commutation rule (8.59). These follow from the definition of the β's (9.40) and the commutation rules of the γ's. In a similar way one can proceed to spinors of higher rank.

10. The Irreducible Representations of the Restricted Lorentz Group

When dealing with tensors we have already pointed out that the various representations obtained with them are in general reducible. Irreducible representations are obtained under symmetry restrictions. The situation is similar when dealing with transformations of spinors. For instance, from a spinor of second rank ϕ_{AB} we can form a symmetrical spinor $\phi_{AB}+\phi_{BA}$ and an anti-symmetrical spinor $\phi_{AB}-\phi_{BA}$. The latter has only two non-vanishing components which have, however, the same absolute value $\phi_{12}-\phi_{21} = -(\phi_{21}-\phi_{12})$. Such a spinor transforms like a scalar. On the other hand, the symmetrical spinor has three different components and yields an irreducible three-dimensional faithful representation of the Lorentz group.

In this section we shall briefly describe a method of getting irreducible representations. We are interested in all the faithful and irreducible representations of the unimodular group, which itself is a two-valued representation of the restricted Lorentz group.

It will be convenient to start with the rotational group. A straightforward method of obtaining all the irreducible representations of the unitary group which represents the group of rotations is to consider the forms

$$V_m = \frac{\psi_1^r \psi_2^s}{\sqrt{(r!\,s!)}} \quad (r+s = 2j,\ r-s = 2m), \tag{10.1}$$

where ψ_1 and ψ_2 are components of a spinor and j is a fixed number, integral or half-integral. r and s run from 0 to $2j$ and m from $-j$ to $+j$. Hence we have $2j+1$ homogeneous V_m's (i.e. with the same total power $r+s = 2j$). All the linear combinations $\sum a_m V_m$ may be looked upon as constituting a vector space with the V_m forming a set of basic vectors and the a_m as components. The number of dimensions in this space is $2j+1$. For $j = \frac{1}{2}$ this number is 2, for $j = 1$ it is 3, etc.

If a unitary transformation $(\psi_1, \psi_2) \to (\psi_1', \psi_2')$ is taking place, the V_m change into some V_m' of the same form as (10.1) and the components a_m change into a_m' where $\sum a_m V_m = \sum a_m' V_m'$ is the same vector. The a_m' depend entirely on the a_m, and the transformations $a_m \to a_m'$ induced in this way in the $(2j+1)$-dimensional space are a representation of the unitary transformations $(\psi_1, \psi_2) \to (\psi_1', \psi_2')$ which is, moreover, nonreducible. The irreducibility will not be proved here. Intuitively we know that the forms (10.1) cannot be split into sums of two or more parts each involving transformations only within itself.

To obtain the representation induced in the above vector space we need only consider infinitesimal transformations. In other words, it is sufficient to calculate the matrices which play in the vector space the role played by the $S_{(\sigma\tau)}$ in infinitesimal spinor transformations, given by

$$d\psi_A = \tfrac{1}{2}\omega^{\sigma\tau}S_{(\sigma\tau)A}{}^{B}\psi_B. \tag{10.2}$$

As already mentioned and illustrated as far as $S_{(23)}$ is concerned, $S_{(23)}$, $S_{(31)}$, and $S_{(12)}$ refer to rotations around the x-, y-, and z-axes. Explicitly (10.2) yields

$$S_{(23)} = \tfrac{1}{2}\begin{bmatrix} 0 & -i \\ -i & 0 \end{bmatrix} \rightarrow d\psi_1 = \frac{1}{2i}\psi_2\,\omega_{23},\ d\psi_2 = \frac{1}{2i}\psi_1\,\omega_{23}, \tag{10.3}$$

$$S_{(31)} = \tfrac{1}{2}\begin{bmatrix} 0 & -1 \\ 1 & 0 \end{bmatrix} \rightarrow d\psi_1 = -\tfrac{1}{2}\psi_2\,\omega_{31},\ d\psi_2 = \tfrac{1}{2}\psi_1\,\omega_{31}, \tag{10.4}$$

$$S_{(12)} = \tfrac{1}{2}\begin{bmatrix} -i & 0 \\ 0 & i \end{bmatrix} \rightarrow d\psi_1 = \frac{1}{2i}\psi_1\,\omega_{12},\ d\psi_2 = -\frac{1}{2i}\psi_2\,\omega_{12}. \tag{10.5}$$

From (10.1) and (10.3 to 5) we get the changes dV_m:

$$S_{(23)} \rightarrow dV_m = \frac{\omega_{23}}{2i}(\sqrt{r}\sqrt{(s+1)}V_{m-1} + \sqrt{(r+1)}\sqrt{s}\,V_{m+1}), \tag{10.6}$$

$$S_{(31)} \rightarrow dV_m = \frac{\omega_{31}}{2}(-\sqrt{r}\sqrt{(s+1)}V_{m-1} + \sqrt{(r+1)}\sqrt{s}\,V_{m+1}), \tag{10.7}$$

$$S_{(12)} \rightarrow dV_m = \frac{\omega_{12}}{2i}(r-s)V_m. \tag{10.8}$$

As mentioned already, it is usual to replace $S_{(23)}$, $S_{(31)}$, and $S_{(12)}$ by $\sigma_x = 2iS_{(31)}$, $\sigma_y = 2iS_{(31)}$, $\sigma_z = 2iS_{(12)}$. We then get, after also replacing r, s by j, m ($r = j+m$, $s = j-m$),

$$\sigma_x \rightarrow dV_m = \{[(j+m)(j-m+1)]^{\frac{1}{2}}V_{m-1} + [(j+m+1)(j-m)]^{\frac{1}{2}}V_{m+1}]\omega_{23}, \tag{10.9}$$

$$\sigma_y \rightarrow dV_m = i\{-[(j+m)(j-m+1)]^{\frac{1}{2}}V_{m-1} + [(j+m+1)(j-m)]^{\frac{1}{2}}V_{m+1}\}\omega_{31}, \tag{10.10}$$

$$\sigma_z \rightarrow dV_m = (r-s)V_m\,\omega_{12} = 2mV_m\,\omega_{12}. \tag{10.11}$$

Of these matrices the simplest is the one corresponding to σ_z. In this case the matrix is diagonal:

$$\sigma_z \to (dV_m) \equiv \begin{bmatrix} dV_j \\ dV_{j-1} \\ \vdots \\ dV_{-j} \end{bmatrix} = \omega_{12} \begin{bmatrix} 2j & 0 & 0 & 0 & \cdot & \cdot & \cdot \\ 0 & 2(j-1) & 0 & 0 & \cdot & \cdot & \cdot \\ & \cdot & \cdot & \cdot & \cdot & \cdot & \cdot \\ 0 & \cdot & \cdot & \cdot & 0 & & -2j \end{bmatrix} \begin{bmatrix} V_j \\ V_{j-1} \\ \vdots \\ V_{-j} \end{bmatrix}. \tag{10.12}$$

On the other hand, in the matrices which represent σ_x and σ_y all the non-vanishing elements are along the lines which flank the diagonal:

$$\sigma_x \to [dV_m] \equiv \begin{bmatrix} dV_j \\ \vdots \\ dV_{-j} \end{bmatrix}$$

$$= \omega_{23} \begin{bmatrix} 0 & \sqrt{(2j)} & 0 & & \cdot \\ \sqrt{(2j)} & 0 & \sqrt{2}\sqrt{(2j-1)} & & \cdot \\ 0 & \sqrt{2}\sqrt{(2j-1)} & 0 & & \\ & & & \ddots & \sqrt{(2j)} \\ \cdot & & 0 & \sqrt{(2j)} & 0 \end{bmatrix} \begin{bmatrix} V_j \\ \vdots \\ V_{-j} \end{bmatrix}, \tag{10.13}$$

$$\sigma_y \to [dV_m] \equiv \begin{bmatrix} dV_j \\ \vdots \\ dV_{-j} \end{bmatrix}$$

$$= i\omega_{31} \begin{bmatrix} 0 & -\sqrt{(2j)} & 0 & & \cdot \\ \sqrt{(2j)} & 0 & -\sqrt{2}\sqrt{(2j-1)} & & \cdot \\ 0 & \sqrt{2}\sqrt{(2j-1)} & 0 & & \cdot \\ \cdot & & & \ddots & -\sqrt{(2j)} \\ 0 & 0 & \sqrt{(2j)} & & 0 \end{bmatrix} \begin{bmatrix} V_j \\ \vdots \\ V_{-j} \end{bmatrix}. \tag{10.14}$$

For $j = \frac{1}{2}$ we get the Pauli matrices. For $j = 1$ we get

$$\sigma_x \to \sigma_x^{(1)} = \begin{bmatrix} 0 & \sqrt{2} & 0 \\ \sqrt{2} & 0 & \sqrt{2} \\ 0 & \sqrt{2} & 0 \end{bmatrix}, \qquad \sigma_y \to \sigma_y^{(1)} = i \begin{bmatrix} 0 & -\sqrt{2} & 0 \\ \sqrt{2} & 0 & -\sqrt{2} \\ 0 & \sqrt{2} & 0 \end{bmatrix},$$

$$\sigma_z \to \sigma_z^{(1)} = \begin{bmatrix} 2 & 0 & 0 \\ 0 & 0 & 0 \\ 0 & 0 & -2 \end{bmatrix}. \tag{10.15}$$

In all cases the matrices representing σ_x, σ_y, σ_z obey the same commutation rules as the Pauli matrices.

The representation (10.15) is not two-valued since it is equivalent to the rotational group itself. The equivalence is established by providing a single transformation matrix which transforms the $\sigma_x^{(1)}/2i$, $\sigma_y^{(1)}/2i$, $\sigma_z^{(1)}/2i$

into R_x, R_y, R_z, the matrices which generate the rotations themselves. This matrix is given by

$$T = \begin{bmatrix} \frac{1}{\sqrt{2}} & \frac{i}{\sqrt{2}} & 0 \\ 0 & 0 & 1 \\ -\frac{1}{\sqrt{2}} & \frac{i}{\sqrt{2}} & 0 \end{bmatrix}, \quad T^{-1} = \begin{bmatrix} \frac{1}{\sqrt{2}} & 0 & -\frac{1}{\sqrt{2}} \\ -\frac{i}{\sqrt{2}} & 0 & -\frac{i}{\sqrt{2}} \\ 0 & 1 & 0 \end{bmatrix}, \tag{10.16}$$

and we find

$$\frac{1}{2i}T^{-1}\sigma_x^{(1)}T = R_x, \quad \frac{1}{2i}T^{-1}\sigma_y^{(1)}T = R_y, \quad \frac{1}{2i}T^{-1}\sigma_z^{(1)}T = R_z. \tag{10.17}$$

Geometrically this transformation results from the unitary transformation which replaces x, y, z by $(1/\sqrt{2})(x+iy)$, z, and $(1/\sqrt{2})(-x+iy)$ respectively. (If in any rotation $X' = RX$, X is replaced by TX, X' by TX', we have $TX' = TRX = TRT^{-1}TX$, hence R is replaced by $R' = TRT^{-1}$ or $R = (1/2i)T^{-1}\sigma T$.)

The general rule is that for half-integral j the representation is two-valued, for integral j it is faithful and equivalent to some special tensor transformation, i.e. a tensor with some special symmetry. It is customary to denote these irreducible representations of the group of rotations by $D(j)$.

In the representations described here rotations around the z-axis are represented by diagonal matrices (see 10.12). From these it is easily seen that the column $V_m = (0, 0, \ldots, V_m, \ldots, 0)$ is an eigenvector of $\sigma_z^{(j)}$:

$$\sigma_z^{(j)}V_m = mV_m. \tag{10.18}$$

A general vector $V = \sum a_m V_m$ in the $(2j+1)$-dimensional space may be expressed as

$$V = \sum a^{A_1A_2\ldots A_{2j}}\psi_{A_1}\psi_{A_2}\ldots\psi_{A_{2j}} \quad (A_\kappa = 1, 2). \tag{10.19}$$

Since, algebraically, V is an invariant, $a^{A_1\ldots A_{2j}}$ has the properties of a spinor of rank $2j$. We need consider only spinors which are symmetrical in all the $2j$ indices, because $\psi_{A_1}\ldots\psi_{A_{2j}}$ is a symmetrical spinor. An anti-symmetrical $a^{A_1\ldots A_{2j}}$ even in a single pair of indices can make no contribution to V. The symmetrical spinor $a^{A_1\ldots A_{2j}}$ has $2j+1$ components as may be seen by arranging the indices as follows:

$$\begin{matrix} 1 & 2 & . & . & . & 2j \\ 1 & 1 & . & . & . & 1 \\ 1 & 1 & . & . & 1 & 2 \\ 1 & 1 & . & 1 & 2 & 2 \\ . & . & . & . & . & . \\ 1 & 2 & . & . & . & 2 \\ 2 & 2 & . & . & . & 2 \end{matrix} \tag{10.20}$$

The symmetrical spinor $a^{A_1...A_{2j}}$ has the form of the following direct product:

$$\phi^{A_1...A_{2j}} = a^{A_1}\times a^{A_2}\times ...\times a^{A_{2j}}. \qquad (10.21)$$

When each of the spinors a^{A_κ} undergoes a unitary transformation the above direct product yields an irreducible $(2j+1)$-dimensional representation of the unitary group and hence also of the group of rotations. It will be convenient to denote a symmetrical spinor of rank $2j$ by $\phi(j)$.

We must now extend the above considerations regarding the rotational group to the Lorentz group. To begin with, a spinor of second rank yields an irreducible representation not only of the group of rotations but also of the Lorentz group proper. It follows, therefore, that the above symmetrical spinors $\phi = \phi(0)$, $\phi^{A_1} = \phi(\frac{1}{2})$, $\phi^{A_1A_2} = \phi(1)$, $\phi^{A_1A_2A_3} = \phi(\frac{3}{2})$, etc., yield irreducible representations of the Lorentz group proper. For reasons which will become clear a little further down it will be convenient to denote these symmetrical spinors not by $\phi(j)$ as was done above, but by $\phi(j,0)$. Thus $\phi(0,0)$ transforms as a scalar, $\phi(\frac{1}{2},0)$ as a spinor ϕ_A of first rank, $\phi(1,0)$ as a symmetrical spinor of second rank ϕ_{AB}, $\phi(\frac{3}{2},0)$ as a symmetrical spinor ϕ_{ABC} of third rank, etc. The corresponding contravariant spinors ϕ, ϕ^A, ϕ^{AB}, ϕ^{ABC}, etc., yield equally irreducible representations; these are, however, equivalent to those obtained from the covariant spinors. Similarly, the symmetrical $\phi_{\dot{A}}$, $\phi_{\dot{A}\dot{B}}$, $\phi_{\dot{A}\dot{B}\dot{C}}$, etc., or $\phi^{\dot{A}}$, $\phi^{\dot{A}\dot{B}}$, $\phi^{\dot{A}\dot{B}\dot{C}}$, etc., yield irreducible representations. We shall denote the latter by $\phi(0,\frac{1}{2})$, $\phi(0,1)$, $\phi(0,\frac{3}{2})$, etc.

The transformations obtained from the spinors $\phi(j,0)$ or $\phi(0,j)$ do not constitute all the irreducible representations of the Lorentz group. The representations obtained so far are associated with vector spaces with basic vectors V_m which are characterized by the eigenvalues m of the operator $\sigma_z^{(j)}$ only. We used three non-commuting operators, σ_x, σ_y, and σ_z, and consequently only one of them, say σ_z, could be represented by a diagonal matrix. It is true that out of these matrices we can obtain another one which may take the form of a diagonal matrix, namely,

$$\sigma_x^{(j)2}+\sigma_y^{(j)2}+\sigma_z^{(j)2} = j(j+1)\mathbf{1}.$$

This matrix is, however, not only diagonal, but also a multiple of a unit matrix. The eigenvalue $j(j+1)$ is therefore not required for further specification of the various basic vectors in the space which yields $D(j)$. The Lorentz group proper is, however, characterized by six infinitesimal transformation matrices and not by three. These, as we know, obey the following commutation rule:

$$[T^{(\kappa\lambda)}, T^{(\mu\nu)}] = g^{\lambda\mu}T^{(\kappa\nu)}-g^{\kappa\mu}T^{(\lambda\nu)}+g^{\kappa\nu}T^{(\lambda\mu)}-g^{\lambda\nu}T^{(\kappa\mu)}. \qquad (10.22)$$

On introducing the matrices

$$K_x = \tfrac{1}{2}(iT^{(23)}+T^{(41)}),\quad K_y = \tfrac{1}{2}(iT^{(31)}+T^{(42)}),\quad K_z = \tfrac{1}{2}(iT^{(12)}+T^{(43)}),$$
$$L_x = \tfrac{1}{2}(iT^{(23)}-T^{(41)}),\quad L_y = \tfrac{1}{2}(iT^{(31)}-T^{(42)}),\quad L_z = \tfrac{1}{2}(iT^{(12)}-T^{(43)}),\qquad (10.23)$$

we find

$$[K_x, K_y] = K_z \quad \text{(cyclically)},$$
$$[L_x, L_y] = L_z \quad (\quad ,, \quad),$$
$$[K_x, L_x] = [K_x, L_y], \text{ etc.} = 0. \qquad (10.24)$$

Since $K_{x,y,z}$ and $L_{x,y,z}$ commute and each of these sets of matrices satisfies the same commutation rule as we encountered before with R_x, R_y, and R_z, it follows that we can have both K_z and L_z in a diagonal form (in addition to K^2 and L^2). Moreover, if K_z is a diagonal matrix of $2k+1$ dimensions and L_z a diagonal matrix of $2l+1$ dimensions, both of the same form as (10.12), the simultaneous eigenvectors of K_z and L_z will be $(2k+1)(2l+1)$ in number, each being characterized by two diagonal elements (eigenvalues) $m(k)$ and $m(l)$. $m(k)$ runs from $-k$ to k and $m(l)$ from $-l$ to l. Let us denote these eigenvectors by $V_{m(k)m(l)}$. They form a set of basic vectors in a $(2k+1)(2l+1)$-dimensional space and may be realized by the following forms:

$$V_{m(k)m(l)} = \frac{\psi_1^r \psi_2^s \phi_1^u \phi_2^v}{(r!\,s!\,u!\,v!)^{\frac{1}{2}}} \quad \begin{aligned}&(r+s = 2k,\ u+v = 2l,\\ &r-s = 2m(k),\ u-v = 2m(l)).\end{aligned} \qquad (10.25)$$

These do not lead us back to (10.1) owing to the dots in the second pair of indices. The evaluation of $dV_{m(k)m(l)}$ from (10.25) will yield matrices which represent the various infinitesimal transformations and in particular the diagonal $(K_z)\times(1_{2l+1})$ and $(1_{2k+1})\times L_z$, where K_z and L_z are defined above (10.23).

As before with the rotations, we can say that each vector

$$V = \sum c_{m(k)m(l)} V_{m(k)m(l)}$$

may be associated with a spinor $c_{A_1 \ldots A_{2k}\dot{B}_1 \ldots \dot{B}_{2l}}$ which is symmetrical in the first group of indices and symmetrical in the second group of indices. This spinor has the form

$$c_{A_1 \ldots A_{2k}\dot{B}_1 \ldots \dot{B}_{2l}} = (a_{A_1}\times a_{A_2}\times \ldots \times a_{A_{2k}})\times(b_{\dot{B}_1}\times b_{\dot{B}_2}\times \ldots \times b_{\dot{B}_{2l}}). \qquad (10.26)$$

We shall denote these symmetrical spinors by $\psi(k,l)$ and their transformation matrices by $D(k,l)$. We can now confirm that $\psi(k,0)$ coincides with our former $\psi(k)$ and $\psi(0,l)$ with the symmetrical $\psi_{\dot{B}_1 \dot{B}_2 \ldots \dot{B}_{2l}}$.

In Table III some low-order symmetrical spinors are listed.

TABLE III

$\psi(k, l)$	*Spinor notation*	$(2k+1)(2l+1)$ = *dimension of representation*
$\psi(0, 0)$	scalar ψ	1
$\psi(\frac{1}{2}, 0)$	ψ_A (or ψ^A)	2
$\psi(0, \frac{1}{2})$	$\psi^{\dot{A}}$ (or $\psi_{\dot{A}}$)	2
$\psi(\frac{1}{2}, \frac{1}{2})$	$\psi_A{}^{\dot{B}}$ (or $\psi^A{}_{\dot{B}}$, $\psi_{A\dot{B}}$, $\psi^{A\dot{B}}$)	4
$\psi(1, 0)$	ψ_{AB} (or ψ^{AB})	3
$\psi(0, 1)$	$\psi^{\dot{A}\dot{B}}$ (or $\psi_{\dot{A}\dot{B}}$)	3
$\psi(1, \frac{1}{2})$	$\psi_{AB}{}^{\dot{C}}$ (or $\psi^{AB}{}_{\dot{C}}$)†	6
$\psi(\frac{1}{2}, 1)$	$\psi_A{}^{\dot{B}\dot{C}}$ (or $\psi^A{}_{\dot{B}\dot{C}}$, etc.)	6
$\psi(\frac{3}{2}, 0)$	ψ_{ABC} (or ψ^{ABC})	4
$\psi(0, \frac{3}{2})$	$\psi^{\dot{A}\dot{B}\dot{C}}$ (or $\psi_{\dot{A}\dot{B}\dot{C}}$)	4
$\psi(1, 1)$	$\psi_{AB}{}^{\dot{C}\dot{D}}$ (or $\psi^{AB}{}_{\dot{C}\dot{D}}$, etc.)	9

The list can be continued for increasing values in k, l. We note the four-dimensional representation yielded by $\psi(\frac{1}{2}, \frac{1}{2})$. In this case

$$k+l = \tfrac{1}{2}+\tfrac{1}{2} = 1$$

and consequently, as explained a little further down, the representation is faithful. We have already seen that the spinors of second rank yield a faithful representation of the Lorentz group, which is in fact transformable into the Lorentz group and so equivalent to it. In other words, the representation obtained from $\psi(\frac{1}{2}, \frac{1}{2})$ is transformable into a four-vector representation. Generally, whenever $k+l$ is an integer, the $\psi(k, l)$ transform according to a faithful representation. This can be seen by considering the rotational sub-groups. As far as this sub-group is concerned $\psi(k, l)$ transforms as the direct product $\psi(k, 0) \times \psi(l, 0)$, i.e. according to the representation $D(k) \times D(l)$.‡ Now such a representation can always be reduced according to the Clebsch–Gordan series:

$$D(k) \times D(l) = D(k+l) \oplus D(k+l-1) \oplus \dots \oplus D(|k-l|). \quad (10.27)$$

If $k+l$ is an integer, all the terms on the right-hand side refer to faithful representations of the rotational group with dimensions $2(k+l)+1, \dots, 2|k-l|+1$. It therefore follows further that when $k+l$ is an integer $\psi(k, l)$ transforms according to a faithful representation of the Lorentz group, and when $(k+l)$ is a half-integer the representation is two-valued.

† Also $\psi^{AB\dot{C}}$ and $\psi_{AB\dot{C}}$ but not $\psi^A{}_{B\dot{C}}$. We can raise or lower dotted or undotted indices as a whole but not individually.

‡ The unitary U is equivalent to U^*: $\epsilon U \epsilon^{-1} = U^*$.

We note in the above list that $\psi(1,0)$ and $\psi(0,1)$ also yield faithful representations. These should be equivalent to some tensor representations because the tensors yield all the faithful representations. A tensor having only three components can exist only under very restrictive symmetry conditions. In our present case $\psi(1,0)$ and $\psi(0,1)$ transform equivalently with anti-symmetrical tensors of second rank which are also self-dual. (Anti-symmetry reduces the number of components of a four-tensor of second rank to six, self-duality reduces it further to three components.) The representation yielded by $\psi(1,1)$ is also faithful and is therefore transformable into a certain tensor representation. In this case the tensor in question is symmetric of second rank with zero trace. (The trace of a second-rank tensor is a scalar and transforms as $\psi(0,0)$. A symmetric tensor $S_{\kappa\lambda}$ has ten components and the condition $\sum S_{\kappa\kappa} = 0$ reduces their number to nine independent components in agreement with $(2k+1)(2l+1) = 3.3 = 9$.) The representation obtained from $\psi(\frac{3}{2},0)$ (or $\psi(0,\frac{3}{2})$) is also four-dimensional. However, being two-valued, it is not equivalent to a four-vector representation.

All the representations with $(k+l)$ integral can be extended to represent the full Lorentz group, by simply combining them with faithful representations of the four-group **E**, **P**, **T**, **PT**. To obtain similar extensions with the two-valued representations we must use direct sums $\psi(k,l) \oplus \psi(l,k)$† and combine them with suitable representations of the four-group. We know already that $\psi(\frac{1}{2},0) \oplus \psi(0,\frac{1}{2})$ yield the four-spinors ψ^a with a representation $D(\frac{1}{2},0) \oplus D(0,\frac{1}{2})$ of the proper group, and this can be extended with a suitable representation of the four-group to cover the full Lorentz group. In general, $D(k,l) \oplus D(l,k)$ with $k \neq l$ and $k+l$ half-integral yields a two-valued representation of the proper group which can be extended to cover the full group. The representation has $(2k+1)(2l+1)$ dimensions. When $k = l$ the representation is always faithful and it can be extended only with a faithful representation of the four-group (without forming $D(k,k) \oplus D(k,k)$). All these representations are non-unitary.

11. Spinors and their Relation to Spin

In the introduction to this chapter we have already explained the reason why spinor transformations find their way into quantum mechanics. It remains to explain the origin of the name spinor and its relation to spin. The intrinsic angular momentum of fundamental

† Correctly paired: e.g. $\psi_{AB}{}^{\dot{C}} \oplus \psi_{A}{}^{\dot{B}\dot{C}}$, but not $\psi_{AB}{}^{\dot{C}} \oplus \psi^{AB\dot{C}}$.

particles or of combinations of such particles as nuclei, is called the spin. This angular momentum was first discovered in electrons by Goudsmit and Uhlenbeck, and Pauli was the first to produce a quantum-mechanical theory which included the spin. Pauli's theory was non-relativistic and possessed invariance only with respect to the group of rotations and we shall do well to restrict ourselves at first to this group. For the purpose of this section we may disregard all the properties of the electron except its spin. Following Pauli we then introduce a spin variable s which can take up only two discrete values ± 1, and consider functions $\psi(s)$. Each such function describes a spin state. Since s can take only two values $+1$ and -1, $\psi(s)$ has only two ordinates and the Hilbert space of functions $\psi(s)$ becomes a two-dimensional plane (Fig. 32). In this plane each vector $\mathbf{\Psi}$ is associated with a function $\psi(s)$:

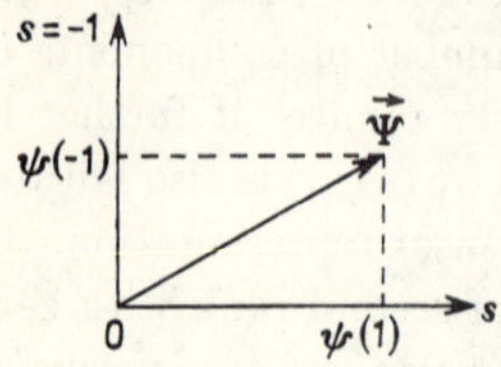

FIG. 32

$$(\mathbf{\Psi})_1 = \Psi_1 = \psi(1), \qquad (\mathbf{\Psi})_{-1} = \Psi_{-1} = \psi(-1). \tag{11.1}$$

According to quantum mechanics we have to consider normalized functions, so that

$$\mathbf{\Psi}^*\mathbf{\Psi} = 1 \quad \text{or} \quad \psi^*(1)\psi(1)+\psi^*(-1)\psi(-1) = 1. \tag{11.2}$$

$|\psi(1)|^2$ represents the probability of finding the spin pointing in a specified direction (say positive z-axis), and $|\psi(-1)|^2$ the probability of finding it pointing in the opposite direction.

We must now consider the transformation of $\mathbf{\Psi}$ when a rotation takes place. Since $\mathbf{\Psi}$ is a state vector it can only undergo a unitary transformation, as this is the only transformation which will leave the transition probability $|\mathbf{\Psi}^*\mathbf{\Phi}|^2$ between two states $\mathbf{\Psi}$ and $\mathbf{\Phi}$ unchanged. We also know that such a unitary representation of the group of rotations exists. It consists of the two-dimensional unitary transformations which represent the group of rotations two-valuedly. If R is a rotation and U_R the unitary matrix which represents it, we have

$$\mathbf{\Psi}^*\mathbf{\Phi} = (U_R\mathbf{\Psi})^*U_R\mathbf{\Phi}. \tag{11.3}$$

Expressed in components and denoting $U_R\mathbf{\Psi}$ by $\mathbf{\Psi}'$ and $U_R\mathbf{\Phi}$ by $\mathbf{\Phi}'$, we have

$$\Psi_1^*\Phi_1+\Psi_{-1}^*\Phi_{-1} = \Psi_1^{*\prime}\Phi_1'+\Psi_{-1}^{*\prime}\Phi_{-1}'. \tag{11.4}$$

An invariance of this form has already been discussed on p. 236 of section 3, and it follows that the introduction of a spin variable which can take up only two values leads directly to state vectors which transform like two-spinors of first rank as far as the group of rotations is concerned.

Let us now consider operators in the two-dimensional Hilbert space. They can only consist of two-dimensional matrices. To represent observables they must be Hermitian. All Hermitian operators in the Hilbert plane can be obtained through linear combinations with real coefficients of the following four Hermitian matrices:

$$\sigma_0 = \begin{bmatrix} 1 & 0 \\ 0 & 1 \end{bmatrix}, \quad \sigma_x = \begin{bmatrix} 0 & 1 \\ 1 & 0 \end{bmatrix}, \quad \sigma_y = \begin{bmatrix} 0 & -i \\ i & 0 \end{bmatrix}, \quad \sigma_z = \begin{bmatrix} 1 & 0 \\ 0 & -1 \end{bmatrix}. \tag{11.5}$$

We have already encountered the Pauli matrices σ_x, σ_y, σ_z and know that they are connected with infinitesimal rotations. In quantum mechanics $\frac{1}{2}i(\sigma_x, \sigma_y, \sigma_z)$ represent the components of the spin angular momentum. We wish now to prove that the σ's remain invariant under a rotation, i.e. under a combined vector transformation and the unitary transformation which represents the rotation. This invariance is a necessary property required of any tensor operator, as was pointed out in the introduction to this chapter. By a tensor we now mean one which belongs to the three-dimensional space since we are considering only the group of rotations. Hence we have to prove that

$$\sigma_k = U_R R_{kl} \sigma_l U_R^{-1} \quad (k, l = x, y, z \text{ or } x', y', z'), \tag{11.6}$$

where (R_{kl}) is the matrix which describes the considered rotation R. By way of an example let us consider a rotation around the z-axis by an angle γ. The matrix R_{kl} is then given by

$$(R_{kl}) = \begin{bmatrix} \cos\gamma & -\sin\gamma & 0 \\ \sin\gamma & \cos\gamma & 0 \\ 0 & 0 & 1 \end{bmatrix}. \tag{11.7}$$

The corresponding unitary transformation U_R is given by (see 3.17)

$$U_R = \begin{bmatrix} e^{-\frac{1}{2}i\gamma} & 0 \\ 0 & e^{\frac{1}{2}i\gamma} \end{bmatrix}. \tag{11.8}$$

According to (11.6), we thus get

$$\begin{aligned} \sigma_{x'} &= \cos\gamma(U_R \sigma_x U_R^{-1}) - \sin\gamma(U_R \sigma_y U_R^{-1}), \\ \sigma_{y'} &= \sin\gamma(U_R \sigma_x U_R^{-1}) + \cos\gamma(U_R \sigma_y U_R^{-1}), \\ \sigma_{z'} &= U_R \sigma_z U_R^{-1}. \end{aligned} \tag{11.9}$$

Using (11.8), we have

$$U_R \sigma_x U_R^{-1} = \begin{bmatrix} e^{-\frac{1}{2}i\gamma} & 0 \\ 0 & e^{\frac{1}{2}i\gamma} \end{bmatrix} \begin{bmatrix} 0 & 1 \\ 1 & 0 \end{bmatrix} \begin{bmatrix} e^{\frac{1}{2}i\gamma} & 0 \\ 0 & e^{-\frac{1}{2}i\gamma} \end{bmatrix} = \begin{bmatrix} 0 & e^{-i\gamma} \\ e^{i\gamma} & 0 \end{bmatrix},$$

$$U_R \sigma_y U_R^{-1} = \qquad = \begin{bmatrix} 0 & -ie^{-i\gamma} \\ ie^{i\gamma} & 0 \end{bmatrix},$$

$$U_R \sigma_z U_R^{-1} = \qquad = \begin{bmatrix} 1 & 0 \\ 0 & -1 \end{bmatrix}. \tag{11.10}$$

Substituting these expressions in (11.9), we find

$$\sigma_{x'} = \sigma_x, \quad \sigma_{y'} = \sigma_y, \quad \sigma_{z'} = \sigma_z. \tag{11.11}$$

The same result is obtained if rotations around the other axes are considered, and hence for any rotation.

It follows that the x-component of the spin angular momentum will always be represented by the first of the Pauli matrices, σ_x, no matter which orientations of axes we use, and similarly for the other components. The representation given in (11.5) is, however, adapted to the z-axis. In other words, it is assumed that the z-component of the spin is a diagonal matrix and that the spin variable s consists of the eigenvalues of σ_z. We could adapt the representation to any other axis, by permuting the Pauli matrices, but this transformation from one representation into another is of no further interest in the present context.

The result obtained in (11.11) is the condition for the operators σ_x, σ_y, σ_z to form a vector operator. We are not free to select any three matrices and attribute to them a vector property, in the same way that we are not free, for instance, to assume that $Q_x = \partial/\partial x$, $Q_y = \partial^2/\partial y^2$, $Q_z = \partial^2/\partial x \partial z$ are components of a vector operator. When transformed like a vector with a rotation matrix R_{kl} and then expressed in terms of x', y', z' (which corresponds to $Q' = U_R Q U_R^{-1}$), we do not get the same operators again (in terms of x', y', z'). On the other hand, the three operators of shift $P_x = \partial/\partial x$, $P_y = \partial/\partial y$, $P_z = \partial/\partial z$ will yield $P_{x'} = \partial/\partial x'$, $P_{y'} = \partial/\partial y'$, $P_{z'} = \partial/\partial z'$, i.e. they yield the same operators again, and consequently they form a vector. From

$$x'_k = R_{kl} x_l, \tag{11.12}$$

we have

$$\frac{\partial}{\partial x_l} = R_{ml} \frac{\partial}{\partial x'_m} = R^{-1}_{lm} \frac{\partial}{\partial x'_m}. \tag{11.13}$$

Transforming P_k as a vector leads to

$$P'_k = R_{kl} P_l, \tag{11.14}$$

where the P'_k are still functions of $\partial/\partial x$, $\partial/\partial y$, $\partial/\partial z$. Using (11.13) to express them in terms of $\partial/\partial x'$, $\partial/\partial y'$, $\partial/\partial z'$, we get

$$P'_k = R_{kl}\,R^{-1}_{lm}\frac{\partial}{\partial x'_m} = \delta_{km}\frac{\partial}{\partial x'_m} = \frac{\partial}{\partial x'_k}. \tag{11.15}$$

Algebraically, we get the same matrices again after transformation, provided the operator is a tensor operator. Analytically we are led back to the same operator expressed in terms of the new variables provided it is a tensor operator. There is no essential difference between the two cases.

So far, we have dealt with the group of rotations only and we have seen that the introduction of a spin variable can be adapted to quantum mechanics, by assuming the state vector to be a spinor. When rotations of axes are performed it undergoes a unitary transformation and transition probabilities remain unchanged. However, we must insist on a theory which allows not only rotations, but any Lorentz transformations. In general, when a Lorentz transformation is performed the vector $\mathbf{\Psi}$ does not undergo a unitary transformation but a unimodular transformation:

$$\left.\begin{aligned}\Psi'_1 &= \alpha\Psi_1+\beta\Psi_{-1}\\ \Psi'_{-1} &= \gamma\Psi_1+\delta\Psi_{-1}\end{aligned}\right\}\quad (\alpha\delta-\beta\gamma = 1). \tag{11.16}$$

Under such circumstances $\mathbf{\Psi}^*\mathbf{\Phi}$ (where $\mathbf{\Phi}$ undergoes the same transformation as $\mathbf{\Psi}$ (11.16)) does not remain invariant, and the general scheme of quantum mechanics cannot be adopted. The difficulty can, however, be removed if we alter the concept of a state in quantum mechanics by changing it from a function to that of a functional. A functional $F[\psi(x,y,z,t)]$ yields for each function $\psi(x,y,z,t)$ a value. Thus $\int \psi(x,y,z,t)\, dxdydzdt$ is a functional because for each function the integral will yield a value. In quantum mechanics we then assume that the functional yields the amplitude of probability of finding a distribution-function of some assembly. This advance from state functions to state functionals is the essential feature of what is known as the second quantization and comes out quite naturally if we deal with the spin. By considering functionals of spin functions and assuming that the Hilbert space is a space of functionals, the spin concept can be extended so as to be in agreement not only with the group of rotations but also with the full Lorentz group.

Functionals of spin functions can be obtained as follows. Let $u_1(s)$ and $u_{-1}(s)$ be two orthogonal functions, for instance,

$$u_1(s) = \delta_{1,s}, \qquad u_1(s) = \delta_{-1,s}. \tag{11.17}$$

Then, any spin function can be expressed in terms of such functions:

$$u_{a_1,a_{-1}}(s) = a_1 u_1(s) + a_{-1} u_{-1}(s), \tag{11.18}$$

where a_1 and a_{-1} are any two numbers. We can say that any pair of such numbers determines a spin function (denoted by $u_{a_1 a_{-1}}$). We can say, further, that any function $\psi(a_1, a_{-1})$ is a functional of the functions $u_{a_1 a_{-1}}$, because any pair a_1, a_{-1} determines a spin function and for any such pair $\psi(a_1, a_{-1})$ yields a value. Hence, the functionals of spin functions are fnnctions of two variables. The Hilbert space in this case is a space of functions of two variables:

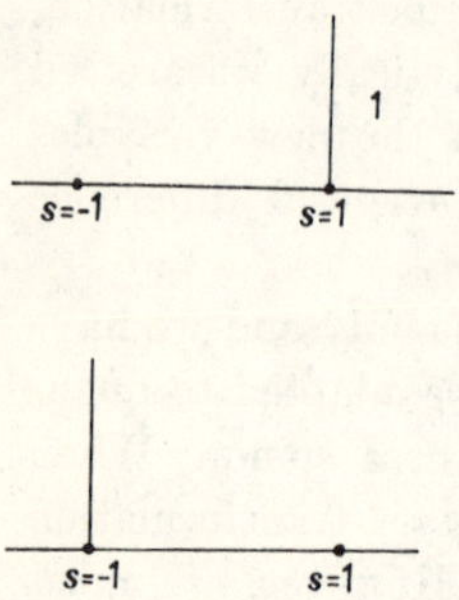

FIG. 33

$$\mathbf{\Psi} \to \psi(a_1, a_{-1}). \tag{11.19}$$

Denoting each axis in the Hilbert space by a pair of values a_1, a_{-1}, and the component of $\mathbf{\Psi}$ along the axis $a_1 a_{-1}$ by $\Psi_{a_1 a_{-1}}$, we have

$$(\mathbf{\Psi})_{a_1 a_{-1}} \equiv \Psi_{a_1 a_{-1}} = \psi(a_1, a_{-1}). \tag{11.20}$$

$\psi(a_1, a_{-1})$ is the probability amplitude of finding a_1 spins in the direction of the positive z-axis, and a_{-1} in the opposite direction. A given state vector provides a probability for each possible distribution. The transition probability from one state $\mathbf{\Psi}$ to another state $\mathbf{\Phi}$ is obtained from

$$\mathbf{\Psi}^*\mathbf{\Phi} = \iint \psi^*(a_1, a_{-1})\phi(a_1, a_{-1})\, d\omega, \tag{11.21}$$

where $d\omega$ is an invariant element of area in the two-dimensional space a_1, a_{-1}. A suitable definition will be given later.

We must prove that the scalar product remains unchanged when any Lorentz transformation is taking place. To this end we note that a_1, a_{-1} transform like a spinor of first rank because they determine the spinor $U_{a_1 a_{-1}}$. Let L be a Lorentz transformation and let the corresponding unimodular transformation be given by

$$\begin{aligned} a'_1 &= \alpha a_1 + \beta a_{-1} \\ a'_{-1} &= \gamma a_1 + \delta a_{-1} \end{aligned} \qquad (\alpha\delta - \beta\gamma = 1). \tag{11.22}$$

Under such transformations all the spin functions (11.18) will undergo a unimodular transformation and hence behave like spinors. Furthermore, the spin functionals, such as $\mathbf{\Psi} \to \psi(a_1, a_{-1})$, $\mathbf{\Phi} \to \phi(a_1, a_{-1})$, will undergo unitary transformations U_L: $\mathbf{\Psi} \to U_L \mathbf{\Psi}$, $\mathbf{\Phi} \to U_L \mathbf{\Phi}$, etc. To prove it we must demonstrate that the scalar product remains unchanged, i.e. that

$$\mathbf{\Psi}^*\mathbf{\Phi} = (U_L \mathbf{\Psi})^* U_L \mathbf{\Phi}, \tag{11.23}$$

where
$$U_L\boldsymbol{\Psi} \to \psi(\alpha a_1+\beta a_{-1}, \gamma a_1+\delta a_{-1}),$$
$$U_L\boldsymbol{\Phi} \to \phi(\alpha a_1+\beta a_{-1}, \gamma a_1+\delta a_{-1}). \tag{11.24}$$

This is achieved by defining the element of area with the help of two infinitesimal spinors $(d_a a_1, d_a a_{-1})$ and $(d_b a_1, d_b a_{-1})$, as follows:
$$d\omega = d_a a_1 d_b a_{-1} - d_a a_{-1} d_b a_1. \tag{11.25}$$

This element of area remains invariant under spinor (unimodular) transformations and so we get, using (11.22),
$$(U_L\boldsymbol{\Psi})(U_L\boldsymbol{\Phi}) = \iint \psi^*(\alpha a_1+\beta a_2, \gamma a_1+\delta a_2)\phi(\alpha a_1+\beta a_2, \gamma a_1+\delta a_2)\times$$
$$\times(d_a a_1 d_b a_2 - d_b a_1 d_a a_2)$$
$$= \iint \psi^*(a_1', a_2')\phi(a_1', a_2')(d_a a_1' d_b a_2' - d_b a_1' d_a a_2'). \tag{11.26}$$

This integral is identical with the one of (11.21) except for a change of letters in the argument, since in both cases we integrate over the same range of the variables.

From the foregoing we see that spin variables which can take up only a discrete set of values can be adopted in the general scheme of quantum mechanics and that they lead necessarily to second quantization. Only in this way can Hilbert spaces of infinite dimensions be obtained which are required to accommodate unitary representations of the Lorentz group.

We know already that entities transforming as spinors (or tensors) can only represent observables and not states. In other words, they must be operators obeying suitable commutation rules, all this being part and parcel of the process of second quantization. In our present case we must attribute operator properties to the spin function $U_{a_1 a_2}(s)$. Let
$$U_{A_1A_{-1}}(s) = A_1 u_1(s)+A_{-1}u_{-1}(s) = A_1\delta_{\frac{1}{2},s}+A_{-1}\delta_{-\frac{1}{2},s}, \tag{11.27}$$
$$U_{B^1B^{-1}}(s) = B^1u_1(s)+B^{-1}u_{-1}(s) = B^1\delta_{\frac{1}{2},s}+B^{-1}\delta_{-\frac{1}{2},s} \tag{11.28}$$
be spinor operators. In (11.18) we have replaced a_1, a_{-1} by A_1, A_{-1}, which are now operators transforming as a covariant spinor. Similarly, we have introduced a contravariant spinor B^{-1}, B^1. For brevity we shall denote $U_{A_1A_{-1}}$ by $U_c(s)$ and $U_{B^1B^{-1}}$ by $U^c(s)$.

We now postulate the following commutation laws:†
$$U_c(s)U^c(s')+U^c(s')U_c(s) = \delta(s-s') = \delta_{1s}\delta_{1s'}+\delta_{-1s}\delta_{-1s'}, \tag{11.29}$$
$$U_c(s)U_c(s')+U_c(s')U_c(s) = 0, \tag{11.30}$$
$$U^c(s)U^c(s')+U^c(s')U^c(s) = 0. \tag{11.31}$$

† $\delta(s-s')$ is a function which is equal to 1 for $s = s' = 1$ and $s = s' = -1$. It is equal to zero for the other pair of points. $\delta_{1,s}\delta_{1,s'}$ is equal to 1 for $s = s' = 1$ and is zero elsewhere. $\delta_{-1,s}\delta_{-1,s'}$ is equal to 1 at $s = s' = -1$ and is zero elsewhere, hence the identity in (11.29).

These commutation rules apply to particles obeying Fermi–Dirac statistics (fermions). At present we are dealing with such particles of spin $\frac{1}{2}$.

Substituting (11.27) and (11.28) in (11.29) to (11.31) we get the following relations:

$$A_N B^M + B^M A_N = \delta^M{}_N \quad (M, N = 1, 2, \text{ replacing from now on } -1, 1), \tag{11.32}$$

$$A_M A_N + A_N A_M = 0, \tag{11.33}$$

$$B^M B^N + B^N B^M = 0. \tag{11.34}$$

The relativistic invariance of these relations is a direct result of their spinor character and the fact that $\delta^M{}_N$ is an invariant spinor of second rank which is given by

$$(\delta^M{}_N) = \begin{bmatrix} 1 & 0 \\ 0 & 1 \end{bmatrix},$$

$$(\delta^{MN}) \equiv (\delta_{MN}) = \begin{bmatrix} 0 & -1 \\ 1 & 0 \end{bmatrix},$$

$$(\delta^M{}_N) = -(\delta_M{}^N). \tag{11.35}$$

B^1 and B^2 are creation operators of up-spin and down-spin respectively. A_1 and A_2 are annihilation operators of up- and down-spin. In the Hilbert space of functions ($\mathbf{\Psi} \rightarrow \psi(a_1 a_2)$, where a_1, a_2 can only be 0 or 1), these operators are represented as follows:

$$B^1\psi(0, a_2) = \psi(1, a_2), \qquad B^2\psi(a_1, 0) = (-1)^{a_1}\psi(a_1, 1), \tag{11.36}$$

$$B^1\psi(1, a_2) = 0, \qquad B^2\psi(a_1, 1) = 0, \tag{11.37}$$

$$A_1\psi(0, a_2) = 0, \qquad A_2\psi(a_1, 0) = 0, \tag{11.38}$$

$$A_1\psi(1, a_2) = \psi(0, a_2), \qquad A_2\psi(a_1, 1) = (-1)^{a_1}\psi(a_1, 0). \tag{11.39}$$

The operator $B^M A_M$ is an invariant and is known as the total number operator. It yields the total number of spins, since its eigenvalues are $a_1 + a_2$:

$$B^M A_M \psi(a_1 a_2) = (a_1 + a_2)\psi(a_1, a_2). \tag{11.40}$$

This relation is obtainable from (11.36) to (11.39) by applying it to the four different cases $\psi(1, 0)$, $\psi(0, 1)$, $\psi(1, 1)$, and $\psi(0, 0)$, with total numbers of spins 1, 1, 2, 0. It is essential to have the number operator expressed as $B^M A_M$ and not as $A_M B^M$. It can also be verified in the same way that $B^1 A_1$ then yields the number of up-spins and $B^2 A_2$ the number of down-spins. In fact, these two operators form the basis of the representation and are therefore in diagonal form. In other words we use the eigenvalues of these two independent and commuting operators as the arguments in the functions.

In the theory of relativity the angular momentum is an anti-symmetrical tensor with six components and not a three-dimensional axial vector. It is closely associated with the six infinitesimal rotations in the x, y, z, t space. In the Hilbert plane of spinors these infinitesimal transformations are determined by the spin-tensor $S_{\kappa\lambda M}{}^{N}$. The six components of the spin angular momentum are given by the spin-tensor

$$\sigma_{\kappa\lambda M}{}^{N} = 2iS_{\kappa\lambda M}{}^{N}. \tag{11.41}$$

However, essentially we are led to the Pauli matrices because $\sigma_{12} = \sigma_z$, $\sigma_{23} = \sigma_x$, $\sigma_{31} = \sigma_y$, and $\sigma_{14} = i\sigma_{23} = i\sigma_x$, $\sigma_{24} = i\sigma_{31} = i\sigma_y$, $\sigma_{34} = i\sigma_{12} = i\sigma_z$. In quantum theory we have to look upon these components as operators in the Hilbert space of functions $\psi(a_1, a_2)$. They are given by:

$$i^{-1}\sigma_{14} = \sigma_{23} = \sigma_x \rightarrow \Sigma_{23} = a_1\frac{\partial}{\partial a_2}+a_2\frac{\partial}{\partial a_1}, \tag{11.42}$$

$$i^{-1}\sigma_{24} = \sigma_{31} = \sigma_y \rightarrow \Sigma_{31} = i^{-1}\left(a_1\frac{\partial}{\partial a_2}+a_2\frac{\partial}{\partial a_1}\right), \tag{11.43}$$

$$i^{-1}\sigma_{34} = \sigma_{12} = \sigma_z \rightarrow \Sigma_{12} = a_1\frac{\partial}{\partial a_1}-a_2\frac{\partial}{\partial a_2}. \tag{11.44}$$

To prove these relations we consider a transformation

$$\psi(a_1, a_2) \rightarrow \psi(a_1+\delta a_1, a_2+\delta a_2) = \psi'(a_1, a_2),$$

therefore
$$\psi'(a_1, a_2) = \psi(a_1, a_2)+\frac{\partial\psi}{\partial a_1}\delta a_1+\frac{\partial\psi}{\partial a_2}\delta a_2$$

and
$$\delta\psi = \left(\delta a_1\frac{\partial}{\partial a_1}+\delta a_2\frac{\partial}{\partial a_2}\right)\psi. \tag{11.45}$$

For a rotation around the x-axis by an angle $\delta\alpha$ we have to use S_{23} which gives

$$\begin{bmatrix}\delta a_1\\ \delta a_2\end{bmatrix} = \frac{1}{2}\begin{bmatrix}-ia_2\\ -ia_1\end{bmatrix}\delta\alpha. \tag{11.46}$$

Inserting $\delta a_1 = -\frac{1}{2}ia_2\,\delta\alpha$ and $\delta a_2 = -\frac{1}{2}ia_1\,\delta\alpha$ in (11.45) and multiplying by $2i$, we get (11.42). In the same way, using S_{31} and S_{12} we get (11.43) and (11.44).

The invariance of the tensor operator $\Sigma_{\kappa\lambda} = -\Sigma_{\lambda\kappa}$ as given by (11.42) to (11.44) can be proved in the same way as was done for instance with $(\partial/\partial x, \partial/\partial y, \partial/\partial z)$ with respect to shifts. We must express the Lorentz transformation in terms of the unimodular transformation which represents it, transform $\Sigma_{\kappa\lambda}$ as a tensor, and then express the operator in terms of a_1', a_2', $\partial/\partial a_1'$, $\partial/\partial a_2'$. However, such a calculation is cumbersome and unnecessary. The operators $\Sigma_{\kappa\lambda}$ obey the same commutation rules as the $\sigma_{\kappa\lambda}$ and consequently they induce a representation of the Lorentz group in the space of functions $\psi(a_1, a_2)$. Such a state of affairs cannot depend on a particular Lorentz frame and consequently the $\Sigma_{\kappa\lambda}$ are invariant tensor operators.

From the $\Sigma_{\kappa\lambda}$ we can form two scalars:

$$S_1 = \Sigma_{\kappa\lambda}\Sigma_{\kappa\lambda}, \tag{11.47}$$

$$S_2 = \Sigma_{\kappa\lambda}\Sigma^d_{\kappa\lambda}, \tag{11.48}$$

where $\Sigma^d_{\kappa\lambda}$ is the dual of $\Sigma_{\kappa\lambda}$ ($\Sigma^d_{\kappa\lambda} = \epsilon_{\kappa\lambda\mu\nu}\Sigma_{\mu\nu}$).

The first of these invariants yields on using (11.42) to (11.44),

$$S_1 = \Sigma^2_{23}+\Sigma^2_{31}+\Sigma^2_{12}+\Sigma^2_{14}+\Sigma^2_{24}+\Sigma^2_{34} \equiv 0. \tag{11.49}$$

The second invariant leads to†

$$S_2 = i(\Sigma_{23}\Sigma_{14}+\Sigma_{31}\Sigma_{24}+\Sigma_{12}\Sigma_{34}), \tag{11.50}$$

therefore
$$S_2 = i\left(a_1\frac{\partial}{\partial a_1}+a_2\frac{\partial}{\partial a_2}\right)\left(a_1\frac{\partial}{\partial a_1}+a_2\frac{\partial}{\partial a_2}+2\right). \tag{11.51}$$

Eqn. (11.50) in terms of the σ's is nothing but

$$(\sigma_x^2+\sigma_y^2+\sigma_z^2) = \sigma^2 = 3 = 4(\tfrac{1}{2})(\tfrac{1}{2}+1).$$

$i^{-1}S_2$ may therefore be denoted more suitably by Σ^2. We have

$$\Sigma^2 = 4\left[\frac{1}{2}\left(a_1\frac{\partial}{\partial a_1}+a_2\frac{\partial}{\partial a_2}\right)\left\{\frac{1}{2}\left(a_1\frac{\partial}{\partial a_1}+a_2\frac{\partial}{\partial a_2}\right)+1\right\}\right]. \tag{11.52}$$

$a_1(\partial/\partial a_1)+a_2(\partial/\partial a_2)$ is an invariant operator because a_1, a_2 transform as a covariant spinor and $\partial/\partial a_1$, $\partial/\partial a_2$ as a contravariant spinor. The 'square' of this operator yields the square of the total spin angular momentum.

All that has been said with regard to spin $\frac{1}{2}$ can be generalized to higher spin. For spin 1 we may use a variable s which can take three values, say, -2, 0, 2. We require the three-dimensional representation of the Lorentz group which is associated with $\psi(1,0)$ or ψ_{AB} and ψ^{AB} (see Table III on p. 294). To make sure that ψ_{AB} is symmetrical we use $\frac{1}{2}(\psi_{AB}+\psi_{BA})$. This spinor has three independent components and yields a faithful three-dimensional representation. For spin $\frac{3}{2}$ ($s = -3$, -1, 1, 3) we require the four-dimensional representation $\psi(\frac{3}{2},0)$. This representation is two-valued. For a spin k when $s = (-2k, -2k+2,..., 2k)$ the required representation is $\psi(k,0)$. The representations $\psi(k,l)$ with $l \neq 0$ are reducible as far as the group of rotations is concerned (see 10.27) and correspond to a spin which is not unique.

† Compare the case of the electromagnetic field when $F_{\kappa\lambda}F^d_{\kappa\lambda}$ yields (**EH**). In both cases we are dealing with anti-symmetrical tensors.

In the case of spin 1 with $s = -2, 0, 2$, the three-dimensional Pauli matrices $\sigma_x^{(1)}$, $\sigma_y^{(1)}$, $\sigma_z^{(1)}$ are those given in (10.15):

$$\sigma_x^{(1)} = \begin{bmatrix} 0 & \sqrt{2} & 0 \\ \sqrt{2} & 0 & \sqrt{2} \\ 0 & \sqrt{2} & 0 \end{bmatrix}, \quad \sigma_y^{(1)} = i\begin{bmatrix} 0 & -\sqrt{2} & 0 \\ \sqrt{2} & 0 & -\sqrt{2} \\ 0 & \sqrt{2} & 0 \end{bmatrix}, \quad \sigma_z^{(1)} = \begin{bmatrix} 2 & 0 & 0 \\ 0 & 0 & 0 \\ 0 & 0 & -2 \end{bmatrix}. \tag{11.53}$$

When we go over to the Hilbert space of functionals we must consider functions of three variables $\phi\{a_{11}, \frac{1}{2}(a_{12}+a_{21}), a_{22}\}$ where $\frac{1}{2}(a_{AB}+a_{BA})$ is a symmetrical spinor of second rank. The corresponding operators $\Sigma_{\kappa\lambda}^{(1)}$ are obtainable in the same way as those of (11.42) to (11.44).

Particles of spin 1 obey Einstein–Bose statistics (bosons) and consequently the creation and annihilation operators obey the commutation rule

$$A_{MK}B^{NL} - B^{NL}A_{MK} = \tfrac{1}{2}(\delta_M{}^N\delta_K{}^L + \delta_K{}^L\delta_M{}^N), \tag{11.54}$$

where it is assumed that the spinors are symmetrized.

In the Hilbert space of functionals these operators are represented as follows:

$$B^{11} \to \psi'(a_{11}, a_{(12)}, a_{22}) = (a_{11}+1)^{\frac{1}{2}}\psi(a_{11}+1, a_{(12)}, a_{22}), \tag{11.55}$$

$$A_{11} \to \psi'(a_{11}, a_{(12)}, a_{22}) = (a_{11})^{\frac{1}{2}}\psi(a_{11}-1, a_{(12)}, a_{22}), \tag{11.56}$$

and similarly for the other components.

12. The Dirac Field

The first to introduce entities which do not transform as four-tensors and yet yield Lorentz covariant relations was Dirac through his wave equation for entities with four components. Dirac argued that the relativistic wave equation

$$(\partial_\mu \partial^\mu + \kappa^2)\psi = 0 \qquad (\kappa = m_0 c/h) \tag{12.1}$$

for a particle of mass m_0 is not acceptable from a quantum-mechanical point of view to deal with a system of one particle, because it is of second order in the time derivative. On the other hand, relativistic covariance demands that all four derivatives ∂_μ should appear with the same power. Consequently, Dirac proceeded to factorize the wave equation into two factors linear in ∂_μ. Such factorization is not possible unless we look upon ψ as an entity of, say, n components and introduce matrices α_ν of n columns and n rows $\alpha_{\nu pq}$ $(p, q = 1,..., n)$ which appear as coefficients in the factorization:

$$\begin{aligned}\partial_\mu \partial^\mu + \kappa^2 &= (\partial_\mu \alpha^\mu + i\kappa)(\partial_\nu \alpha^\nu - i\kappa) \\ &= (\partial^\mu \alpha_\mu + i\kappa)(\partial^\nu \alpha_\nu - i\kappa),\end{aligned} \tag{12.2}$$

where

$$\alpha^\nu = g^{\nu\lambda}\alpha_\lambda. \tag{12.3}$$

The wave equation (12.1) will then be satisfied if any one of the following equations hold:

$$(\partial_\mu \alpha^\mu + i\kappa)\psi = 0, \tag{12.4}$$

$$(\partial^\nu \alpha_\nu - i\kappa)\psi = 0, \tag{12.5}$$

where the α's obey the following conditions:

$$\alpha^\mu \alpha^\nu + \alpha^\nu \alpha^\mu = 2g^{\mu\nu},\dagger \tag{12.6}$$

from which it also follows that

$$\alpha_\mu \alpha_\nu + \alpha_\nu \alpha_\mu = 2g_{\mu\nu}. \tag{12.7}$$

Since a change from α_μ to $-\alpha_\mu$ leaves these relations unchanged, and by such a change (12.5) follows from (12.4), there is no essential difference between these equations and we shall use (12.4).

Eqns. (12.6) or (12.7) cannot be satisfied by ordinary numbers. They can, however, be satisfied by matrices of lowest order four. These form an irreducible set, and any other solution which satisfies (12.6) and is of higher dimensions can only be composed of the irreducible four-dimensional matrices. It can thus have 8, 12, 16,... dimensions, but represents no further interest. It follows, therefore, that ψ is an entity with four components ψ_a and such expressions as $\alpha\psi$ have the following meaning:

$$\alpha\psi = \begin{bmatrix} \alpha_{11} & \alpha_{12} & \alpha_{13} & \alpha_{14} \\ \alpha_{21} & \alpha_{22} & \alpha_{23} & \alpha_{24} \\ \alpha_{31} & \alpha_{32} & \alpha_{33} & \alpha_{34} \\ \alpha_{41} & \alpha_{42} & \alpha_{43} & \alpha_{44} \end{bmatrix} \begin{bmatrix} \psi_1 \\ \psi_2 \\ \psi_3 \\ \psi_4 \end{bmatrix}. \tag{12.8}$$

Dirac proved that the linear factors (12.4) and (12.5) are Lorentz invariant. In other words, for each Lorentz transformation $L_\mu{}^\nu$ there exists a corresponding (i.e. representing) transformation

$$\psi'_a = S_{am}\psi_m, \tag{12.9}$$

which yields
$$\alpha'_{\mu ab} = S_{am} L_\mu{}^\nu \alpha_{\nu mn} S^{-1}_{nb} = \alpha_{\mu ab}, \tag{12.10}$$

and therefore leaves (12.4) unchanged, with the same α's.

On the basis of the spinor analysis which was proposed by L. van d. Waerden after Dirac's theory was published and which has been treated in this chapter, we recognize ψ as a four-spinor and the invariant α's

† On carrying out the multiplication in (12.2) we get $\partial_\mu \alpha^\mu \partial_\nu \alpha^\nu + \kappa^2 + i\kappa(\partial_\nu \alpha^\nu - \partial_\mu \alpha^\mu)$ where $\partial_\nu \alpha^\nu = \partial_\mu \alpha^\mu$. Hence we get further $\partial_\mu \partial_\nu \alpha^\mu \alpha^\nu + \kappa^2$. In this expression we can exchange μ and ν, and because of the symmetry of $\partial_\mu \partial_\nu$ this gives $\frac{1}{2}\partial_\mu \partial_\nu(\alpha^\mu\alpha^\nu + \alpha^\nu\alpha^\mu) + \kappa^2$. If (12.6) is satisfied this yields

$$\partial_\mu \partial_\nu g^{\mu\nu} + \kappa^2 \equiv \partial_\mu \partial^\mu + \kappa^2.$$

as the γ's. Hence, using the γ's of (8.25) and (8.26) we can express Dirac's equation in the following forms:

$$\partial_\mu \gamma^\mu{}_a{}^b \Psi_b + i\kappa \Psi_a = 0, \tag{12.11}$$

$$\partial_\mu \gamma^{\mu a}{}_b \Psi^b - i\kappa \Psi^a = 0. \tag{12.12}$$

In (12.11) we assume that Ψ is a covariant four-spinor and in (12.12) it is assumed to be a contravariant four-spinor. Each of these equations follows from the other. Thus, changing Ψ_a into Ψ^a involves a transformation in the spin frame which is given by

$$\Psi^a = \epsilon^{ab}\Psi_b, \tag{12.13}$$

where ϵ^{ab} is regarded as a matrix and Ψ_b as a column; ϵ^{ab} is unitary. To obtain the transformation of $\gamma^\mu{}_a{}^b$ which goes with this change in frame, it is best to look upon γ as an operator (a matrix) in the spin space, which must therefore transform according to

$$(\gamma^\mu{}_a{}^b)' = (\epsilon^{am})(\gamma^\mu{}_m{}^n)(\epsilon^{-1nb}). \tag{12.14}$$

When this transformation is carried out we obtain $-\gamma^{\mu a}{}_b$ and hence (12.12).†

As pointed out in the introduction to this chapter, entities which represent observables must transform according to a faithful representation of the Lorentz group. Our next objective, therefore, is to obtain tensor entities from the Ψ_a. The transformation of the Ψ_a themselves represents the Lorentz group two-valuedly. Let us begin with a scalar. If we take two four-spinors Ψ_a and Φ^a, then $\Psi_a \Phi^a$ is a scalar. However, in general it will be a complex number (or from a quantum-mechanical point of view it will not be a Hermitian operator). In the following we shall assume the Ψ to be numbers and not operators, as we shall not deal with the quantization of Dirac's equation. Now if $\Psi_a \Phi^a$ is to be real there must exist a special relationship between Ψ_a and Φ^a. It is easily seen that the following relationship satisfies this demand:

$$(\Psi_a) = \begin{bmatrix} \psi_1 \\ \psi_2 \\ \phi^{\dot 1} \\ \phi^{\dot 2} \end{bmatrix}, \qquad (\Phi^a) = \begin{bmatrix} \phi^1 \\ \phi^2 \\ \psi_{\dot 1} \\ \psi_{\dot 2} \end{bmatrix} \equiv (\Psi^{+a}). \tag{12.15}$$

Thus we get for the scalar which we denote by J:

$$\begin{aligned} J = \Psi_a \Phi^a &= \psi_1 \phi^1 + \psi_2 \phi^2 + \phi^{\dot 1} \psi_{\dot 1} + \phi^{\dot 2} \psi_{\dot 2} \\ &\equiv \psi_1 \phi^1 + \psi_2 \phi^2 + (\phi^1 \psi_1 + \phi^2 \psi_2)^*. \end{aligned} \tag{12.16}$$

† (12.12) follows from (12.11) also on using the general formula $\phi^a \psi_a = -\phi_a \psi^a$.

This is a real number since the sum of the last two terms is the complex conjugate of the sum of the first two terms. The spinor Φ^a as given in (12.15) is called the adjoint of Ψ_a. This relationship is reciprocal. As indicated in (12.15) we denote the adjoint by Ψ^{+a} and so we have

$$\Psi^{++a} = \Psi_a. \tag{12.17}$$

It is customary to introduce a matrix formalism and express the scalar product as between two matrices. We must then look upon one of the spinors, say Ψ_a, as a column, and transpose Ψ^{+a}, since in (12.15) we have already assumed both to be columns. The horizontal row must be written in front of the vertical column. Hence

$$J = \Psi_a \Psi^{+a} = (\widetilde{\Psi^{+a}})(\Psi_a). \tag{12.18}$$

Usually only the matrix formalism is used and it is understood that Ψ^{+a} is a horizontal row and the scalar is expressed simply as $\Psi^+\Psi$. However, as we wish to treat Dirac's equation mainly from a spinor point of view, and use the matrix formalism only where it helps the calculation, we have adopted a more detailed notation as above.

Next we form the four-vector

$$J^\mu = \gamma^{\mu a}{}_b \Psi_a \Psi^{+b}. \tag{12.19}$$

Here again it is convenient to use the matrix formalism:

$$J^\mu = (\widetilde{\Psi^{+b}})(\widetilde{\gamma^{\mu b}}{}_a)(\Psi_a). \tag{12.20}$$

Using (8.26), this yields

$$\begin{aligned} J^1 &= \psi_1\psi_{\dot 2}+\psi_{\dot 1}\psi_2-(\phi^2\phi^{\dot 1}+\phi^1\phi^{\dot 2}), \\ J^2 &= i[(\psi_1\psi_{\dot 2}-\psi_{\dot 1}\psi_2)-(\phi^2\phi^{\dot 1}-\phi^1\phi^{\dot 2})], \\ J^3 &= -\psi_1\psi_{\dot 1}+\psi_2\psi_{\dot 2}+\phi^1\phi^{\dot 1}-\phi^2\phi^{\dot 2}, \\ J^4 &= \psi_1\psi_{\dot 1}+\psi_2\psi_{\dot 2}+\phi^1\phi^{\dot 1}+\phi^2\phi^{\dot 2}. \end{aligned} \tag{12.21}$$

All the quantities are real. In addition, J^4 is positive definite. This enables us to define J^4 as a density and to look upon J^μ as a four-current density. The property of J^4 as a positive definite presents one of the essential features of the Dirac theory and from Dirac's equation we shall derive further down the continuity equation:

$$\partial_\mu J^\mu = 0. \tag{12.22}$$

In a similar way we can form the tensors

$$J^{\mu\nu} = \gamma^{\mu a}{}_m \gamma^{\nu m}{}_b \Psi_a \Psi^{+b}, \tag{12.23}$$

$$J^{\lambda\mu\nu} = \gamma^{\lambda a}{}_m \gamma^{\mu m}{}_n \gamma^{\nu n}{}_b \Psi_a \Psi^{+b}, \tag{12.24}$$

$$J^{\kappa\lambda\mu\nu} = \gamma^{\kappa a}{}_m \gamma^{\lambda m}{}_n \gamma^{\mu n}{}_s \gamma^{\nu s}{}_b \Psi_a \Psi^{+b}, \tag{12.25}$$

and subsequently their anti-symmetrized expressions. We get

$$J^{\mu\nu} = (\widetilde{\Psi^{+b}})(\widetilde{\gamma^{\nu b}}_m)(\widetilde{\gamma^{\mu m}}_a)(\Psi_a), \tag{12.26}$$

and similarly for the other quantities. $J^{\kappa\lambda\mu\nu}$ we can express in terms of the matrix γ^5, after changing to the dual $\epsilon_{\kappa\lambda\mu\nu}J^{\kappa\lambda\mu\nu}$:

$$\epsilon_{\kappa\lambda\mu\nu}J^{\kappa\lambda\mu\nu} = (\widetilde{\Psi^{+a}})(\widetilde{\gamma^{5a}}_b)(\Psi_b). \tag{12.27}$$

$J^{[\mu\nu]}$ is anti-symmetrical and has only six independent components. $J^{[\lambda\mu\nu]}$ is anti-symmetrical in all indices and has only four independent components, while $J^{[\kappa\lambda\mu\nu]}$, which is anti-symmetrical in all four indices, has only one independent component; its dual is a pseudo-scalar.

For any Ψ_a which obeys Dirac's equation, the adjoint Ψ^{+a} is obtainable as shown in (12.15). We shall now prove that it obeys the differential equation

$$\partial_\mu \gamma^{\mu a}{}_b \Psi^{+b} - i\kappa\Psi^{+a} = 0. \tag{12.28}$$

This equation is called the adjoint Dirac equation. That Ψ^{+a} satisfies it we prove as follows. Ψ^{+a} is obtainable from Ψ_a through the following transformation in the spinor frame:

$$(\Psi^{+\dot{a}}) = (\gamma^{4a}{}_b)(\Psi_b), \qquad (\gamma^{4a}{}_b) = \begin{bmatrix} 0 & 0 & 1 & 0 \\ 0 & 0 & 0 & 1 \\ 1 & 0 & 0 & 0 \\ 0 & 1 & 0 & 0 \end{bmatrix}. \tag{12.29}$$

As a consequence, $(\gamma^\mu{}_a{}^b)$ in Dirac's equation, when regarded as an operator (a matrix) in the spinor space, changes into

$$(\gamma^{4a}{}_m)(\gamma^\mu{}_m{}^n)(\gamma^{4n}{}_b)^{-1}. \tag{12.30}$$

When this transformation is carried out by using (8.25) for $\gamma^\mu{}_m{}^n$ we get the matrix (see 8.26)

$$\gamma^{\mu *a}{}_b = \gamma^{\mu\dot{a}}{}_{\dot{b}}. \tag{12.31}$$

Hence Dirac's equation transforms into

$$\partial_\mu \gamma^{\mu\dot{a}}{}_{\dot{b}} \Psi^{+\dot{b}} + i\kappa\Psi^{+\dot{a}} = 0. \tag{12.32}$$

Taking the complex conjugate, (12.28) follows.

We can now prove the equation of continuity (12.22). For this purpose we multiply (12.11) by Ψ^{+a} and (12.28) by Ψ_a. On adding the resulting relations we get

$$\partial_\mu(\gamma^\mu{}_a{}^b\Psi_b)\Psi^{+a} + \partial_\mu(\gamma^{\mu a}{}_b \Psi^{+b})\Psi_a = 0. \tag{12.33}$$

Consequently,

$$\partial_\mu(\gamma^\mu{}_a{}^b\Psi_b)\Psi^{+a} + \partial_\mu(\gamma^{\mu b}{}_a \Psi^{+a})\Psi_b = 0. \tag{12.34}$$

Now, from (8.25) and (8.26) we have

$$\gamma^\mu{}_a{}^b = \gamma^{\mu b}{}_a. \tag{12.35}$$

Hence (12.33) becomes

$$\partial_\mu(\gamma^\mu{}_a{}^b\Psi^{+a}\Psi_b) = 0 \tag{12.36}$$

and so (12.22) is proved, where

$$J^\mu = \gamma^\mu{}_a{}^b\Psi^{+a}\Psi_b = \gamma^{\mu a}{}_b\Psi^{+b}\Psi_a \tag{12.37}$$

in accordance with (12.19).

It may be added that had we used the γ's as given in (8.21) and (8.22), then in view of the commutation relations (8.30) and (8.31) which differ in the sign of $g^{\kappa\lambda}$ as compared with (12.7), Dirac's equation would take the form

$$\partial_\mu\gamma^\mu{}_a{}^b\Psi_b + \kappa\Psi_a = 0. \tag{12.38}$$

The adjoint is now defined by

$$\Psi_a = \begin{bmatrix}\psi_1\\ \psi_2\\ \phi^{\dot 1}\\ \phi^{\dot 2}\end{bmatrix}, \qquad \Psi^{+a} = i\begin{bmatrix}\phi^1\\ \phi^2\\ -\psi_{\dot 1}\\ -\psi_{\dot 2}\end{bmatrix} \tag{12.39}$$

and we have on using $i\gamma^{4a}{}_b$ (using (8.22) for the γ's)

$$\Psi^{+\dot a} = i\begin{bmatrix}0 & 0 & -1 & 0\\ 0 & 0 & 0 & -1\\ 1 & 0 & 0 & 0\\ 0 & 1 & 0 & 0\end{bmatrix}\begin{bmatrix}\psi_1\\ \psi_2\\ \phi^{\dot 1}\\ \phi^{\dot 2}\end{bmatrix} = i\begin{bmatrix}-\phi^{\dot 1}\\ -\phi^{\dot 2}\\ \psi_1\\ \psi_2\end{bmatrix} \tag{12.40}$$

For the γ's we have again

$$i\gamma^{4a}{}_m(\gamma^\mu{}_m{}^n)(i\gamma^{4n}{}_b)^{-1} = \gamma^{\mu*a}{}_b. \tag{12.41}$$

Hence (12.38) leads to

$$\partial_\mu\gamma^{\mu a}{}_b\Psi^{+b} + \kappa\Psi^{+a} = 0. \tag{12.42}$$

From (12.38) and (12.42) we get after multiplication with Ψ^{+a} and Ψ_a respectively and subtracting the results:

$$\partial_\mu(\gamma^\mu{}_a{}^b\Psi_b)\Psi^{+a} - \partial_\mu(\gamma^{\mu a}{}_b\Psi^{+a})\Psi_b = 0. \tag{12.43}$$

However, in our present case we find from (8.21) and (8.22),

$$\gamma^\mu{}_a{}^b = -\gamma^{\mu b}{}_a. \tag{12.44}$$

Consequently, this leads again to the equation of continuity. The four-current J^μ is now defined by

$$J^\mu = i\gamma^{\mu a}{}_b\Psi_a\Psi^{+b}. \tag{12.45}$$

The four-spinor field Ψ_a corresponds to spin $\frac{1}{2}$. This can be seen if we consider the infinitesimal transformations induced in this space through rotation with which we have already dealt in section 9. Using (9.37) and (8.25) we find

$$\mathscr{S}_{(12)} = \tfrac{1}{2}\begin{bmatrix} i & 0 & 0 & 0 \\ 0 & -i & 0 & 0 \\ 0 & 0 & i & 0 \\ 0 & 0 & 0 & -i \end{bmatrix}, \qquad \mathscr{S}_{(23)} = \tfrac{1}{2}\begin{bmatrix} 0 & i & 0 & 0 \\ i & 0 & 0 & 0 \\ 0 & 0 & 0 & i \\ 0 & 0 & i & 0 \end{bmatrix},$$

$$\mathscr{S}_{(31)} = \begin{bmatrix} 0 & -1 & 0 & 0 \\ 1 & 0 & 0 & 0 \\ 0 & 0 & 0 & -1 \\ 0 & 0 & 1 & 0 \end{bmatrix}; \tag{12.46}$$

therefore

$$-(\mathscr{S}^2_{(12)}+\mathscr{S}^2_{(23)}+\mathscr{S}^2_{(31)}) = \tfrac{3}{4}\begin{bmatrix} 1 & 0 & 0 & 0 \\ 0 & 1 & 0 & 0 \\ 0 & 0 & 1 & 0 \\ 0 & 0 & 0 & 1 \end{bmatrix} = \tfrac{1}{2}(\tfrac{1}{2}+1).\mathbf{1}. \tag{12.47}$$

Dirac's equation was obtained without any reference to spinor calculus, which was developed later. The algebraic method, introduced by Dirac through his matrices which obey certain commutation rules, is even now more popular with theoreticians than the spinor calculus in the strict sense of the word. This applied not only to Dirac's original equation but also for instance to the Duffin–Kemmer equation:

$$\partial_\mu \beta^{\mu ab}{}_{mn} \Psi^{mn} + i\kappa \Psi^{ab} = 0, \tag{12.48}$$

where Ψ^{ab} is a four-spinor of second rank and the β's are sixteen-dimensional matrices obeying the Duffin–Kemmer commutation rules (8.59).

On the basis of the spinor calculus the invariance of (12.48) is assured. From an algebraic point of view it is possible to derive this equation by pursuing a different factorization of the wave equation

$$\partial_\mu \partial^\mu \psi + \kappa^2 \psi = 0$$

from the one which was first adopted by Dirac. A systematic treatment of this question of factorization of the wave equation can be found in Umezawas's book on quantum theory of fields. Here we only remark that the β-matrices of Duffin–Kemmer are not irreducible. With regard to the full Lorentz group they reduce into three irreducible parts of one, five, and ten dimensions respectively. Four-spinors of second rank yield a faithful representation of the Lorentz group and consequently the irreducible representation which they contain can refer only to particles of integral spin.

APPENDIX

1. General Remarks

In a way, the theory of relativity was the first to be confronted with a certain duality in the nature of light, which was later to play a fundamental role in the quantum theory of photons (light-quanta). On the one hand, light exhibits interference and diffraction properties and has a velocity which is independent of the motion of the source, all this pointing to a wave motion in some medium. On the other hand, the experimental evidence also indicates the capacity of light to travel in vacuum, in the same way as any body or particle which has been imparted some momentum and is left to itself. A full synthesis of these apparently contradictory wave and particle aspects is achieved in quantum theory. From a purely relativistic point of view, the difficulty which had to be explained was confined to the independence of the velocity of light on the motion of the source, once the property of light of travelling in vacuum and not in some hypothetical medium had been accepted. This state of affairs was created by the Michelson experiment (section 3 of Appendix). Let us assume that in a given frame x, y, z in which light always travels in straight lines irrespective of direction we make use of light signals emitted from standard sources at rest in x, y, z to synchronize clocks, in the same way as can be done with guns and bullets. Next let us consider a second frame x', y', z' which moves with constant velocity and along a straight line relative to x, y, z. From the fact that light can travel in vacuum it follows that both systems are physically equivalent as far as propagation of light is concerned, and we can synchronize clocks in x', y', z' in the same way as was done in x, y, z. Both in x, y, z and in x', y', z' we thus make use of the inherent isotropy of space and obtain a time definition which is in accord with it. Moreover, if identical clocks are used in both systems, we should also find that a standard source (at rest in each frame) emits light with the same velocity.

So far the 'particle' properties of light have appeared, i.e. among others, its ability to travel in vacuum. The difficulty arises when some of the wave properties are exhibited, namely when it is found that the velocity of light is independent of the nature of the source and its state of motion. We have then to speak of the velocity of light as such and not of the velocity of light emitted from some prescribed source under some prescribed conditions. As this is true in every inertial frame of reference this means that the velocity of a given signal of light has the same value no matter from which frame of reference it is measured. Thus, from the equivalence of all inertial frames of reference (principle of relativity) and from the independence of the velocity of light on the motion of the source (law of constancy of light velocity) it follows that the velocity of light has the same value in all inertial frames. We referred to this property as the invariance of light velocity.

The invariance of light velocity constitutes a grave difficulty from a classical point of view. If a Galilei transformation is applied, a velocity relative to K' is obtained from a velocity relative to K by adding to the latter the relative velocity between K and K'. It follows that a velocity along a given direction can never have the same value in two frames which move relative to each other along the considered direction. The invariance of light velocity and the principle of relativity (as conceived classically) thus appear incompatible. Einstein realized, however, that there can be no question of a conflict between such independent laws as the principle of relativity (or the equivalence of all inertial frames both in dynamics and optics or any other branch of physics) and the independence of the velocity of light on the motion of the source. (As we have seen from these two laws, the invariance of light velocity follows.) The position is rather such that the principle of relativity and the invariance of light velocity put us in a position to determine the interrelations which exist between space and time coordinates of moving systems without postulating certain properties which are very difficult to check with sufficient accuracy by direct experiments. In principle we could, of course, have obtained these relations from direct observations after having synchronized the clocks in each inertial frame as already described on several occasions. In principle we could also verify directly that whether we synchronize clocks mechanically (e.g. with bullets) or optically, we get the same result in each inertial frame. However, such direct observations, with present-day experimental technique, and with the small relative velocities at our disposal, cannot reveal the transformation formulae with the required accuracy. On the other hand, the law of constancy of light velocity and the applicability of the principle of relativity in optics can be verified experimentally with a high degree of accuracy. The relevant experimental evidence is discussed in the next two sections.

The invariance of the velocity of light provides us with the information which is required in order to derive the transformation formula for space and time coordinates. We have already seen that on the basis of the principle of relativity a general relation can be derived, in which an invariant velocity is figuring, but it is not possible to determine the value of the velocity on the basis of this principle alone, nor to tell whether it is finite or infinite. By identifying this velocity with the velocity of light, we thus obtain the relation which must exist between space and time coordinates of moving systems. The relation derived in this way, although not through direct observations, is nevertheless free from arbitrary assumptions since it is based on relevant measurements.

2. The Independence of the Velocity of Light on the Motion of the Source

This result can be derived in a convincing way from observations on spectroscopic double-stars. These stars have components which, although close to each other, possess sufficiently large orbital velocities to give rise to an observable Doppler effect with a characteristic time dependence. The latter is a function of the shape of the orbit and the velocity of the components.

For simplicity let us assume that the components S_1 and S_2 move in a circle about their common centre of gravity. At time t_1 let S_1 be at A and move in a direction towards the earth (Fig. 34). If S_1 imparts a velocity v to the light which it emits the velocity of the latter will be $c+v$. Light emitted from the position A would then reach us at a time

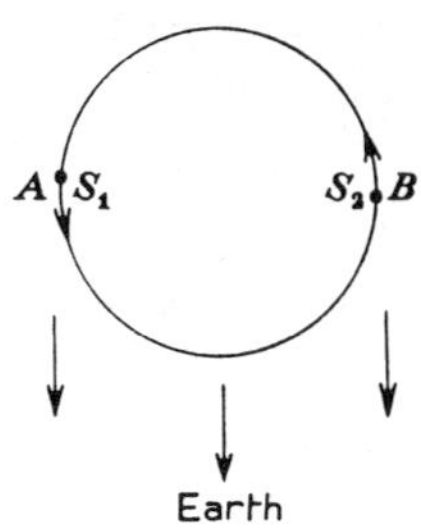

FIG. 34

$$t_1' = t_1 + \frac{\Delta}{c+v}, \tag{2.1}$$

where Δ is the distance of the earth from the double-star. Let T be the time it takes S_1 to reach the point B. Light emitted from this position would reach us at

$$t_1'' = t_1 + T + \frac{\Delta}{c-v}. \tag{2.2}$$

We have assumed again that the motion of the source affects the velocity of propagation, this time, since the source moves away from us, by subtracting the quantity v. From (2.1) and (2.2) it follows that the half period of S_1 will appear to us as having a duration

$$T_1' = t_1'' - t_1' = T + \frac{\Delta}{c-v} - \frac{\Delta}{c+v} = T + \tau, \tag{2.3}$$

where
$$\tau = \frac{\Delta}{c-v} - \frac{\Delta}{c+v} = \frac{2v\Delta}{c^2-v^2} \cong \frac{2v\Delta}{c^2}. \tag{2.4}$$

Let us now consider the component S_2. Suppose it is at B at a time t_2. This we shall observe at a time

$$t_2' = t_2 + \frac{\Delta}{c-v}. \tag{2.5}$$

On the other hand, the arrival of S_2 at A will be observed at a time

$$t_2'' = t_2 + T + \frac{\Delta}{c+v}. \tag{2.6}$$

Hence this half period will appear to us to have a duration

$$T_2' = t_2'' - t_2' = T + \frac{\Delta}{c+v} - \frac{\Delta}{c-v} = T - \tau. \tag{2.7}$$

It would thus turn out that for each component of the double-star one half period is $T+\tau$ and the other $T-\tau$. The difference 2τ is proportional to Δ, the distance between the double-star and the earth. de Sitter, who first called attention to this possible effect, pointed out that with many double stars τ would be of the order of T (assuming that v is equal to the velocity of the stars in their orbit). To explain then the actually observed time dependence of the Doppler effect (or the orbital velocity) we should have to assume that the motion of the double-star does not proceed according to Kepler's laws as they follow from Newton's theory.

It may be added that when Einstein published his first paper (1905) de Sitter's argument (1913) was not yet known. However, all experience with light as a wave propagation indicated that the velocity must depend on the properties of some medium and not on the motion of the source in the medium.

3. The Michelson Experiment

We shall now turn our attention to the second property of the velocity of light, its independence of the direction of propagation in each inertial frame. If this turns out to be the case, we can say that the law of emission of light satisfies the principle of relativity. This question is closely connected with that of the ether, a hypothetical medium which was thought to fill the whole of space and to carry the light waves. In a frame of reference which is at rest relative to the ether the propagation is isotropic. However, in a frame of reference which moves relative to the ether, the velocity will appear non-isotropic. From the observation of this effect it should then be possible to measure the velocity of the frame relative to the ether. On the basis of this argument, A. A. Michelson and E. W. Morley set themselves the task, in 1887, to measure the velocity of the earth relative to the ether.

Briefly, the experiment was carried out as follows. A Michelson interferometer was set on a turntable, which was floating on a pool of mercury and the interference fringes observed. Then the whole apparatus, interferometer and source, was rotated through 90°. The ether theory predicted that this rotation would cause a shift in the position of the fringes. It is this shift which was found to be non-existent.

Let us consider a ray of light which travels to and fro between two points P and M_1 a distance l_1 apart. P and M_1 are assumed to be points on a rigid body which travels with a velocity v relative to the ether. Let t' be the time it takes

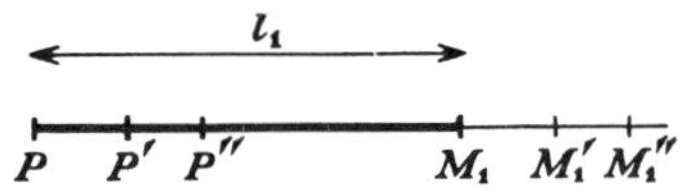

FIG. 35

the ray to get to M_1 and t'' the time it takes to get back to P after reflection at M_1. These events, reflection and return of the ray, occur in the ether at M_1' and P'' (see Fig. 35).

The following relations will then hold:

$$ct' = l_1+vt', \tag{3.1}$$

$$ct'' = l_1-vt'', \tag{3.2}$$

$$t' = \frac{l_1}{c-v}, \qquad t'' = \frac{l_1}{c+v}. \tag{3.3}$$

Hence

$$t_1 = t_1'+t_1'' = \frac{2l_1}{c}\frac{1}{1-\beta^2}, \tag{3.4}$$

where

$$\beta = \frac{v}{c}. \tag{3.5}$$

The same result can also be obtained as follows. Relative to the ether the velocity of light is c. Again, relative to the ether, the velocity of the body carrying the points P and M_1 is v. Therefore, as observed in the ether, the velocity of the ray relative to the body is $c-v$ before reflection and $c+v$ after reflection. Now if l is the distance between P and M_1 (as measured in the ether) the times it takes the ray to travel from P to M_1 and back are $l(c-v)^{-1}$ and $l(c+v)^{-1}$ in agreement with (3.3).

From (3.4) we see that the time t_1 differs from $2l_1/c$ which would have been the required time had the body been at rest relative to the ether.

Next let us consider a ray which travels almost perpendicularly to the velocity of the rigid body and is reflected by the mirror M_2. The velocity of the ray relative to the ether from P to M_2' is c and this is also the case after it has been reflected at M_2' to get to P'' (Fig. 36). Hence the time it takes the ray to get from P to M_2' is equal to the time it takes the ray to travel back. Let the total time be t_2, so that $PP'' = vt_2$. From the triangle $PM_2'P'$ we get

$$(\tfrac{1}{2}vt_2)^2+l_2^2 = (\tfrac{1}{2}ct_2)^2, \tag{3.6}$$

$$t_2 = \frac{2l_2}{c}\frac{1}{(1-\beta^2)^{\frac{1}{2}}}. \tag{3.7}$$

From this calculation it follows that if a ray arriving from S (see Fig. 37) is divided by a half-silvered plate P, into two rays, one travelling towards the mirror M_1 and the other towards the mirror M_2, the difference in times to get back to the half-silvered plate is given by

$$\tau_1 = t_1 - t_2 = \frac{2l_1}{c}\frac{1}{1-\beta^2} - \frac{2l_2}{c}\frac{1}{\sqrt{(1-\beta^2)}}. \tag{3.8}$$

As long as τ_1 is not too large the capacity of the partial beams to interfere with each other will be maintained and we shall be able to observe interference fringes.

In case $v = 0$, we get from (3.8)

$$\tau_0 = \frac{2l_1}{c} - \frac{2l_2}{c}. \tag{3.9}$$

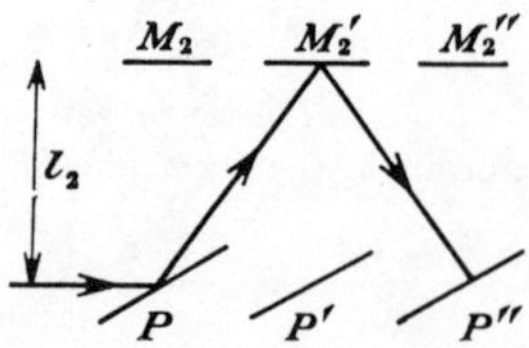

FIG. 36

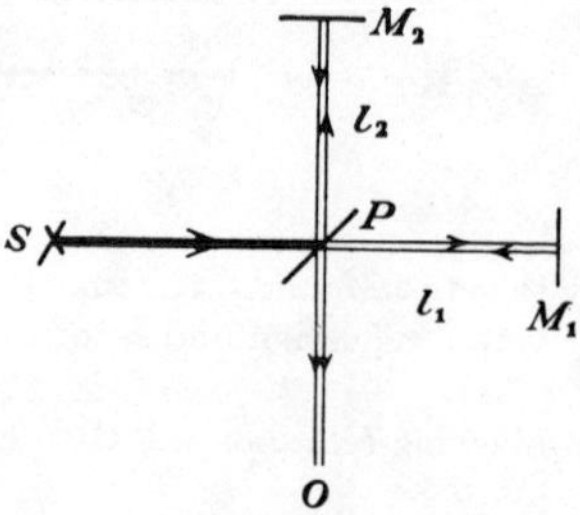

FIG. 37

Michelson's interferometer serves to measure wavelengths by observing the number of fringes which pass through the cross-wire of the eye-piece when either l_1 or l_2 is shortened by a known amount. If l_1 is changed the circular fringes close in, and if l_2 is changed (causing a negative time delay, see (3.9)) they open up (or vice versa, depending on the relative position between source and eye). Suppose we change l_1 by $\frac{1}{2}\lambda_0$, which exactly corresponds to a shift of one fringe. This will correspond to a delay which is equal to T, where T^{-1} is the frequency of the light used. When (3.9) is applicable we have

$$\tau_0 + T = \frac{2(l_1+\frac{1}{2}\lambda_0)}{c} - \frac{2l_2}{c}. \tag{3.10}$$

From (3.9) and (3.10) we get

$$cT = \lambda_0. \tag{3.11}$$

cT is the wavelength of the light when the source is at rest in the ether. The same result is obtained if we change l_2 by $-\frac{1}{2}\lambda_0$.

Matters are, however, different when (3.8) applies, i.e. when the whole apparatus moves in the ether. In this case we get different results when we alter l_1 and l_2 respectively. We have to alter these arms by different amounts to get a shift of one fringe, which in every case corresponds to T. Eqn. (3.8) may be written to a sufficient degree of accuracy as follows:

$$\tau_1 = \frac{2l_1}{c}(1+\beta^2) - \frac{2l_2}{c}(1+\tfrac{1}{2}\beta^2). \tag{3.12}$$

Therefore, if l_1 is altered we have to change the length by an amount $\frac{1}{2}\lambda_1$, where

$$cT = \lambda_1(1+\beta^2), \tag{3.13}$$

whereas to get a shift of one fringe by altering l_2 we have to alter the length by $\frac{1}{2}\lambda_2$, where

$$cT = \lambda_2(1+\tfrac{1}{2}\beta^2). \tag{3.14}$$

Thus we shall get a different wavelength according to whether we change one arm of the interferometer or the other. Equating (3.13) and (3.14) we have

$$\frac{\lambda_1}{\lambda_2} = \frac{1+\frac{1}{2}\beta^2}{1+\beta^2} \cong 1-\tfrac{1}{2}\beta^2. \tag{3.15}$$

This equation should enable us to calculate v/c from observations of λ_1 and λ_2. For the earth which moves in its orbit with a velocity of about 30 km/sec we may expect v/c to be at least of the order 10^{-4} and β^2 of the order 10^{-8}. The measurements of such a small effect naturally puts high demands on skill and precautions. At best we can shorten one of the arms, say l_1, by a length of l which is nearly equal to l_1. We shall then observe a certain number n_1 of fringes passing by, where

$$cTn_1 = l(1+\beta^2). \tag{3.16}$$

Next, we also shorten l_2 by the same amount l and compare the corresponding number n_2 with n_1. For n_2 we have

$$cTn_2 = l(1+\tfrac{1}{2}\beta^2). \tag{3.17}$$

Hence,

$$\frac{n_1}{n_2} = \frac{1+\beta^2}{1+\frac{1}{2}\beta^2} \cong 1+\tfrac{1}{2}\beta^2. \tag{3.18}$$

Instead of shortening the arms, Michelson and Morley adopted a different and more useful method. As already mentioned they had the apparatus floating on mercury so that it was easily rotatable. When the apparatus is rotated through 90 degrees we have to exchange l_1 and l_2 in (3.8) and reverse the order of subtraction. This will yield the new time delay τ_2,

$$\tau_2 = -\frac{2l_2}{c}\frac{1}{1-\beta^2}+\frac{2l_1}{c}\frac{1}{\sqrt{(1-\beta^2)}}. \tag{3.19}$$

If τ_2 differs from τ_1, the rotation will show a shift of fringes. The difference between τ_1 and τ_2 is given by

$$\tau = \tau_1-\tau_2 = \frac{2l_1}{c}\left(\frac{1}{1-\beta^2}-\frac{1}{\sqrt{(1-\beta^2)}}\right)+\frac{2l_2}{c}\left(\frac{1}{1-\beta^2}-\frac{1}{\sqrt{(1-\beta^2)}}\right)$$

$$\cong \frac{2l_1}{c}\tfrac{1}{2}\beta^2+\frac{2l_2}{c}\tfrac{1}{2}\beta^2 = \frac{l_1+l_2}{c}\beta^2. \tag{3.20}$$

The method adopted by Michelson and Morley resulted in a simplification of the apparatus. In addition, by using several mirrors and multiple reflections they further increased its range. The expected shift amounted to 0·4 of a fringe spacing and they were in a position of observing it to an accuracy of 1 per cent. However, no shift was found. The negative result obtained by Michelson and Morley proves that the ether hypothesis must be abandoned. The most natural view to adopt as a result of this experiment is to assume that light travels relative to the apparatus with the same velocity in all directions, as if the apparatus was at rest relative to the ether. If this is the case we have to use (3.9) to calculate the phase delay between the two beams. On rotating the apparatus the phase shift is given by

$$\tau_0' = -\frac{2l_2}{c}+\frac{2l_1}{c} = \tau_0. \tag{3.21}$$

Consequently no change should take place when the apparatus is rotated.

4. Plane Electromagnetic Waves. Doppler Effect and Aberration. de Broglie Waves

Maxwell's equations *in vacuo* (see (2.4) and (2.6), pp. 95 and 96)

$$\frac{\partial F_{\kappa\lambda}}{\partial x_\lambda} = 0, \qquad \frac{\partial G_{\kappa\lambda}}{\partial x_\lambda} = 0 \tag{4.1}$$

allow a solution of the following form:

$$F_{\kappa\lambda} = F^0_{\kappa\lambda} e^{ik_\mu x_\mu}, \qquad G_{\kappa\lambda} = G^0_{\kappa\lambda} e^{ik_\mu x_\mu}, \tag{4.2}$$

where $F^0_{\kappa\lambda}$ and $G^0_{\kappa\lambda}$ are constant tensors, i.e. they are independent of x, y, z, t. It follows that the exponential factor is necessarily a scalar and hence k_μ is a four-vector. If (4.2) is to form a solution of Maxwell's equations we must assume k_μ to be independent of x, y, z, t. We get from (4.1) and (4.2) the following relations for the amplitudes $F^0_{\kappa\lambda}$ and $G^0_{\kappa\lambda}$ and the vector k_μ:

$$F^0_{\kappa\lambda} k_\lambda = 0, \qquad G^0_{\kappa\lambda} k_\lambda = 0. \tag{4.3}$$

They represent eight relations for the six components $F^0_{\kappa\lambda}$ since the $G^0_{\kappa\lambda}$ are the duals of $F_{\kappa\lambda}$. Of these relations, two express the orthogonality of **E** and **H** to **k** which is the space part of the four-vector k_μ ($k_\mu = (\mathbf{k}, ik_0)$). The equations in question are those which pertain to $\kappa = 4$:

$$F^0_{4\lambda} k_\lambda = 0, \qquad G^0_{4\lambda} k_\lambda = 0. \tag{4.4}$$

These equations are equivalent to (see pp. 95 and 96):

$$\begin{aligned} E_x k_x + E_y k_y + E_z k_z &\equiv \mathbf{E}.\mathbf{k} = 0, \\ H_x k_x + H_y k_y + H_z k_z &\equiv \mathbf{H}.\mathbf{k} = 0. \end{aligned} \tag{4.5}$$

The remaining six equations in (4.3), which are homogeneous in the six $F^0_{\kappa\lambda}$ (they still contain all these components), can be solved only if the determinant vanishes and this imposes a condition on the four-vector k_μ. It will be convenient to introduce now a system of coordinates in which $k_y = k_z = 0$. Then from (4.5) it follows that

$$E_x = H_x = 0 \quad (\text{for } k_y = k_z = 0) \tag{4.6}$$

and (4.3) yields apart from (4.6):

$$\begin{aligned} -k_x H_z + k_0 E_y &= 0, \qquad & k_x E_z + k_0 H_y &= 0, \\ k_x H_y + k_0 E_z &= 0, \qquad & -k_x E_y + k_0 H_z &= 0. \end{aligned} \tag{4.7}$$

These equations may be regrouped into the following two pairs:

$$\begin{aligned} -k_x H_z + k_0 E_y &= 0, \qquad & k_x E_z + k_0 H_y &= 0, \\ k_0 H_z - k_x E_y &= 0, \qquad & k_0 E_z + k_x H_y &= 0. \end{aligned} \tag{4.8}$$

Both these pairs of equations lead to

$$k_x^2 - k_0^2 = 0. \tag{4.9}$$

We can now disengage ourselves from the condition $k_y = k_z = 0$ and generalize (4.9) into

$$k_x^2 + k_y^2 + k_z^2 - k_0^2 = k^2 - k_0^2 = 0, \tag{4.10}$$

or

$$k_\mu^2 = 0. \tag{4.11}$$

Such a condition is invariant under Lorentz transformations. We have to distinguish between two cases

$$\text{(i)} \quad k_0 = k = |\mathbf{k}|, \qquad \text{(ii)} \quad k_0 = -k. \tag{4.12}$$

They yield

$$F_{\kappa\lambda} = F^0_{\kappa\lambda} e^{i(k_x x + k_y y + k_z z \mp kct)}. \tag{4.13}$$

In the first case we get plane waves which travel in the direction of **k**, in the second case they travel in the opposite direction. By selecting the first of the alternatives in (4.12) we secure that the waves shall always travel in the direction of **k**. This vector is called the propagation vector.

(4.11) expresses the fact that electromagnetic waves travel with the velocity of light. In the four-dimensional Minkowski world (x, y, z, ct) they travel along the light cones. k_μ is a null-vector.

Let
$$\omega = k_0 c = kc = |\mathbf{k}|c. \tag{4.14}$$
(4.13) can then be written
$$F_{\kappa\lambda} = F^0_{\kappa\lambda}\, e^{i(\mathbf{k}.\mathbf{r}-\omega t)}, \tag{4.15}$$
where **r** is the radius vector (x, y, z). The phase ϕ of the wave is given by
$$\phi = \mathbf{k}.\mathbf{r}-\omega t. \tag{4.16}$$
It is useful to introduce the concept of phase velocity v_p. This is defined as the velocity of an observer who moves relative to the wave along a straight line in such a way as to see constantly the same phase. Hence, the phase velocity is obtained from the condition
$$d\phi = 0. \tag{4.17}$$
Since **k** is a constant vector, (4.16) and (4.17) yield
$$\mathbf{k}.d\mathbf{r} = \omega\, dt. \tag{4.18}$$
If we move along a radius vector **r** which makes an angle α with the propagation vector, we have $k\,dr\cos\alpha = \omega\,dt$ so that
$$v_p = dr/dt = \omega/k\cos\alpha = c/\cos\alpha, \tag{4.19}$$
where (4.14) has been used. Hence, if we move along a radius vector which is parallel to the propagation vector ($\alpha = 0$) we get
$$v_p = dr/dt = \omega/k = c. \tag{4.20}$$
In this case the phase velocity is equal to that of light. According to (4.19) if we move along a direction which is different from that of **k**, we must move with a velocity which is greater than c, if we wish to see the same phase all along our way. We could, of course, never move with a velocity greater than c, and to observe the same phase, the best we could do is to move with a velocity c parallel to the propagation vector. This direction is perpendicular to the planes of equal phase. The equation which determines such a plane at any time t = const and for a given phase ϕ = const is
$$\phi+\omega t = \mathbf{k}.\mathbf{r} = \text{const}. \tag{4.21}$$

The phase velocity must be distinguished from the group-velocity. This is obtained if we consider a group of waves, containing a range of angular frequencies ω, all travelling in the same direction and constituting a pulse of electromagnetic waves. Electromagnetic energy is transmitted with the group velocity, although in this particular case which we are considering now the group velocity v_g is equal to the phase velocity c. It is well known from optics that the group velocity is given by
$$v_g = \frac{d\omega}{dk}. \tag{4.22}$$

The group velocity will always differ from the phase velocity when there is dispersion, i.e. when the phase velocity depends on ω. In the case of electromagnetic waves in free space there is no dispersion (see (4.20)) and (4.14) yields
$$v_g = c. \tag{4.23}$$

That phase velocities of waves in general can be greater than c presents no difficulty to the principle that light velocity is the greatest possible velocity.

This statement refers only to the velocity of matter or of transmission of energy. The phase of a harmonic motion is neither material nor related to energy. In the case of electromagnetic waves in free space, if we consider the phase velocity along a direction which is different from that of **k**, it is greater than c. However, electromagnetic energy travels along **k** and no energy travels in a direction perpendicular to it. In wave guides electromagnetic waves are propagated in such a way that the phase velocity along the tube is greater than c. The group velocity is, however, smaller than c. In the case of wave guides the phase velocity is frequency-dependent.†

The relations (4.5) and (4.11) do not exhaust all the properties of a plane electromagnetic wave. From (4.7) looked upon as equations for k_x and k_0, we find that a solution is possible only if

$$E_y H_y + E_z H_z = 0. \tag{4.24}$$

Taking into account (4.6) we get

$$\mathbf{E}.\mathbf{H} = 0. \tag{4.25}$$

This relation will also hold in any other system of coordinates, because a rotation of axes so as to make $E_x = H_x = 0$ cannot affect the scalar product of two vectors (we know already that both these components can be made to vanish). Apart from this from (4.4), p. 98, we know already that $\mathbf{E}.\mathbf{H}$ is a relativistic invariant, so that if it vanishes in one Lorentz frame it vanishes in any other frame as well, which may differ not only in orientation but may also be moving with some velocity relative to the first frame. Since (4.5) was obtained for an arbitrary frame, we come to the conclusion that in a plane electromagnetic wave the three vectors **k**, **E**, **H** are orthogonal to each other in every Lorentz frame.

Finally, from (4.6) and (4.8), we get

$$E_x^2 + E_y^2 + E_z^2 = H_x^2 + H_y^2 + H_z^2. \tag{4.26}$$

This relation also is independent of the particular Lorentz frame used, because we know already from (4.3), p. 98, that $\mathbf{E}^2 - \mathbf{H}^2$ is an invariant. For plane waves this invariant vanishes. From (4.26) we thus get generally, for any plane wave

$$|\mathbf{E}| = |\mathbf{H}|. \tag{4.27}$$

We shall now turn our attention to the problem of transforming a given plane wave in a frame K into an electromagnetic wave in a frame K'. It is evident from (4.2) that in K' we shall get again a plane wave

$$F'_{\kappa\lambda} = F'^{0}_{\kappa\lambda}\, e^{i(k'_\mu x'_\mu)} \tag{4.28}$$

and because of the invariance of the relations (4.3) all the characteristic properties of a plane wave will be confirmed in K'. (k'_μ is a null vector, $\mathbf{k}'$, $\mathbf{E}'$, $\mathbf{H}'$ are mutually orthogonal, $|\mathbf{E}'| = |\mathbf{H}'|$.)

For simplicity let us consider two parallel frames K and K' which move along the x-axis with relative velocity v and let us begin with the null-vector k_μ. Let the wave be propagated in K with a propagation vector

$$\mathbf{k} = (k\cos\alpha,\, k\cos\beta,\, k\cos\gamma) = ((\omega/c)\cos\alpha,\, (\omega/c)\cos\beta,\, (\omega/c)\cos\gamma), \tag{4.29}$$

so that

$$ck_\mu = (\omega l,\, \omega m,\, \omega n,\, i\omega), \tag{4.30}$$

† As a further illustration, consider a series of pendulums a distance s apart swinging independently with the same angular frequency ω, but with a phase delay ϕ from one pendulum to the next. The motion as a whole has the appearance of a wave. To see the same phase we have to move along the lines of pendulums with a velocity $s\omega/\phi$. By making ϕ sufficiently small, we can achieve phase velocities which are bigger than any given value. No energy is transmitted in this case.

where l, m, and n are the direction-cosines $\cos\alpha$, $\cos\beta$, and $\cos\gamma$ respectively. Like k_μ, ck_μ is also a four-vector. On performing a four-vector transformation we get for ck'_μ:

$$\begin{aligned} \omega' l' &= \gamma\omega l + (\gamma v/ic)i\omega = \gamma\omega(l+v/c), \\ \omega' m' &= \omega m, \\ \omega' n' &= \omega n, \\ i\omega' &= (i\gamma v/c)\omega l + \gamma(i\omega) = i\gamma\omega(lv/c+1). \end{aligned} \tag{4.31}$$

From the fourth of these equations we get

$$\omega'/\omega = \gamma(lv/c+1), \tag{4.32}$$

and this yields for l', m', n'

$$l' = \frac{l+v/c}{lv/c+1}, \qquad m' = \frac{m}{\gamma(lv/c+1)}, \qquad n' = \frac{n}{\gamma(lv/c+1)}. \tag{4.33}$$

As direction-cosines of $\mathbf{k}$ relative to K, l, m, and n satisfy the relation

$$l^2+m^2+n^2 = 1. \tag{4.34}$$

Using (4.33) and (4.34) we find that l', m', n' are direction-cosines of $\mathbf{k}'$ relative to K because we find that

$$l'^2+m'^2+n'^2 = 1. \tag{4.35}$$

Eqn. (4.32) expresses the relativistic Doppler effect. For $l = 1$ ($\alpha = 90°$ and as a consequence $m = n = 0$) we have a plane wave with a propagation vector parallel to the x-axis. In this case

$$\omega'/\omega = \gamma(1+v/c) = \frac{(1+v/c)^{\frac{1}{2}}}{(1-v/c)^{\frac{1}{2}}}. \tag{4.36}$$

In our present case K' moves relative to K in a direction which is opposite to that of $\mathbf{k}$ so that $\omega' > \omega$. To the first order in v/c we confirm the non-relativistic formula

$$\omega'/\omega \approx 1+v/c. \tag{4.37}$$

The relativistic formula (4.36) displays the following symmetry: if we exchange ω' with ω and change v into $-v$ it yields a result which is consistent with (4.36). This is not the case with the classical formula. Here we get a difference of the order v^2/c^2. However this does not apply to (4.32). Since it contains l, it is not a relation between ω and ω' only.†

The relativistic formula (4.32) contains also a transverse Doppler effect. For $l = 0$, $\mathbf{k}$ is perpendicular to the x-axis and we get

$$\omega'/\omega = \gamma. \tag{4.38}$$

This effect is independent of the sign of v.

The transformation formulae (4.33) for the direction-cosines contain all the relevant relations with regard to the aberration of light. Let $\mathbf{k}$ lie in the x, z plane, so that $m = 0$, $l = \cos\alpha$, $n = \sin\alpha$. (4.33) leads to

$$\cot\alpha' = \gamma(\cos\alpha+v/c)/\sin\alpha, \qquad \cot\alpha = \gamma(\cos\alpha'-v/c)/\sin\alpha', \tag{4.39}$$

where the second formula has been obtained from the first by exchanging α and α' and reversing the sign of v.‡ From these equations we get after some manipulation

$$2\sin\tfrac{1}{2}(\alpha-\alpha') = \frac{\gamma v\sin\alpha\sin\alpha'}{c\sin\tfrac{1}{2}(\alpha+\alpha')}. \tag{4.40}$$

† The principle of relativity demands in general only that the whole complex of equations shall display this symmetry.

‡ This is possible in this case because the formulae (4.33) for l', n' involve only l, n. The relations are symmetrical in this case provided v is changed into $-v$.

The difference $\Delta\alpha = \alpha - \alpha'$ is called the aberration. If this is small ($\alpha \approx \alpha'$) (4.40) gives

$$\Delta\alpha = \gamma v \sin\alpha/c. \tag{4.41}$$

It follows that if a star is observed continuously through a telescope, as a result of the quasi-circular motion of the earth around the sun and the aberration, the telescope must be continuously reoriented. In the course of a year the telescope generates a cone with an opening which corresponds to an angle $2\Delta\alpha$. This amounts to an apparent circular motion of each star which is a replica of our own motion around the sun (the circular motion is superimposed on any relative motion between the stars and the sun). The first to observe this effect was J. Bradley in 1727. For small velocities, $\gamma \approx 1$, the formula (4.41) becomes identical with the one obtained classically either through the emission theory of light or through the ether theory provided it is assumed that Fresnel's theory of drag through optical media holds. According to this theory the aberration is independent of the properties of the media filling the telescope in accordance with observations made by Airy who filled a telescope with water and found that it had no effect on the aberration. According to the theory of relativity the aberration is of purely kinematical nature and the properties of the telescope do not enter into the explanation of the effect. The telescope must be directed along the propagation vector both in K and K' and the transformation of $\mathbf{k}$ determines the aberration. The aberration due to the earth's motion around the sun is too small to enable detection of the difference of order v^2/c^2 between the relativistic and classical formulae.

Let us also consider the formula (4.22) for the group velocity and see whether as a result of $(\mathbf{k}, i\omega/c)$ being a four-vector v_g transforms correctly as a velocity. Let v_g' be the group velocity with respect to K': $v_g' = d\omega'/dk'$. Let us also assume for simplicity that the wave travels along the x-axis

$$(l = l' = 1,\ m = m' = n = n' = 0).$$

(4.31) can be rewritten as follows (we need consider only the first and fourth equations):

$$\begin{aligned} k_x' &= \gamma k_x + \gamma v\omega/c^2, \\ \omega' &= \gamma v k_x + \gamma\omega. \end{aligned} \tag{4.42}$$

This yields

$$\begin{aligned} dk_x' &= \gamma\, dk_x + \gamma v\, d\omega/c^2, \\ d\omega' &= \gamma v\, dk_x + \gamma\, d\omega. \end{aligned} \tag{4.43}$$

Consequently

$$v_g' = d\omega'/dk' = \frac{v + d\omega/dk_x}{1 + v\, d\omega/c^2\, dk_x} = \frac{v + v_g}{1 + vv_g/c^2}. \tag{4.44}$$

A similar result also holds for the phase velocity $v_p = \omega/k$. In this calculation we have only assumed that $(\mathbf{k}, i\omega/c)$ is a four-vector but did not assume that it is a null-vector. We can say quite generally that if the fourth component of a four-vector is divided by the length of the space vector $(\mathbf{k})$ we get a dimensionless quantity which on multiplication with c transforms like a velocity. It is essential to multiply the quotient by c and not by any other velocity.

Next let us consider the intensity of the wave in K' and compare it with the intensity in K. The intensity is given by Poynting's vector (we are considering intensity of radiation per unit area). Again we assume the wave to travel along the x-axis.† In K' we have

$$P_x' = c[\mathbf{E}', \mathbf{H}']_x = c(E_y' H_z' - E_z' H_y'). \tag{4.45}$$

† In this case a cross-section transforms like a scalar $yz = y'z'$. Hence if a beam of light has a unit cross-section in K it has also a unit cross-section relative to K'.

Using the formula (3.1), p. 97, we get

$$P'_x = c\gamma^2(E_y+(v/c)H_z)(H_z+(v/c)E_y)-c\gamma^2(E_z-(v/c)H_y)(H_y-(v/c)E_z). \quad (4.46)$$

Now from (4.7) with $k_x = k_0$ (see (4.12), where we use the first alternative) we get

$$E_y = H_z, \qquad E_z = -H_y. \quad (4.47)$$

Hence, (4.46) gives

$$P'_x = c\gamma^2(1+v/c)^2E_y H_z - c\gamma^2(1+v/c)^2E_z H_y = \gamma^2(1+v/c)^2P_x. \quad (4.48)$$

Therefore
$$P'_x = \frac{1+v/c}{1-v/c}P_x. \quad (4.49)$$

It is interesting to compare also the amount of energy flowing per cycle in K with the corresponding quantity in K'. This will also yield a comparison of the energies of a finite wave train as measured in K and K'. We must use now real parts of the complex quantities (4.2) and consider the integrals

$$\int\limits_0^{2\pi/\omega} \cos^2\omega t\, dt = \pi/\omega \quad \text{and} \quad \int\limits_0^{2\pi/\omega'} \cos^2\omega' t'\, dt' = \pi/\omega'. \quad (4.50)$$

The energies per cycle will be given by

$$E = (\pi/\omega)P_x^0 \quad \text{and} \quad E' = (\pi/\omega')P_x'^0, \quad (4.51)$$

where
$$P_x^0 = [\mathbf{E}^0, \mathbf{H}^0]_x \quad \text{and} \quad P_x'^0 = [\mathbf{E}'^0, \mathbf{H}'^0]_x. \quad (4.52)$$

We thus get
$$E'/E = \omega P_x'^0/\omega' P_x^0. \quad (4.53)$$

Therefore, using (4.36) and (4.49) we obtain

$$\frac{E'}{E} = \frac{(1-v/c)^{\frac{1}{2}}(1+v/c)}{(1+v/c)^{\frac{1}{2}}(1-v/c)} = \frac{(1+v/c)^{\frac{1}{2}}}{(1-v/c)^{\frac{1}{2}}}. \quad (4.54)$$

Thus we find
$$E'/E = \omega'/\omega. \quad (4.55)$$

It is important to realize that to obtain this result (and thus arrive at a comparison between total energy in a wave as observed in K and K' respectively) we have to calculate $P_x \Delta t$, where Δt is the duration of the pulse in K, and $P'_x\, \Delta t'$, where $\Delta t'$ is the duration of the pulse in K'. Δt and $\Delta t'$ are, however, not corresponding time elements which are related to the same events in K and K'. Let us assume that at $t = t' = 0$ the wave train just begins to enter the planes $x = 0$ and $x' = 0$. After $\Delta t'$ seconds the end of the wave train will pass the plane $x' = 0$, but at this time, which will appear as of duration $\gamma\Delta t'$ in K, the end of the wave train will not have reached yet the plane $x = 0$. Hence K and K' refer to different events when they observe Δt and $\Delta t'$. For this reason we have evaluated energies per cycle in each frame. The total number of cycles is an invariant: it has the same value for both observers. All this may also serve to explain why the period of a wave in K ($T = 2\pi/\omega$) does not relate to the period of this wave in K' ($T' = 2\pi/\omega'$) in the same way as a period dt of a clock in K to the duration dt' of this period as observed in K'. Here both K and K' refer to the same pair of events: two positions of the hand of a given clock.

The relation (4.55) shows that Planck's formula $E = h\nu$ or $E = h\omega/2\pi$, where h is a universal constant (a scalar) is relativistically invariant.

A similar result is obtained if we consider the amount of momentum contained in a volume (of unit cross-section) of length λ, where λ is the wavelength

$$\lambda = 2\pi/k = 2\pi c/\omega = c/\nu. \quad (4.56)$$

Now we must consider the integrals

$$\int_0^\lambda \cos^2 kx\,dx = \pi/k \quad \text{and} \quad \int_0^{\lambda'} \cos^2 k'x'\,dx' = \pi/k'. \tag{4.57}$$

The density of electromagnetic momentum G_x is proportional to Poynting's vector ($= c^{-2}P_x$); consequently the total momentum M_x per wavelength is given by

$$M_x = G_x^0\pi/k = c^{-2}P_x^0\pi/k = c^{-1}P_x^0\pi/\omega = c^{-1}E \tag{4.58}$$

and we get

$$M_x'/M_x = E'/E. \tag{4.59}$$

What has been said above with regard to the period $T = 2\pi/\omega$ applies in a similar way to $\lambda = 2\pi/k$. From $k = \omega/c$ we have

$$k'/k = \omega'/\omega = \lambda/\lambda'. \tag{4.60}$$

Wavelength does not transform as the length of a rod.

From (4.58) we find an invariant relation between total energy and total momentum of a wave. If $E = h\nu$, we get $M_x = h\nu/c$. This result enabled Einstein to develop Planck's hypothesis with regard to emission and absorption of energy $h\nu$, into a corpuscular theory of light in which the energy E of a photon is necessarily associated with a momentum E/c.

In this connexion it is also of interest to recall the essential result of de Broglie's argument in attributing wave properties to particles. On the one hand we consider a wave, the wavelength and frequency of which are determined by k_μ,

$$k_\mu = 2\pi(\cos\alpha/\lambda,\ \cos\beta/\lambda,\ \cos\gamma/\lambda,\ i\nu/c), \tag{4.61}$$

where, however, we are not considering now an electromagnetic wave, for which $k_\mu^2 = 0$, but any wave for which $k_\mu^2 \neq 0$. On the other hand, if we consider the energy-momentum vector of a particle, we have

$$P_\mu = (P_x, P_y, P_z, iW/c). \tag{4.62}$$

Consequently, a relativistically invariant relation $W = h\nu$ necessitates a corresponding relation between P_x, P_y, P_z, and $\lambda^{-1}(\cos\alpha, \cos\beta, \cos\gamma)$. It yields

$$|\mathbf{P}| = h/\lambda. \tag{4.63}$$

Assuming that the particle has a mass m_0 and moves with a velocity v, we get

$$\frac{m_0 v}{(1-v^2/c^2)^{\frac{1}{2}}} = \frac{h}{\lambda}, \qquad \frac{m_0 c^2}{(1-v^2/c^2)^{\frac{1}{2}}} = h\nu. \tag{4.64}$$

For the phase velocity we get

$$v_p \equiv \nu\lambda = c^2/v. \tag{4.65}$$

Since $v < c$, the phase velocity of the de Broglie waves is always greater than c. To calculate the group velocity we must evaluate $d\omega/dk$ or $d\nu/d(\lambda^{-1})$. It will be more convenient to evaluate v_g^{-1}. From (4.64) we have

$$h/\lambda = m_0 c(\gamma^2-1)^{\frac{1}{2}}, \qquad \gamma = h\nu/m_0 c^2. \tag{4.66}$$

Consequently

$$-\frac{d\lambda}{\lambda^2} = \left(\frac{h^2\nu^2}{m_0 c^4}-1\right)^{-\frac{1}{2}} \frac{h\nu\,d\nu}{m_0 c^3} = (\gamma^2-1)^{-\frac{1}{2}}\gamma\,d\nu/c \tag{4.67}$$

and

$$v_g^{-1} = -\frac{1}{\lambda^2}\frac{d\lambda}{d\nu} = (\gamma^2-1)^{-\frac{1}{2}}\gamma/c = v^{-1}. \tag{4.68}$$

Therefore

$$v_g = v. \tag{4.69}$$

This result expresses the most attractive feature of de Broglie's original ideas.

Another way of deriving (4.69) is to make use of the relation between energy and momentum of a particle (see (2.13), p. 152):

$$W^2 = P^2c^2 + m_0^2 c^4. \tag{4.70}$$

From the proportionality between energy and frequency on the one hand and momentum and wave-number λ^{-1} on the other hand, where the proportionality factor is Planck's constant in both cases, we have

$$v_g = \frac{d\nu}{d(\lambda^{-1})} = \frac{dW}{dP}. \tag{4.71}$$

On using (4.70) this yields

$$v_g = c^2P/W. \tag{4.72}$$

Since $W = m_0 c^2\gamma$ and $P = m_0 v\gamma$, we get (4.69).

Both for electromagnetic waves as well as for de Broglie waves we get the following relation between phase velocity and group velocity:

$$v_p v_g = c^2. \tag{4.73}$$

This also applies to electromagnetic waves which travel in wave guides made of perfect conductors. That this should be so is satisfactory since in all these cases we are dealing with waves in vacuum with its characteristic velocity c.

The relation (4.73) will hold for any particle irrespective of its nature (i.e. its spin, mass, charge, and parity). In general the waves will have tensor or spinor character. The relation

$$P_\mu^2 = -m_0^2 c^2 \tag{4.74}$$

will yield on changing P_μ into $i^{-1}\hbar\partial/\partial x_\mu$, the wave equation

$$\Box\psi_{\kappa\lambda\ldots} = m_0^2 c^2/\hbar^2\psi_{\kappa\lambda\ldots}. \tag{4.75}$$

To obtain particles of definite spin we shall need additional conditions as was illustrated in the case of the Proca field.

REFERENCES

AHARONI, J., 'Spin Functionals', *Nature*, **181** (1958),1385–6.

—— *Antennae*, at the Clarendon Press, Oxford, 1946.

BADE, W. L., and JEHLE, H., 'An Introduction to Spinors', *Rev. Mod. Phys.* **25** (1953), 714.

CORSON, E. M., *Introduction to Tensors, Spinors and Relativistic Wave Equations*, Blackie & Son Ltd., 1953.

COSTA DE BEAUREGARD, O., *La Théorie de la relativité restreinte*, Mason & Co., 1949.

DANCOFF, S., and INGLIS, D. R., 'On the Thomas Precession of Accelerated Axes', *Phys. Rev.* **50** (1936), 784.

DIRAC, P. A. M., 'Relativistic Quantum-mechanics', *Proc. Roy. Soc.* A, **136** (1932), 453.

—— 'Unitary Representations of the Lorentz Group', *Proc. Roy. Soc.* A, **183** (1945), 284.

—— 'Classical Theory of Radiating Electrons', *Proc. Roy. Soc.* A, **167** (1938), 148.

EINSTEIN, A., 'Zur Elektrodynamik bewegter Körper', *Ann. d. Physik*, **17** (1905).

—— Princeton lectures, 1921.

FERMI, E., 'Über einen Wiederspruch zwischen der elektrodynamischen und der relativistischen Theorie der elektrodynamischen Masse', *Phys. Zeitschrift*, **23** (1922), 340.

HARISH-CHANDRA, 'Infinite Irreducible Representations of the Lorentz Group', *Proc. Roy. Soc.* A, **372** (1947).

HILGEVOORD, J., and WOUTHUYSEN, S. A., 'On the Spin Angular Momentum of the Dirac Particle', *Nuclear Phys.* **40** (1963), 1.

HOGARTH, J. E., 'Cosmological Considerations of the Absorber Theory of Radiation', *Proc. Roy. Soc.* A, **267** (1962), 365.

HOYLE, F., and NARLIKAR, J. V., 'Time Symmetrical Electrodynamics and the Arrow of Time in Cosmology', *Proc. Roy. Soc.* A, **277** (1964), 1.

IWANENKO, D., and SOKOLOW, A., *Klassische Feldtheorie*, Akademie-Verlag, Berlin.

JAUCH, J. M., and ROHRLICH, F., *The Theory of Photons and Electrons*, Addison-Wesley Publ. Co., 1955.

KEMMER, N., 'The Particle Aspect of Meson Theory', *Proc. Roy. Soc.* A, **173** (1939), 91.

KWAL, B., 'Les expressions de l'énergie et de l'impulsion du champ électromagnétique propre de l'électron en mouvement', *J. Phys. et Rad.* **10** (1949), 103.

LANDAU, L., and LIFSHITZ, E., *The Classical Theory of Fields*, Addison-Wesley Publ. Co., 1951.

LAPORTE, O,. and UHLENBECK, G., 'Application of Spinor Analysis to the Maxwell and Dirac Equations', *Phys. Rev.* **37** (1931), 1380.

MCLENNAN, J. A., Jr., 'Parity Nonconservation and the Theory of the Neutrino', *Phys. Rev.* **106** (1957), 821.

MØLLER, C., *The Theory of Relativity*, Clarendon Press, 1952.

PAULI, W., 'Relativitätstheorie', *Encyklopädie der mathematischen Wissenschaften*, V, 2, 1921. (English translation, Pergamon Press, 1958.)

ROHRLICH, F., 'The Classical Electron, Lectures in Theoretical Physics', *Boulder* 1959, vol. 2 (Interscience, 1960).

ROHRLICH, F., 'Self-energy and Stability of the Classical Electron', *Am. J. of Phys.* **28** (1960), 639.

RZEWUSKI, J., *Feldtheorie*, Part 1. Polish Academy of Sciences, Physical Monographs. Warszawa, 1958.

SYNGE, J. L., *Relativity; the Special Theory*, North-Holland Publ. Co., 1956.

UMEZAWA, H., *Quantum Field Theory*, North-Holland Publ. Co., 1956.

WEYSSENHOFF, J., and RAABE, A., 'Relativistic Dynamics of Spin Fluids and Spin-particles', *Acta Phys. Polonica*, **9** (1947).

WHEELER, J. A., and FEYNMAN, R. P., 'Interaction with the Absorber as the Mechanism of Radiation', *Rev. Mod. Phys.* **17** (1945), 157.

INDEX

PRINTED IN GREAT BRITAIN
AT THE UNIVERSITY PRESS, OXFORD
BY VIVIAN RIDLER
PRINTER TO THE UNIVERSITY

A CATALOGUE OF SELECTED DOVER BOOKS IN ALL FIELDS OF INTEREST

A CATALOGUE OF SELECTED DOVER BOOKS IN ALL FIELDS OF INTEREST

CONDITIONED REFLEXES, Ivan P. Pavlov. Full translation of most complete statement of Pavlov's work; cerebral damage, conditioned reflex, experiments with dogs, sleep, similar topics of great importance. 430pp. 5⅜ x 8½. 60614-7 Pa. $4.50

NOTES ON NURSING: WHAT IT IS, AND WHAT IT IS NOT, Florence Nightingale. Outspoken writings by founder of modern nursing. When first published (1860) it played an important role in much needed revolution in nursing. Still stimulating. 140pp. 5⅜ x 8½. 22340-X Pa. $3.00

HARTER'S PICTURE ARCHIVE FOR COLLAGE AND ILLUSTRATION, Jim Harter. Over 300 authentic, rare 19th-century engravings selected by noted collagist for artists, designers, decoupeurs, etc. Machines, people, animals, etc., printed one side of page. 25 scene plates for backgrounds. 6 collages by Harter, Satty, Singer, Evans. Introduction. 192pp. 8⅞ x 11¾. 23659-5 Pa. $5.00

MANUAL OF TRADITIONAL WOOD CARVING, edited by Paul N. Hasluck. Possibly the best book in English on the craft of wood carving. Practical instructions, along with 1,146 working drawings and photographic illustrations. Formerly titled *Cassell's Wood Carving*. 576pp. 6½ x 9¼. 23489-4 Pa. $7.95

THE PRINCIPLES AND PRACTICE OF HAND OR SIMPLE TURNING, John Jacob Holtzapffel. Full coverage of basic lathe techniques—history and development, special apparatus, softwood turning, hardwood turning, metal turning. Many projects—billiard ball, works formed within a sphere, egg cups, ash trays, vases, jardiniers, others—included. 1881 edition. 800 illustrations. 592pp. 6⅛ x 9¼. 23365-0 Clothbd. $15.00

THE JOY OF HANDWEAVING, Osma Tod. Only book you need for hand weaving. Fundamentals, threads, weaves, plus numerous projects for small board-loom, two-harness, tapestry, laid-in, four-harness weaving and more. Over 160 illustrations. 2nd revised edition. 352pp. 6½ x 9¼. 23458-4 Pa. $6.00

THE BOOK OF WOOD CARVING, Charles Marshall Sayers. Still finest book for beginning student in wood sculpture. Noted teacher, craftsman discusses fundamentals, technique; gives 34 designs, over 34 projects for panels, bookends, mirrors, etc. "Absolutely first-rate"—E. J. Tangerman. 33 photos. 118pp. 7¾ x 10⅝. 23654-4 Pa. $3.50

THE AMERICAN SENATOR, Anthony Trollope. Little known, long unavailable Trollope novel on a grand scale. Here are humorous comment on American vs. English culture, and stunning portrayal of a heroine/villainess. Superb evocation of Victorian village life. 561pp. 5⅜ x 8½. 23801-6 Pa. $6.00

WAS IT MURDER? James Hilton. The author of *Lost Horizon* and *Goodbye, Mr. Chips* wrote one detective novel (under a pen-name) which was quickly forgotten and virtually lost, even at the height of Hilton's fame. This edition brings it back—a finely crafted public school puzzle resplendent with Hilton's stylish atmosphere. A thoroughly English thriller by the creator of Shangri-la. 252pp. 5⅜ x 8. (Available in U.S. only) 23774-5 Pa. $3.00

CENTRAL PARK: A PHOTOGRAPHIC GUIDE, Victor Laredo and Henry Hope Reed. 121 superb photographs show dramatic views of Central Park: Bethesda Fountain, Cleopatra's Needle, Sheep Meadow, the Blockhouse, plus people engaged in many park activities: ice skating, bike riding, etc. Captions by former Curator of Central Park, Henry Hope Reed, provide historical view, changes, etc. Also photos of N.Y. landmarks on park's periphery. 96pp. 8½ x 11. 23750-8 Pa. $4.50

NANTUCKET IN THE NINETEENTH CENTURY, Clay Lancaster. 180 rare photographs, stereographs, maps, drawings and floor plans recreate unique American island society. Authentic scenes of shipwreck, lighthouses, streets, homes are arranged in geographic sequence to provide walking-tour guide to old Nantucket existing today. Introduction, captions. 160pp. 8⅞ x 11¾. 23747-8 Pa. $6.95

STONE AND MAN: A PHOTOGRAPHIC EXPLORATION, Andreas Feininger. 106 photographs by *Life* photographer Feininger portray man's deep passion for stone through the ages. Stonehenge-like megaliths, fortified towns, sculpted marble and crumbling tenements show textures, beauties, fascination. 128pp. 9¼ x 10¾. 23756-7 Pa. $5.95

CIRCLES, A MATHEMATICAL VIEW, D. Pedoe. Fundamental aspects of college geometry, non-Euclidean geometry, and other branches of mathematics: representing circle by point. Poincare model, isoperimetric property, etc. Stimulating recreational reading. 66 figures. 96pp. 5⅝ x 8¼. 63698-4 Pa. $2.75

THE DISCOVERY OF NEPTUNE, Morton Grosser. Dramatic scientific history of the investigations leading up to the actual discovery of the eighth planet of our solar system. Lucid, well-researched book by well-known historian of science. 172pp. 5⅜ x 8½. 23726-5 Pa. $3.50

THE DEVIL'S DICTIONARY. Ambrose Bierce. Barbed, bitter, brilliant witticisms in the form of a dictionary. Best, most ferocious satire America has produced. 145pp. 5⅜ x 8½. 20487-1 Pa. $2.25

THE EARLY WORK OF AUBREY BEARDSLEY, Aubrey Beardsley. 157 plates, 2 in color: *Manon Lescaut, Madame Bovary, Morte Darthur, Salome,* other. Introduction by H. Marillier. 182pp. 8⅛ x 11. 21816-3 Pa. $4.50

THE LATER WORK OF AUBREY BEARDSLEY, Aubrey Beardsley. Exotic masterpieces of full maturity: *Venus and Tannhauser, Lysistrata, Rape of the Lock, Volpone,* Savoy material, etc. 174 plates, 2 in color. 186pp. 8⅛ x 11. 21817-1 Pa. $5.95

THOMAS NAST'S CHRISTMAS DRAWINGS, Thomas Nast. Almost all Christmas drawings by creator of image of Santa Claus as we know it, and one of America's foremost illustrators and political cartoonists. 66 illustrations. 3 illustrations in color on covers. 96pp. 8⅜ x 11¼. 23660-9 Pa. $3.50

THE DORÉ ILLUSTRATIONS FOR DANTE'S DIVINE COMEDY, Gustave Doré. All 135 plates from Inferno, Purgatory, Paradise; fantastic tortures, infernal landscapes, celestial wonders. Each plate with appropriate (translated) verses. 141pp. 9 x 12. 23231-X Pa. $4.50

DORÉ'S ILLUSTRATIONS FOR RABELAIS, Gustave Doré. 252 striking illustrations of *Gargantua and Pantagruel* books by foremost 19th-century illustrator. Including 60 plates, 192 delightful smaller illustrations. 153pp. 9 x 12. 23656-0 Pa. $5.00

LONDON: A PILGRIMAGE, Gustave Doré, Blanchard Jerrold. Squalor, riches, misery, beauty of mid-Victorian metropolis; 55 wonderful plates, 125 other illustrations, full social, cultural text by Jerrold. 191pp. of text. 9⅜ x 12¼. 22306-X Pa. $7.00

THE RIME OF THE ANCIENT MARINER, Gustave Doré, S. T. Coleridge. Dore's finest work, 34 plates capture moods, subtleties of poem. Full text. Introduction by Millicent Rose. 77pp. 9¼ x 12. 22305-1 Pa. $3.50

THE DORE BIBLE ILLUSTRATIONS, Gustave Doré. All wonderful, detailed plates: Adam and Eve, Flood, Babylon, Life of Jesus, etc. Brief King James text with each plate. Introduction by Millicent Rose. 241 plates. 241pp. 9 x 12. 23004-X Pa. $6.00

THE COMPLETE ENGRAVINGS, ETCHINGS AND DRYPOINTS OF ALBRECHT DURER. "Knight, Death and Devil"; "Melencolia," and more—all Dürer's known works in all three media, including 6 works formerly attributed to him. 120 plates. 235pp. 8⅜ x 11¼. 22851-7 Pa. $6.50

MECHANICK EXERCISES ON THE WHOLE ART OF PRINTING, Joseph Moxon. First complete book (1683-4) ever written about typography, a compendium of everything known about printing at the latter part of 17th century. Reprint of 2nd (1962) Oxford Univ. Press edition. 74 illustrations. Total of 550pp. 6⅛ x 9¼. 23617-X Pa. $7.95